D0216933

STATISTICS
AND
SOCIETY

STATISTICS: Textbooks and Monographs

A Series Edited by

D. B. Owen, Coordinating Editor
Department of Statistics
Southern Methodist University
Dallas, Texas

R. G. Cornell, Associate Editor
for Biostatistics
University of Michigan

W. J. Kennedy, Associate Editor
for Statistical Computing
Iowa State University

A. M. Kshirsagar, Associate Editor
for Multivariate Analysis and
Experimental Design
University of Michigan

E. G. Schilling, Associate Editor
for Statistical Quality Control
Rochester Institute of Technology

ADDITIONAL VOLUMES IN PREPARATION

STATISTICS AND SOCIETY

Data Collection and Interpretation

Second Edition, Revised and Expanded

WALTER T. FEDERER

Biometrics Unit and
Mathematical Sciences Institute
Cornell University
Ithaca, New York

Marcel Dekker, Inc.   New York • Basel • Hong Kong

Library of Congress Cataloging--in--Publication Data

Federer, Walter Theodore
  Statistics and society : data collection and interpretation / by
  Walter T. Federer. -- -- 2nd ed.
      p.   cm. -- -- (Statistics, textbooks and monographs; v.117)
  Includes bibliographical references and index.
  ISBN 0-8247-8249-6
  1. Statistics. I. Title. II. Series.
HA29.F34   1991
001.4'22--dc20   91-11596
                                                         CIP

This book is printed on acid-free paper.

Copyright © 1991 by MARCEL DEKKER, INC. All Rights Reserved

Neither this book nor any part may be reproduced or transmitted in any form
or by any means, electronic or mechanical, including photocopying, micro-
filming, and recording, or by any information storage and retrieval system,
without permission in writing from the publisher.

MARCEL DEKKER, INC.
270 Madison Avenue, New York, New York, 10016

Current printing (last digit):
10 9 8 7 6 5 4 3 2 1

PRINTED IN THE UNITED STATES OF AMERICA

TO MY WIFE, EDNA, WITH APPRECIATION AND LOVE

PREFACE TO SECOND EDITION

The second edition represents a considerably revised and expanded version of the original book. This has resulted from advances in Statistics, changes in student populations over the years, and experiences with teaching the course entitled "Statistics and the World We Live In", many times. Several instructors have taught this course, Statistics and Biometry 200, since its inception in 1966. Some taught it as a baby probability course, some as a baby statistical methods course, and some as a topic of interest to them, e.g., interval estimation. Thus, it should be obvious that the students received quite different material from each of the instructors. If this course is to be an "artsy" type course on the general concepts, ideas, and philosophies of the subject of Statistics, it should be broadly based to cover a majority of the areas of Statistics and not some particular sub-set. This is why a broad range of topics are included in this book.

The following correspondence exists between chapters in the first and second editions. Roman numerals were used for chapter numbers in the first edition and arabic numbers for the chapters in the second edition: I =1, II = 2, III = 3, IV = 4, V = 5, 6 is new, VI = 7, VII = 8, VIII = 9, IX = 11, X = 10, XI = 12, XII = 13, XIII = 15, and XIV = 14. The content of the chapters was changed to reflect what was being taught in "Statistics and the World We Live In" by the author at the present time. The coverage of topics was expanded. For example, sampling in an analytical laboratory and for catastrophic events can be used to greatly increase the efficiency of a laboratory and provide needed information in coping with catastrophes; population structures for various experiment and treatment designs are discussed, a topic that is omitted from statistical texts. The topics of "product improvement", quality control, acceptance sampling, clinical trials, and others are presented to demonstrate the depth and broadness of the subject of Statistics.

The assistance of Mrs. Colleen Bushnell in preparing a number of the figures and graphs is greatly appreciated.

Walter T. Federer

PREFACE TO FIRST EDITION

The initial impetus for this book came from a meeting of the author with T. C. Watkins, past Director of Resident Instruction, and R. P. Murphy, past Head of the Department of Plant Breeding and Biometry, of Cornell University, where it was decided that a different approach of presenting Statistics was required. The introductory methods and theory courses did not appear to be satisfying the needs of all students. A course presenting the ideas, concepts, and philosophies associated with the procurement of numbers with a minimum emphasis on statistical methodology or on mathematical theory was required; the course was to be concerned with numbers and the use of numbers as related to the many aspects of Society. The course was to be complementary to and non-repetitive of other introductory statistics courses. It was to be an "artsy" type course in statistics in that the student would learn about the subject of Statistics but would not attain competence in the technical aspects of the subject.

Such a course was first attempted in the spring term of 1966 and was repeated during the spring terms of 1967, 1968, 1969, 1970, 1971, and 1972. The course was given at 8:00 a. m. with no announced instructor and as an elective course. Despite the early hour and other adversities, the enrollments in the first seven years were 57, 101, 99, 75, 70, 85, and 100. (In 1972 the time was changed.) In addition to the students registered for credit, several visitors have attended the course each year. The author taught the course during the first three years and the sixth year; Professor B. L. Raktoe, University of Guelph, Professor Carl Marshall, Oklahoma State University, and Professor F. B. Cady, Cornell University, taught the course during the fourth, fifth, and seventh years respectively. These experiences were utilized in the preparation of the present text.

The material presented in the text together with the associated problems, three term quizzes, and a final examination has been considered sufficient for three semester hours of credit at the sophomore level in a university. For the first six years, three lectures were given each week and weekly discussion periods with optional attendance were held. The material has not been difficult for the upper one-fourth of the class but has been for the remaining three-fourths. The concept of orthogonality in experimental designs has been the most difficult concept but without it several students have indicated that the course would not be challenging enough. Student evaluations have been obtained repeatedly but these have produced only minor changes in content or presentation. Any instructor using this text in a course should make certain that most of the students really understand what has been presented before proceeding to the next topic.

Statistics is defined to be the development and application of techniques involved with (1) the design of the investigation, (2) the summarization of results from an investigation, and (3) inference from the results of the investigation to the population from which the results were derived. Hence, a prime concern should be with the origins of the data and how to obtain reliable information in any investigation; unfortunately methodology and mathematical statistics books define away these important problems related to real world situations. The emphasis in this text is on the relationship of the sample and the population, statistic and parameter, and on the underlying variability of all observations in the real world. The universality of variability and the various sources of variation represent the underlying theme throughout the book. One designs to take account of certain types of variation. Consequently, this text emphasizes design concepts and techniques mostly with some attention being devoted to summarization concepts and techniques, and with little attention being devoted to inference. This is the reverse of the emphasis in current statistics texts.

A search was made for a suitable text for this course. In the course of the search, the author was amazed to discover that methods books contain mostly computational procedures and techniques of computation with little or no space being devoted to methods of procuring meaningful and accurate data; it appears that many statistics books are concerned mainly with the manipulation of numbers and algebraic symbols or with mathematical manipulations. Although many students are interested in these manipulations, many are not. This latter group would form the nucleus of students for the material covered in this text. An instructor using the present text will find the books by E. B. Wilson, "*An Introduction to Scientific Research*", by D. Huff, "*How to Lie With Statistics*", and by M. J. Slonim, "*Sampling in a Nutshell*", useful for supplementary reading material, especially with chapters I to V.

Several types of courses may be taught from this text by changing the emphasis and sequence of chapters and by adding material in selected areas.

During the first four presentations of the material, the chapter sequence followed was I to IV, VI, VII, V, VIII, X, XI, and XIII; most of the material in chapters IX, XII, and XIV was omitted. The present sequence of chapters is considered to be better pedagogically because the sample survey concepts presented in chapter V are easier to comprehend than the experiment and treatment design concepts in chapters VI and VII. During the fifth and sixth presentations of the course the chapter sequence was essentially that presented in the text except that some of the material in chapters IX to XII and XIV was included with the material in chapters V to VII; also, sample survey designs received more emphasis and experimental and treatment designs less emphasis than given in the text. A third type of course that could be taught from the text would emphasize sample survey design and analysis concepts using the material in chapters I to V, parts of VIII to XII, XIV, with either more emphasis on summary statistics or with a considerably expanded version of chapter XIII on the type of statistics gathered and the methods used to procure the data for the United States, for the United Nations, or for some selected country like India. This text is designed for a one-semester course but could easily be expanded into a two-semester course by supplementing, for example, the material in chapters V, VIII, XIV, IX, and/or XI.

I am indebted to the Literary Executor of the late Sir Ronald A. Fisher, F. R. S., and to Oliver & Boyd, Edinburgh, for their permission to reprint tables VIII.5, XI.1, XI.2, of t-values in section XI.5, and of chi-square values in section XIV.5 from their book *"Statistical Methods for Research Workers"*.

The author wishes to acknowledge the interest shown by and the comments received from Prof. B. L. Raktoe, Prof. Carl Marshall, Prof. Foster B. Cady, and his other colleagues at Cornell University in the preparation of the text. Prof. Oscar Kempthorne devoted a considerable amount of time and effort and offered valuable suggestions; his efforts are greatly appreciated. Sincere gratitude is expressed for the editorial assistance of my wife, Lillian (deceased in 1978). Also, gratitude and thanks are acknowledged to Mrs. Donna Van Order, Mrs. Anne White, Mrs. Norma Phalen, and Mrs. Helen Seamon for their efforts and skills exhibited in typing the manuscript.

Walter T. Federer
1972

TABLE OF CONTENTS

Contents

CHAPTER 1. INTRODUCTION AND DEFINITIONS

1.1 THE TITLE

Statistics and Society--what do we mean by these two words? The term **Society** means the entire structure of a self-contained group of people. The United States of America is a Society. Likewise, Canada and Mexico are Societies. There are professional societies of doctors, lawyers, and statisticians, and religious societies of Christians, Moslems, and Jews. Habits, customs, fashions, daily living, culture, recreational activity, pastimes, politics, religion, and so forth, all go to make up a Society. The American Heritage Dictionary defines this word as follows: "The totality of social relationships among human beings. A group of human beings broadly distinguished from other groups by mutual interests, participation in characteristic relationships, shared institutions, and a common culture. The institutions and culture of a distinct self-perpetuating group." The above conception of the term Society in one form or another is probably fairly universal among all people. There would normally be only small differences in conception of the term.

A populist conception of the term **Statistics** is very diverse and often far from the appropriate conception. Even among those professionally trained in the discipline of Statistics, there appears to be no single unique conceptual definition. Hence, it is imperative that all writers of statistical treatises precisely define what they mean by the term Statistics. A historical perspective will be useful in forming ideas about the word and whether it is a subject (proper noun), a noun, or an adjective. The following is an adaptation of material from the book entitled *An Introduction to the Theory of Statistics* by Yule and Kendall (1950). The words statist, statistics, and statistical all appear to be derived from the Latin word "status" in the sense of a political state. The term "statist" is the earliest of the three, appearing in *Hamlet* in 1602, in *Cymbeline* in 1610 or 1611, and in *Paradise Regained* in 1671. The word "statistics" first appeared in *The Elements of Erudition* by Baron J. F. von Bielfeld in 1787; this book was translated into the English by W. Hooper, M. D. "Statistics" was the title of one of the chapters of this book. The definition of the subject Statistics was given as "The science that teaches us what is the political arrangement of all the modern states of the known world". In the book, he states, "It is about forty years ago that the branch of political knowledge, which has for its object the actual and relative power of the several modern states, the power arising from their natural advantages, the industry and civilization of their inhabitants, and the wisdom of their governments, has been formed, chiefly by German writers, into a separate science.... By the more convenient form it has now received...this science, distinguished by the new-coined name of 'statistics', is become a favorite study in Germany... To the several articles contained in the work, some respectable 'statistical' writers have added a view of the principal epochas of the history of each country." Note that both the noun "statistics" and the adjective "statistical" appear in this quotation. In German the word for statistics is "statistik", and "statisticus" is the Latin word for statistical.

Sir John Sinclair, editor and organizer of the first "Statistical Account of Scotland" in the late 1700s, did much to assure the acceptance of the new words "statistics" and "statistical". In a circular letter to the Clergy of the Church of Scotland in 1790, he states that in Germany, "'Statistical Inquiries', as they are called, have been carried on to a very great extent," and he explains this as "Statistical inquiries - or inquiries respecting the population, the political circumstances, the productions of a country, and other matters of state". His hopes that these words would be accepted were realized and their meanings under-went rapid development during the next half century. Statistics or Statistik as used by the writers of the eighteenth century meant the exposition of the noteworthy characteristics of a state or government. The expositions were mostly of the verbal rather than the numerical type. However, during the beginning of the nineteenth century, numerical data began to displace verbal descriptions of the characteristics of the state; this represented a narrowing of the meaning of Statistics to the exposition of the characteristics of a State by "numerical" methods. As pointed out by the following statement in the first volume of *The Journal of the Royal Statistical Society* in 1838-9, "Statistics may be said, in the words of the prospectus of this Society, to be the ascertaining and bringing together of those facts which are calculated to illustrate the condition and prospects of society...the statist commonly prefers to employ figures and tabular exhibitions"; the adoption of the narrower definition was not universal. By the end of the nineteenth century, the use of the words Statistics as a subject and science, statistical as an adjective, and statistics as a noun were in common usage. The person whose profession was Statistics was called a statistician. During the first half of the twentieth century, terms like social and economic statistics, biological statistics, medical statistics, genetic statistics, physical and engineering statistics, geological statistics, educational statistics, and others were used to indicate the collection and/or application of statistical procedures in biology, medicine, etc.. These terms are misfortunate in that statistics can hardly be biological, social, or genetic. A much better terminology would have been to use statistics, or statistical procedures, for studies in biology, medicine, or genetics. However, these terms are now well entrenched and would be difficult if not impossible to dislodge despite their ambiguity.

Many people in the general public think of Statistics in very limited terms and more in the nature of the nineteenth century. To them, all statisticians work with numbers in the manner of a hockey statistician, a baseball statistician, a basketball statistician, etc., as commonly heard in the wrap-up of a television sportscast. It is likely that their "statistician" never completed a course in Statistics but they simply found some person to keep numerical track of saves, goals scored by each team, shots on goal, and other numerical items that the sportscaster believes interests listeners. A glance through the pages of the Sports Section of any newspaper provides ample evidence of the heavy reliance of sports writers on numerical facts. Some people think of Statistics as computing, some as mathematics, some as preparing tables and graphs, some as working for the Census Bureau, some as working for an insurance company, some as hoodwinking the public, and some as working for a sportscaster. Others may think of the vast number of numerical details appearing in publications such as *The Wall Street Journal* and other business and economic oriented publications as statistical publications since there are numerous statistics appearing in these publications. The subject of Statistics encompasses all of these and more. Undoubtedly there are more concepts of what constitutes the subject of Statistics and about what a statistician does. Regardless of what one's conception of Statistics is, a moment of reflection is all that is needed to know that Statistics and Society are very much intertwined and that it behooves any individual who professes to have intellectual curiosity to learn something about the subject. For many individuals, it may not be necessary to learn the methodology of Statistics, but it is important to learn the concepts. We shall attempt to present a number of the concepts with a minimum of effort being spent on methodology. Most so-called introductory textbooks in Statistics devote a majority of their effort to manipulations of numbers (statistical methodology) or of symbols (mathematical statistics) with little effort being devoted to concepts. The writer of textbooks or the teacher of a Statistics course finds it relatively simple to write or to teach numerical or symbolic arithmetic but in order to present the concepts in a straight-forward and logical manner, the concepts must be thoroughly and completely understood by the writer and the teacher.

1.2 THE GOALS

All books should have objectives and goals. Those for this conceptual and introductory text are:
(a) to develop ideas and concepts related to measurements and measuring instruments, to investigations, to data collection, and to experimentation,
(b) to define and describe various aspects of the subject of Statistics,
(c) to study some uses and some abuses of statistics and statistical procedures, and
(d) to develop a constructively critical attitude toward investigations, numbers, and users of numbers in our Society.

To fulfill these goals, a number of topics, ideas, and procedures will be covered. In order to obtain a fuller understanding and a more complete orientation for the material presented, it is helpful to read the references listed at the end of each chapter. To comprehend the concepts and techniques of Statistics more fully, it is highly desirable for the reader to work the problems given at the end of each of the chapters. Statistical ideas are usually not completely absorbed solely by listening and reading. It is necessary to work a large number of problems of many and varied types. As one student remarked when he finally got around to working all the problems given in Snedecor (1946), "My knowledge of Statistics has increased forty-fold". Note that he had taken the course for credit once before and he had "sat through or audited" the course twice before. He finally realized that he was not going to learn statistical concepts and procedures by osmosis but that he had to work the problems.

1.3 SOME STATISTICAL TERMINOLOGY

In order to eliminate some of the abstractness associated with a particular set of results from an investigation, it is suggested that the height, weight, eye color, and age in months of each student in the class be obtained as a class project (See Problem 2.4.). The numbers and adjectives obtained would have meaning to class members and would be less impersonal than other sets. A characteristic like height is called a **variable** and sometimes a **factor** in statistical jargon. Heights vary from individual to individual and hence the term variable. For a particular variable, we can think of the totality of individuals having this characteristic. This totality, or universe, is denoted as a **population**. Any subset of the population is denoted as a **sample**. To illustrate, for a variable, say income per year, the population could be all the individuals residing in the United States, in the State of New York, or all registered students at Cornell University in a given year.

An individual, an item, or a member of a population is denoted as a **sampling unit**. In many populations, the sampling units are obvious but in others they are not. For example, for a variable like height of members of the freshman class of a particular college, the sampling unit is the individual freshman. If the universe is the land area of County X of State Y, what is the sampling unit? It could be an acre, a hectare, a square meter, a farm, or some other size of unit but there is no obvious and unique sampling unit. In all cases, the investigator should precisely and clearly define the sampling unit and the population. Otherwise, these items will be conceptually vague and imprecise and open to misinterpretation.

A **survey** or **sample survey** is a collection of facts on one or more characteristics of members of a sample. Such a study or investigation as a survey is of common occurrence in our Society. The particular sample used for the survey may or may not be a fair and unbiased representation of the population under consideration. For example, the population could be all of the registered students at College X in a particular term with the sample being members of a freshman biology class. This particular sample would not likely be representative of the population for most characteristics of interest. There are many ways in which the members of a particular class could be non-representative, e.g., all class members are the same sex, all are seniors, all are political science majors, all have the same ethnic background, etc.. Many surveys used in our Society are representative and many are unrepresentative even though the investigator may like for us to believe that the survey is representative. This topic is discussed in more detail in later chapters.

A **census** is a 100% sample, that is, all the members of the universe are included in the sample. Many surveys described as censuses may not represent a 100% sampling but only an attempt to obtain a 100% sample. The U. S. Census, which is taken every ten years, fell under considerable criticism in 1980 because many cities were believed to be undercounted. These cities stood to lose Federal funds because the 1980 Census indicated that their population had declined since the last Census. The U. S. Census Bureau takes elaborate and costly steps in attempting to obtain a 100% sample. In many situations, a census is impractical and a sample, often a small one, will be used. For example, to determine how a cake, the universe, tastes, a person would normally cut one small piece, the sample, and taste the piece of cake, the sample survey. This person would be assuming that the single piece of cake was representative of the entire cake. To test the temperature of the water in a swimming pool, the population, one does not test the temperature of the water in the entire pool. Instead, the testing is made at one spot in the pool, and it is believed that the sample spot is representative of the entire pool. Sampling is a way of life in all societies but in most cases, little thought is given as to whether or not the sample is representative of the population. Remember, an assumption that a sample is representative does not make it representative.

Another frequently used term in Statistics is parameter. A **parameter** is a characteristic of the population and is obtained from all members of the population; it does not vary but is a constant and fixed value for a particular population. If the value for a particular characteristic for each and every member of a population is summed and divided by the number of members in the population, this is the well-known **mean** or **arithmetic mean** and is a parameter of the population. Furthermore, the value that divides all the sampling unit values into two equal parts such that 50% of the values are below and 50% are above this value is denoted as the **median** and is a parameter of the population. The largest value for any sampling unit in the population minus the smallest value of any sampling unit is known as the **range** of the population values and is a parameter. No matter how many times a parameter is obtained for a specified universe, its value is the same. Parameters are facts about a population whereas facts from less than a 100% sample of a population, are known as **statistics**, the noun. The mean, median, and range obtained from a sample are statistics which vary from sample to sample, representing uncertainty about parameters. Statistics relate to sample facts and parameters relate to population facts.

1.4 STATISTICS, THE SUBJECT

The subject of Statistics is considered to pertain to the development, the application, and the characterization of procedures for

    (a) planning for any type of study or investigation involving the procurement of numerical or non-numerical facts about members of a sample,

    (b) summarizing the results of the investigation in (a), and

    (c) from the sample design and sample statistics, relating or making inferences and implications about the parameters of the population.

Step (a) is denoted as statistical design, step (b) as statistical methodology, and step (c) as statistical inference. It would be incorrect to say that Statistics is only one of these three items, rather it is all three items and none can be divorced from the others. They are all members of one subject. Sometimes writers of statistical treatises become careless and imply or state that Statistics is, say, statistical inference only but this is incorrect. Often a researcher works on a particular statistical procedure which pertains to only one of the above three items. If they work solely in one area, there is a tendency for them to think that this area is all that there is to Statistics. This causes little difficulty in the developmental and characterization aspects; but for applications, all three items must be considered.

Each of the above three parts of Statistics contains many sub-parts and items. Some of these are described below. Statistical design encompasses:

    (a) a complete description of the population and variables under consideration in the study, including a description of the sampling units, the nature of the responses for the variables, the nature of the investigation, i.e., a sample survey, an experiment (a study and design under the control of the investigator), or an observational study (record keeping and similar recordings of facts such as police records, hospital records, student course grades, etc.),

    (b) complete description, validation, and standardization of measurements, measuring instruments, and measuring procedures used in the study,

    (c) treatment design, i.e., the selection of the entities, treatments, to be studied in an experiment as well as the selection of adequate and appropriate controls and standards,

    (d) experiment design, the arrangement of the treatments in an experiment and the conduct of the experiment,

    (e) survey design and conduct of the survey,

    (f) sample size considerations and the nature of the units in an experiment or survey, and

(g) size, shape, and allocation of sampling units in the study.

Statistical design is not just one but is all of the above and more. Many and diverse statistical methods have been devised to meet the needs of investigators. Study summarization requirements often varies from subject matter field to field. Procedures required for investigations in the field of Education may be and often are completely different from those in, say, Engineering or Genetics. As knowledge in any field advances, the need for new, amended, or different statistical methodology frequently arises. This provides a rich area for research for the statistician in developing new statistical methodology to meet the needs of investigators. Because of the complexities and multitude of needs, many of them are yet to be fulfilled. By far the greatest number of statistical textbooks is on statistical methodology. A part of the methodology relates to computing and computing procedures. Presently there is a flourishing branch of statistical methodology that is known as statistical computing. Owing to the recent and rapid development of PC computers over the past seven to ten years, statistical computing has been selected by many statisticians as their area of research and teaching. The development of software packages for PC computers related to statistical procedures occupies the attention of growing numbers of statisticians and computer scientists.

Procedures for relating sample facts to population parameters, or statistical inference, involves everything done in the first two parts of Statistics. Unless the statistical design and the statistical methodology used in an investigation is clearly understood and correctly used, no meaningful statistical inferences can be made. Statistical inference is the newest part of Statistics and is being used and studied in many ways. It involves meaningful statements about the relative odds or probability of various outcomes from an investigation.

1.5 SOME THINGS STATISTICIANS DO

A statistician performs one or more of the following:
(a) teaching courses in Statistics at various levels and on diverse topics,
(b) conducting research on statistical topics,
(c) consulting with investigators about the statistical and research aspects of their investigations, and
(d) doing or conducting the computations, compiling tables, preparing graphs, and performing other steps associated with the summarization

of results from an investigation.

As with any other field of study or science, there is a need for elementary and introductory courses for students entering the field and for those wishing to learn something about the subject. There may be as many as a dozen beginning courses in Statistics at many universities. The content, level, and material in these courses may be very diverse. Some may have a calculus requirement, some only high school algebra, and others no requirement other than enrollment. One course (and sometimes more than one) in Statistics is often a requirement for many areas of specialization. These requirements from other fields provide the basis for large numbers of undergraduates in the introductory courses in Statistics.

Intermediate and advanced level courses are needed to provide the necessary statistical training and expertise required by various areas of study as well as for students specializing in Statistics. These courses may cover one or more of the three parts of Statistics, i.e., design, methodology, and inference, but most will be on methodology and a fair number will be on mathematical statistics and/or probability. Occasionally, courses on survey design and the conduct of a survey and on experiment and treatment design and the conduct of an experiment, will be presented. Too often courses purported to be in this last area devote most of their attention to statistical methodology rather than design. Aspects of planning and conducting investigations, part (a) of Statistics, should receive more attention than now given in Statistics curricula and in statistical textbooks. Unless the results obtained are reliable, representative, and accurate, statistical computations and inferences may have no practical usefulness. The procurement of good data is a requirement for any worthwhile investigation.

As described earlier, there are many unsolved statistical problems and there always will be. These problems arise from many sources. This means that a wide variety of interests and expertise can be accommodated in attempting to find solutions. There are many categories of expertise and interest among statisticians. Some individuals may work entirely in the realm of Mathematics, Probability, or Computer Science in an abstract manner and isolated from investigators in other fields. Or, they may work closely with an investigator in all aspects of the investigation from its planning stages to its completion and writing the report. The statistical expertise for the latter may involve only a sophisticated use of known statistical theory or it may involve the development of new statistical and/or computing procedures.

Statistical consulting is an activity occupying numerous statisticians in various degrees and areas from hospitals, city governments, pharmaceutical firms, government bureaus, etc. to insurance companies. Some statisticians are full-time consultants, some may consult only a few hours per week or month, and still others may not consult at all. This activity, or service, for investigators has proven to be useful to and in demand by investigators. An investigator cannot be expected to acquire all the necessary statistical expertise for the many and varied investigations that they ordinarily conduct. Hence, it often is necessary for them to seek advice and consultation on the statistical aspects of an investigation. This should be done in the planning stages as it is too late to amend plans after they have been carried out. If consultation is done at the planning stage, a statistician is often able to act as a research consultant as well as a statistical consultant. Suggestions are often made in the course of consulting which change the direction and/or allows the research to be performed in a more efficient manner. Many research publications are jointly authored with a statistician, as a member of the research team.

Statisticians become involved in a variety of activities. They are active in the development, testing, and approval stages for new medical practices, medicines, and drugs. In marketing studies, they are involved in the planning and study analysis stages, as well as in the decision stage as to whether or not to market a product or service. Desirable locations for new enterprises are found by using statistical procedures, market surveys, and consultations. Opinion polls, especially election polls, utilize the services of statisticians. The author was a member of the NBC National Election Team under the leadership of John W. Tukey and David Wallace from 1968 to 1980. As the voting results were being reported across the United States, predictions of winners were being made throughout Election Day night after the voting booths had closed. After 1980, polls were taken of voters exiting the voting booths, and predictions of winners could often be made long before the voting had ceased. In a large dietary-cancer mortality study conducted jointly by the Beijing and Cornell Universities, hundreds of biochemical analyses and other measurements were proposed. Since it was not possible to obtain a large enough blood sample from each individual person for all proposed analyses and since the number of required analyses would be time, space, and cost prohibitive, the author suggested that pools of 25 blood, food, and other samples be used. This procedure was adopted and cut the number of biochemical analyses to 4% of the originally proposed number, provided a large enough blood sample to run all the proposed analyses, and the desired results would be obtained. Also, in a large plant breeding screening program, the author was able to provide procedures which reduced the work by one-third and it was possible to obtain better and more reliable information than with the old procedure. B. R. Murty (personal communication) was able to formulate two new biological findings through use of a statistical procedure known as multivariate analysis. These findings would not have been made without the use of this procedure. Over the years, class projects by members of a class using this text, have resulted in changes in Cornell Campus life in library hours, library procedures, graduation requirements in the College of Agriculture and Life Sciences, minority student programs, sex education, dining hall practices, and various other activities. The day-to-day consulting activity of a statistician provides many ideas for statistical research topics and a broad background in many areas of study. These, in addition to a desire to help people and to be a member of a research team, are the pay-offs for the service aspects of statistical consulting.

Many Federal government and international organizations employ relatively large numbers of statisticians to perform various statistical tasks. The Bureaus of Census and Labor Statistics, the U. S. Department of Agriculture Crops and Markets, the World Health Organization (WHO), and others utilize the services of many statisticians. A large part of their work involves sample surveys and statistical computing.

1.6 STATISTICS, MATHEMATICS, AND COMPUTER SCIENCE

Outside of a language like English, French, German, Russian, Spanish, etc., no other language has a broader application over all fields of Science than Statistics. It has been called the language of Science. Statistics and statistical procedures are used in all fields of Science in one way or another. Statistical procedures can be and are invaluable in planning, summarizing, and interpreting results from an investigation. However, when they are used by the novice or the deceitful, they may be misused and/or misinterpreted. Examples of this will be given later in the text. Since statistical procedures are used universally in investigations, it is to be expected that positions held by statisticians and methods utilized are diverse and varied.

The research and teaching tools of a statistician are mathematical procedures and theory, computers, and computing procedures and theory. Mathematics and a language like English are the languages of Statistics, even though the former is used extensively. Most new theoretical results in Statistics are formulated and/or obtained via mathematical procedures. Note that Statistics is not considered to be a branch of Mathematics or of Computer Science but a science in its own right. Two statistical journals that are highly mathematical in nature are *The Annals of Probability* and *The Annals of Statistics*. For the most part, papers in these journals use complex and advanced mathematical procedures. Elementary mathematics for the most part would not be tolerated in these journals. Many other statistical journals such as *Applied Statistics, Biometrics, Biometrika, Communications In Statistics (Series A and B), Journal of the American Statistical Association, Journal of the Royal Statistical Society (Series A and B), Journal of Economic and Business Statistics, Journal of Statistical Planning and Inference, Sankhya, Technometrics*, etc., utilize the language of Mathematics to a considerable degree. Hence, mathematical procedures are a necessary and much used tool in statistical research and theory and in teaching courses on the mathematical aspects of Statistics. Utilizing mathematical procedures to derive or to explain statistical procedures could be described as the arithmetic of symbols.

Computing procedures, on the other hand, involve the arithmetic of numbers. The manipulation of numbers or symbols is denoted herein as arithmetic. Quite often the manipulation of symbols is described as theory and the manipulation of numbers as applied or conceptual. Although the preceding statement can be partly true, it is not the whole story. A set of formulas does not in itself provide theory and a set of computations does not necessarily describe or conceptualize a statistical procedure. Much more is required; and here is where a language like English is a necessary supplement to the mathematical and computing languages. Knowing computing languages and procedures is necessary to obtain correct and efficient methods for summarizing the information in a set of numbers, especially large sets.

When mathematical derivations become overly complex and difficult, the statistician often turns to the computer and approximates the underlying mathematical result by a process known as simulation. Here many special cases may be combined to obtain a mathematically complex or difficult numerical solution. The experimental results obtained with a computer then give the investigator some insight about the true nature of the process. In other situations, the statistician may suspect that a result is true but the mathematics required to solve the problem is difficult; he may run a number of specific examples on a computer and if these substantiate his guess, he will then attempt to solve the problem analytically. In other cases, the numerical results obtained via the computer may give an indication of how to prove a result.

The computer is useful as a teaching tool in providing much more diversity, depth, and computing experience for the student. Since results can be obtained quickly, many things can be accomplished. For example, a particular assignment in Physics required eight hours on a slide rule whereas the same assignment required only two hours of computing on a pocket calculator. The author worked very hard for three weeks on a computing problem which was then done by one of his students in three seconds on a high speed computer! Prior to the advent of Personal Computers (PCs), students could only be assigned a limited amount of computing because of the time required for computations. With a PC, a student can often do 50 to a 1,000 different problems in the same time previously required to do one. Also, many different procedures can be tried. The student will be able to visualize the performance of a statistical procedure on a wide variety and number of numerical examples. In the future, symbolic arithmetic on a PC will be commonplace. The full use of the PC and future computers as a teaching tool is in the process of development and should be commonplace by the beginning of the 21st century.

1.7 A CAREER IN STATISTICS

Undergraduate and graduate degrees in Statistics may be obtained at most major universities in the United States as well as in other countries. A list of schools, colleges, and universities offering degrees in Statistics in the United States and Canada is published in *Amstat News*, a publication of the American Statistical Association, in the November issue each year. Material related to the listing as well as careers in Statistics, may be obtained by writing to the American Statistical Association, 1429 Duke Street, Alexandria, Virginia 22314-3402.

Students aspiring to have a career in Statistics should possess interest and some competence in mathematical and computing procedures. If the student possesses a certain amount of missionary spirit and a desire to aid investigators with their projects, the individual may wish to specialize in statistical consulting. If the preceding is not an area of interest, then the individual may wish to specialize in teaching and/or research. To perform well in any of these areas, creativity and ingenuity are necessary ingredients for adapting and developing statistical procedures for new situations and needs.

The course load demands for most Statistics curricula are not heavy. This allows the student time to concentrate in a second area such as Biology, Business, Computer Science, Mathematics, Sociology, or other areas. With proper planning, it may be possible to specialize in two or three areas. In a course of study with two or three majors, a student's education is broadened and enhanced and the job opportunities are expanded. There will be many more doors open to students trained in more than one area than if concentration of study is solely in one area.

The job opportunities for students majoring in Statistics have been and are excellent. The numbers entering the field have not been sufficient to meet the demands. Quite often there are so many job opportunities that not only do students have a choice of jobs but even of the area where they would like to live. Students receiving advanced degrees command relatively high salaries and benefits. There are opportunities for establishing consulting businesses on a full-time or part-time basis. Since a relatively small amount of capital for computing equipment and office space is required, it is relatively simple to start a statistical and computing business. Often the business can be set up in a home, which solves the office space problem. Statistics graduates are employed for the most part by academic, governmental, and business organizations. The pharmaceutical industry is a big employer of statisticians. As world competition in industry and manufacturing becomes more intense, the demand for statisticians trained in aspects of quality control and product improvement will increase. Businesses must remain competitive in order to stay in operation (See Mann, 1989).

1.8 AN OVERVIEW

As pointed out previously, this text attempts to stress the conceptual rather than the mathematical or computational aspects of various procedures associated with the planning for and the collection of facts from an investigation, summarizing the results with various methods, and interpreting the results and making inferences about the parameters of a population. Aspects of planning for and collecting data from an investigation should be discussed with a statistician prior to conducting the study. Too frequently the advice of a statistician is sought after the data have been collected, and the information sought is how to salvage something after the fact. It is not efficient investigating if the end result is a salvage operation to obtain something of value from a poorly planned study. Many consulting situations have been encountered wherein the first time the statistician was consulted was after the paper had been rejected for publication on statistical grounds. Any study can benefit from consultation with other scientists, especially if done prior to conducting the study. A well-planned study will usually contain valuable information whereas a poorly planned one may be useless and be fit only for the well-known circular file. Too many investigations fall in the latter category and time spent on them represents misspent effort.

In the following chapter, a discussion of measuring devices, measurements, calibration of devices and measurements, and some problems encountered in obtaining certain results is given. Types of variation are described and are put into two categories, assignable causes and non-assignable causes. A measurement for a single sampling unit is composed of assignable causes of variation and non-assignable causes. The response equation used throughout this text is the sum of these two sources of variation even though other forms are briefly mentioned. The value for any statistic will be decomposed into these two sources of variation. Many examples of "phony statistics" are presented to illustrate misconceptions and fallacies. In the third chapter, data collection details and the conduct of an investigation are described. In the next chapter, these items are described in the more formal context of the Principles of Scientific Investigation; relationships between the principles and statistical procedures are discussed. One aspect of statistical design, survey design, is presented in the fifth chapter. Many types of survey designs are discussed in the context of probability and non-probability designs. In discussing properties of various designs, the ideas of representativeness and unbiasedness are considered. It is shown how to design surveys to obtain the desired kind of information, i.e., how to collect good data. The population structure for a particular survey and sample as well as the response model are discussed. Interviewing techniques and procedures for obtaining answers to sensitive questions are presented. In the sixth chapter, sampling procedures in the laboratory are considered and the ideas related to pooling samples before conducting laboratory analyses are presented. It is demonstrated how to effect considerable savings in time and money by using these ideas. Another aspect of statistical design, experiment design, is studied in the seventh chapter. Population structures, response model equations, and the nature of the units in an experiment are described for several of the more commonly used and simpler experiment designs. This chapter is followed by one on treatment design. The needs for and types of controls are discussed as well as are several types of treatment designs. Types of treatment effects are described and illustrated with examples.

The first eight chapters are on Statistics, scientific investigation, and statistical design, and methods for designing investigations are discussed. Well-planned and conducted investigations lead to the procurement of desired and reliable results. After the investigation has been conducted and the data obtained, it will be necessary to summarize the results as the raw data in themselves are not sufficient to meet the goals of most studies. Some useful summary statistics are the arithmetic mean, median, and range as well as many others. The form of summarization depends on the nature of the study and the goals of the investigation and should be such as to transmit the desired information to the consumer of the results, whether it be a casual reader, the investigator, or another investigator. The communication of results is a necessary and sometimes difficult part of the successful completion of a study. Statistical graphics, charts, pictures, and illustrations represent a useful format for summarizing the information from a study and conveying it to others "at a glance". These methods, properly used, allow for quick assimilation of the results by the consumer. It is recommended that they be utilized in connection with other statistical methods for summarization purposes. The ninth chapter is concerned with these aspects. In the tenth chapter, a number of statistical procedures for summarizing facts from a study are presented. Measures of central tendency such as the arithmetic mean, the median, and others are illustrated for the various survey and experiment designs discussed in previous chapters. Several measures of variation, for example the range, are described and illustrated. These measures of central tendency and of variation have proved to be very useful summary statistics for investigators in all fields. A reason for including these summary statistics all in one chapter is to demonstrate the relationships of the statistics from design to design and to illustrate the concepts. Other textbooks treat each design as a separate entity, and students often fail to comprehend the relationships and concepts. All the statistics presented to this point are called point estimates of a parameter. In order to construct interval estimates of a parameter, it will be necessary to discuss some elementary ideas of probability along with some associated concepts. Without the probability concepts and procedures, it is not possible to go much further with statistical procedures. The eleventh chapter contains several elementary ideas and procedures of probability. In the next chapter, a number of frequently occurring patterns for the measurements under consideration are described. When the patterns of measurements are known, much more can be said about the results from an investigation. Knowledge of these patterns and concepts from probability are made use of in constructing interval estimates of a parameter. Additional statistical procedures for summarizing results can also be constructed utilizing this information. Many of these are used for statistical inferences. Since the question of appropriate sample size frequently arises in planning an investigation, a chapter is devoted to this topic. Since many users of this textbook will not take an additional course in Statistics, or even not purchase another book of this nature, the fourteenth chapter contains a number of frequently used statistical procedures. The concepts and results presented in previous chapters are used to present these methods. Tests of significance as summary and weight of evidence statistics are presented but hypothesis testing has been omitted. The latter concept is difficult for beginners in courses in Statistics, and it is felt that this topic should be relegated to a second course in Statistics. The last chapter deals with statistics and statistical publications of governmental and international organizations. Problem sets are given at the end of each chapter in order to illustrate the material in the text in more detail. Sample examinations with answers are given at the end of the book.

1.9 REFERENCES AND SUGGESTED READING

Brook, R. J., G. C. Arnold, R. M. Pringle, and T. H. Hassard (1986). *The Fascination of Statistics, volume 4 in Popular Statistics*, Marcel Dekker, New York, 456 pp..

Kotz, S. and D. F. Stroup (1983). *Educated Guessing, Volume 2 in Popular Statistics*, Marcel Dekker, New York, 200 pp.

Mann, N. R. (1989). *The Keys to Excellence. The Story of the Deming Philosophy, third edition*, Prestwick Books, Los Angelos, xviii + 196 pp..

Moore, D. S. (1979). *Statistics: Concepts and Controversies*, W. H. Freeman and Company, San Francisco, xviii +313 pp..

Snedecor, G. W. (1946) *Statistical Methods, 4th edition*, The Iowa State University Press, Ames, Iowa, xvi + 485 pp.. (Note that several later editions of this book have been printed.)

Tanur, J. M., Editor (1972). *Statistics: A Guide to the Unknown*, Holden-Day, San Francisco, xxiv + 430 pp..

Yule, G. Undy and M. G. Kendall (1950). *An Introduction to the Theory of Statistics, 14th edition*, Hafner Publishing Company, New York, xxiv + 701 pp..

1.10 PROBLEMS

Problem 1.1 Describe in detail ten different universes or populations with which you are associated or come in contact in everyday life. What are the sampling units? Are you in contact with all or only a sample of the elements of the population?

Problem 1.2   Obtain three examples of everyday usage of statistical procedures from the news media. Describe the use as you comprehend it.

Problem 1.3 What are the key words in this chapter? Make a diagram with connecting lines to demonstrate the relationships among the key words.

Problem 1.4 A table, shape unspecified, of unknown length is to be measured to a high degree of accuracy, say .01 inch. The only measuring instrument available to you is a long tape calibrated in feet to a high degree of accuracy. You are not allowed to recalibrate the tape, the measuring device, although this might be the reasonable thing to do. Give a design procedure using this measuring device and determining the length of the table to the desired degree of accuracy. (Do not spend long periods of time on this problem but think about it occasionally as you proceed through this book. The solution will require careful thought and some ingenuity in setting up the investigational procedure, the design, and to obtain the desired result. Many measurements may be required to attain the specified degree of accuracy.)

Problem 1.5. What was your definition of Statistics, statistics, parameter, sample, and population prior to reading Chapter 1?

CHAPTER 2. MEASUREMENTS AND VARIABILITY

2.1 INTRODUCTION

Man is continually attempting to quantify more and more characteristics about himself and his environment. This quantification is arrived at by measuring characteristics of an aggregate of individuals or items and it is expressed on a number, letter, or a word scale; our present number system was devised for this purpose. Measures describing certain characteristics about an individual or group of individuals are, for example, the number of stone axes owned by members of village X, the number of fish caught out of Cayuga Lake last year, the average quantity of milk consumed by ten year old boys in New York last year, the percent of protest rallies attended by one or more Cornell students in 1970, the number of redheads dated during one's senior year in high school, a numerical grade one receives in a freshman chemistry class, the yearly incomes by household in Tompkins County in New York, etc..

In order to obtain meaningful and consistent numbers to characterize certain aspects of a population, an appropriate **measuring device** is necessary. To have repeatable or reproducible **measurements** (numbers or symbols obtained by the measuring device), it is necessary to have a measuring device with a known and a measurable margin of error. Note that we do not say that the measuring instrument or device must be error-free, but only that the error of measurement must fall within prescribed limits. Having established the limits of error of measurement, we are then in a position to determine whether or not we can measure a characteristic of the individuals in a sample or in the population with the desired accuracy.

There are situations wherein it has not been possible to devise a method of determining the limits of error; this is true for certain areas of the Social Sciences as well as for various areas of all sciences utilizing experimentation. To overcome this deficiency, investigators may need to exhibit considerable ingenuity in devising procedures which yield repeatable measurements within the prescribed limits of error and perhaps in constructing the measuring device. An example of this situation is illustrated by Problem 1.4.

2.2 MEASURING DEVICES

Many diverse measuring devices are utilized by experimenters, teachers, administrators, consumers, etc.. We shall list some of the more common measurements and measuring devices with some comments on their use. However, before doing this let us look at the various types of measurements or responses that can arise in investigations. To do this we need to know the scale used for measurements made during the course of an investigation. The type of statistical procedure for summarizing responses will depend upon the scale. The kinds of scales used are:

Nominal scale. A measurement of a characteristic has a **nominal scale** if the measurement indicates only what class a response falls in without any ordering of the classes. A nominal scale may also be designated as a **categorical scale**, i.e., the measurement falls in a specific category. The term nominal is derived from the French word *nom* meaning name. Brands of a product, varieties of plants, breeds of animals, races of people, sex designation, organizations, and many other such groups in our Society represent examples of nominal or categorical responses.

Ordinal scale. A measurement of a characteristic has an **ordinal scale** if in addition to designating the class or category, it also indicates when one measurement has more, or less, of the characteristic than another measurement, i.e., there is an *ordering* of the classes. For example, good-better-best, gold-silver-bronze medal winners, winner-first runner up-second runner up-fourth place-fifth place in the Miss America Beauty Pageant, year in college, rank in class standing, rank of applicants for a position, and many other similar examples represent measurements on an ordinal scale. Even though ten races of people are numbered from one to ten, the scale would be nominal and not ordinal. The difference in amount between two measurements on an ordinal scale is not specified.

Interval scale. A measurement of a characteristic has an **interval scale** if in addition to the ordering of measurements, the difference in amount between two measurements is known. Dress sizes, shoe sizes, temperature measurements, intelligence quotients, and other such measurements without a specified starting point represent examples of interval scales. The size of the difference between two measurements is known. The difference between brand one and brand two is unknown as is the difference between two ranks whereas the difference between two shirt sizes is known.

Ratio scale. A measurement of a characteristic has a **ratio scale** if the ratio of two measurements is designated and is meaningful. A measurement on the ratio scale has all the properties of the previous scales plus the property that the ratio of two measurement is meaningful. For example, a 100 pound person weighs one-half as much as a 200 pound person and in the same ratio as a ten pound baby does to a 20 pound baby; a forty year old person is twice as old as a twenty year old just as a ten year old is twice as old as a five year old child. A 20 degree temperature relative to a 40 degree temperature is not in the same ratio as 60 degrees is to 120 degrees as the starting point of zero is not meaningful. Note that zero degrees temperature has different meanings on the Celsius and Fahrenheit scales.

Interval and ratio measurements may be discrete or continuous. **Continuous responses** are those which may take on *any* possible value and can vary by infinitely small amounts whereas **discrete responses** can only have specified values which are separated by intervals. The nature of the measurement and not the method of measuring determines whether the measurements are discrete or continuous. Weight measurements theoretically can take on any possible value whereas they may be read and recorded in pounds. Temperatures may have any possible value but we report them in degrees such as 20 degrees Celsius and not as 20.10000009 degrees, for example. Foot sizes are continuous whereas shoe sizes as available in stores are discrete and used as an approximation for actual foot sizes. Counts represent a good example of discrete measurements

2.2.1 Counts as Measurements

Throughout the centuries man has utilized counts to measure his wealth and worldly goods and other items of interest to him. From past records, we note that the earliest of the human race conceived of none and some. Many individuals still use this form of counting, such as for example, sick or not sick, wealthy or poor, for or against, and many other forms of dichotomous measurements. In fact, some individuals live in a dichotomous, sometimes called the black and white, world. The fact that man has two eyes, two ears, two nostrils, two arms, two legs, two kidneys, two sides of the brain, etc., may have something to do with this manner of categorizing items. Many examples of this kind of measurement include the categories all-or-none, liberal-or-conservative, racist-or-nonracist, student-or-nonstudent, mature-or-immature, drunk-or-sober, speeding or non-speeding, and so forth. At some time later, man counted to three on the joints of his finger. Then, man learned to count to ten using the fingers and thumbs of his hands. This method is still utilized by man today! Our standard arithmetic base is ten. As time went on, various number systems evolved including our present number system, and man was able to count his money, wives, children, herds of cattle, flocks of sheep and goats, fleets of ships, members in an organization, men in an army, etc.. Numbers of people in various states or sovereignities were and are important items for the governing bodies in making plans for various activities. Present-day man measures many characteristics about himself and his environment by counts, for example, pollen count, number of dresses or suits, number of courses or credit hours, number of "close" girl-friends or boy-friends, number of marriages and divorces, number of days until Christmas, number of snow-days, number of vacation days, and many other characteristics. Man is the measuring device and numbers represent the measurements. Usually there is little variation in a count but humans do make errors in counting.

2.2.2 Orders or Ranks as Measurements

The ranking of individuals with respect to a particular characteristic is a form of measurement with a long and useful history, and it is still relevant and popular in many present-day contexts. For example, a girl who has a number of boy-friends, say $b > 1$, may, and usually does, have preferences. She may rank the set of b boys she considers to be her boy-friends with respect to dancing ability, conversational ability, masculine appearance, good looks, or general over-all like-ability, where these characteristics have definite meaning in her mind and are specific to her. For any given trait, the b boys are ranked from 1 to b with allowance for ties whenever she does not rank two or more boys differently. Such rankings are useful to her in making decisions as to which boy to date for a particular activity. Likewise, most of us rank our recreational activities from lowest to highest and participate in those most interesting to and liked by us. Some may prefer a hard game of ice hockey or of field hockey, a game of bridge or chess, or simply watching Soap Operas on the Boob Tube.

In many competitive events, the contestants are ranked first (gold medal), second (silver medal), third (bronze medal), etc.. All of the events in the Olympics use this system which has been popularized by television. Contestants in beauty, race (horse, dog, auto, sailing, etc.), cooking (pie, cake, chili, etc.), picture or snapshot, sculpture, plant, decorating, advertising layouts, and many, many other types of contests or events may be ranked from first to last. Note that the actual differences between first and second, between second and third, etc. may be considerably different. For example, the majority of advertising contracts go to gold medal winners with little attention being paid to the other finishers in an event. This is a form of only two ranks, gold medal winner and non-gold medal winner. In today's world, a first-place winner gets the glory and the other finishers are also-rans which is a reversion to only counting to two.

In many cases actual counts or continuous measurements are taken but then these scores are used to rank the list of measurements. In scholastic aptitude tests, individuals are ranked in percentiles. For entrance into college, it may be that only the upper 20% are selected. For income levels, an economist may rank all individuals and then find the income for which 50% of all incomes fall below and for which 50% of all incomes above this number, that is, the median income. He may further find the values for the upper and lower quartiles or even the ten percentiles. This form of ranking has proven rather useful over the years for many situations. For example, a manufacturer or distributor may be interested in the number of hours that 90%, say, of light bulbs will last before they burn out (fail to light again). The actual number of hours for each individual bulb is not of interest except for finding the point for which 90% of the bulbs are still burning. Another test on light bulbs might be the number of times of turning a light bulb on and off that is required before the bulb burns out. Then, these measurements for a large number of bulbs could be ranked and a number like a 90%-tile could be obtained. This would give the number of times for which 90% of the bulbs could be turned on before they burned out.

The measuring instrument is generally the person performing the ranking. The ranking by an individual may be exact or subjective. In the latter case, rankings of one individual may not be repeatable by other individuals but are usually repeatable by the same person. The degree of arbitrariness or subjectiveness varies with the trait being ranked and with the individual involved. Most placements in a track event are repeatable by other individuals, despite the occasional "rhubarbs" encountered, whereas the ranking of two girl-friends with the respect to "like-ability" certainly varies with each individual concerned, fortunately for the survival of *Homo sapiens.* Despite the arbitrariness of some rankings, ranking is and will continue to be a useful form of measurement for all people.

2.2.3 Linear or Length, Square, and Cubic Measurements

Perhaps the first measuring device for lengths or linear measurements that comes to mind is a ruler, a yardstick, or a meter stick. These instruments may be calibrated in feet, inches, 1/16 inches, centimeters, millimeters, or other units of measurement. It has been implied that these units are fixed units that never vary, that is, the centimeter divisions on a ruler are all the exact same length and never vary. There is an implication that this is true even if not explicitly stated in the classroom from kindergarten through college. The idea that measuring instruments and measurements can vary receives very limited attention in our high school and college classrooms except for physics classes and even here it receives limited attention. Did you ever stop to question how much variation there is in the distance between the calibration marks on the rulers manufactured as brand X? Our experience leads us to believe that commercially manufactured rulers, yardsticks, and metersticks are calibrated accurately enough so that we do not have to worry about this in everyday life. Unfortunately when this sense of security is carried over into scientific investigations requiring very precise measurements, less than desired results may be obtained. To illustrate, let us suppose that a difference of one centimeter is of importance in an investigation. When four commercially manufactured metersticks were taken out of a package, it was noted that the 100 centimeter marks were all at the same length but that all four metersticks were different in the middle. Suppose that measurements of length 50 centimeters were of interest. Depending upon which one of the four metersticks was used, measurements of 49, 50, or 51 centimeters would have been obtained. Such variations would be intolerable for certain studies but of no consequence in most every-day life situations. Most of us would never recognize the fact that a "12-inch" ruler was actually only eleven and 31/32 inches long. A yardstick that was 35 inches long could add considerable profit for a merchandiser of cloth goods sold by the yard. It could also add dismay to a seamstress making garments as there would a shortage of material. How many of us ever bother to check the lineal measuring devices we use? Probably very few because we rely heavily on our Federal Bureau of Weights and Measurements to ferret out unscrupulous merchandisers and to have the problem corrected. With regard to heights, do any of you know how tall you are to the nearest centimeter when you arise in the morning and how tall you are just before retiring? Did you know that height measurements can vary by two inches from morning to night? One person desiring to enlist in the Marines was only five feet five inches tall during the day; he laid down and stretched out for a period of time before being officially measured by the Marines as five feet six inches tall, thereby qualifying for enlistment. Another person has official U. S. Air Force height measurements of 6 feet 2 inches and 6 feet 4 inches with the former being obtained late in the afternoon and the latter very early in the morning. Also, in measuring heights of individuals, was the height obtained with or without shoes? When using the measuring instrument, was it held vertically or was it at an angle? Heights for many purposes must be obtained under standard conditions and with accurate measuring devices. For other purposes, the requirements may not be so stringent and only height within an inch, say, is suitable. Even the latter type, despite its impreciseness, is much more descriptive of heights than such terms as midget, short, medium, tall, and "how's the weather up there you pro-basketball prospect"?

What is an inch and was an inch always an inch? We note from the dictionary that the word inch is derived from the Latin word *unica* meaning twelfth part. Historical records show that the inch has been used to measure items in the Anglo-Saxon world since the days of William the Conqueror. It was first defined as the width of a man's thumb, and later as the length of the thumb from the joint to the tip, both variable standards for determining an inch. The first real effort to establish a precise basis for the inch was made under Edward I during the last part of the thirteenth century, when it was defined as being equal to three round, dry barley corns laid end to end. Thus, if one knew what barley corns were and had three of the desired type, one would have a fairly precise standard for calibrating a ruler into 12 or 15 inches. Only recently has the inch been standardized in the precise form needed to manufacture and use present-day high-precision instruments.

In addition to the inch, other units of distance have been used through the ages. Once upon a time a "short piece" and a "far piece" sufficed as did "one day's journey". Today, "one day's journey" on foot or horseback is radically different from one by the Concord or the Space Shuttle Columbia. A term like "about this long" may be sufficient for describing the length of a fish but is useless for precise determinations of length. At one time, a cubit (length from elbow to fingertip), a foot length, a stride or pace, etc. were used to measure distances, but these units were too variable for any type of precision work and greatly bothered the early map-makers. Today we have very precise units for measuring distances. Not only do we have the English system of the inch, foot, yard, rod, mile, and league but also the metric system of micron, millimeter, centimeter, decimeter, meter, decameter, hectometer, and kilometer. A nautical system of measurements involving a fathom, cable, nautical mile, marine league, knot, and degree of a great circle of the earth is in use today.

Systems of square and cubic measures are in common use today. For land measures we have a square meter, an acre, a hectare, a square mile, etc.. The square footage of a house or yard is a popular and frequently used measure of squared distances. Cubic measures are also frequently used as, for example, the cubic footage of house is used in establishing heating and ventilating requirements. Systems of circular or angular measurements are readily available in the form of areas of a circle or the volume of a sphere or cone. As civilization became more complex and detailed, man required more diverse, varied, and precise units of measure and measuring. Methods of assessing the quality of measurements is required for most situations.

2.2.4 Volume Measurements

Many culinary artists tend to utilize very imprecise units for both dry and liquid measure such as a "pinch of salt", "whisker of paprika", "dash of salt", "sprig of mint", "dab of flour", "few drops of vanilla", and many other similarly vague terms with directions "to stir until the right consistency is achieved" or "until it feels right". This makes it impossible to reproduce an epicurian masterpiece of a chef. Fortunately for the less gifted chefs, recipe books refrain from using such terms in recipes given, for example, by Betty Crocker, Pillsbury, and Duncan Hines. We do have a precise set of dry and liquid measurements in the English system, volume and capacity measurements in the metric system, and an apothecaries liquid measuring system that allow determination of measurements with the degree of repeatability required for today's technology. The units of the system have been precisely defined and the precision of the measuring devices determined.

In the English system of liquid, we have the teaspoon, tablespoon, cup, pint, quart, gallon, barrel, and hogshead, as well as several others such as a fifth, number 2 can, etc.. In dry measure, the English system has the pint, quart, peck, and bushel as well as the British dry quart. In the metric system, the cubic millimeter, cubic centimeter, cubic decimeter, deciliter, liter, decaliter, hectoliter, and kiloliter represent the units of measure. Measuring devices for dry and liquid measurements consist of containers graduated to the desired degree of accuracy. There are international standards for volume measurements just as there are for the measurements described in the pevious sub-section.

2.2.5 Time Measurements

Early man may have been satisfied with day and night, with the four phases of the moon, or with the four seasons. These measurements of time satisfy many purposes. The sundial, the graduated candle, and the hourglass were early devices to obtain still more finely graduated measurements of time. We now have time graduations of a second, a minute, an hour, a day, a week, a month, a year, a decade, a century, and even a light year. For certain events such sports events, the times are recorded in thousands of a second. In the computer world, an event may be timed in milliseconds or even finer. Elaborate, diverse, complex, and even simple timing devices have emerged to obtain time measurements required for individuals today. A stop-watch graduated in one-hundreths of a second is common-place. We have calendar watches and clocks, transistorized timepieces, and atomic clocks. Our daily routines are affected or even controlled by clocks beginning when the radio clock alarm goes off in the morning. Airplane, bus and train schedules are based on time measurements. Passengers become upset when their mode of transportation is behind or ahead of the scheduled time. Sports events, science, business, academic institutions, etc. are dependent and even sub-servient to the timepiece in every-day activities. Certain areas of science could not function without time measurements with a case in point being the astronomers. Distances to far-off celestial bodies are measured in terms of light-years and number of years for these bodies to rotate around the sun.

2.2.6 Weight Measurements

A measuring device that is very much on the minds of weight conscious Americans is the bathroom scale, or more generally the scale, calibrated in pounds and ounces. We seemingly put a lot of trust in the scale's accuracy but do we ever bother to ascertain this characteristic for the scales we use? Why do we always weigh more on the doctor's scale than on our home bathroom scale? Is it because we do not disrobe at the doctor's office but are relatively unclothed when we take our weights the first thing in the morning? Is the difference in the weights in the way the scales are read, in the amount of clothing worn, and/or in the time of day? Can we talk about which scale gives "correct" readings until the method of weighing and the weight of an individual are precisely defined? If a person cared enough about this they could take their scale to the doctor's office and compare it with the doctor's scale.

A few years ago, a research organization checked a scale used for weighing heavy objects; the scale was found to weigh high for relatively light objects and to weigh low for relatively heavy objects. This meant that the difference in weight between heavy and medium, heavy and light, and medium and light objects was smaller than it should have been. The error in measurement of weights could have led to erroneous conclusions. A simple check would have revealed this error which had gone undetected for an unknown length of time. If a scale is utilized for precise weights with important consequences, as for instance, in certain research investigations, it should be calibrated against a known standard throughout the total range of weights employed on the scale.

When a package of meat is purchased in a grocery store, and the label states that the weight is two pounds four ounces, what is our concept of this weight?   Do you take this to be 2 lbs. 4 oz. of meat at this moment, of meat at the time of packaging, of the meat and paper, or of meat, paper and the butcher's heavy thumb?   Do you ever bother to re-weigh the meat at home or to have the butcher re-weigh the meat at the store?   It was reported recently that a particular butcher had sold over 30 pounds of paper per week at the price of steak, and the public was warned to have the meat re-weighed at the store.   One can be fairly certain that the number of overweight packages is smaller than the number that are underweight.

In addition to avoirdupois weight for the jeweler, there is apothecary weight for the pharmacist. There is also a troy weight system which has the same basic unit as avoirdupois. The units of avoirdupois are the grain, dram, ounce, and pound. The units of apothecary weight are the scruple dram, ounce, and pound.   How many of us ever use avoirdupois weight to check if our one-carat diamond actually weighs 3.086 grains = 200 milligrams?

For scientific investigations, the weighing device, whether balance, spring, electronic, or atomic, is calibrated according to the metric system.

It is necessary to ascertain the accuracy of the scales used in order that the measurements can be repeated by other scientists. Many investigations require extremely fine calibrations in weight in order to weigh, for example, a single feather from a bird, a hair from an animal, or the air in one cubic centimeter.

2.2.7 Measurements by a Judge

Another very common measuring device is the human judge; he or she serves as the measuring device for sports events, beauty contests, talent contests, taste panels, reading other measuring devices, scoring plant strains for disease infection, palatibility of food or drink, and a host of other items. One of the key criteria for a judge is the ability to discriminate and to differentiate between varying levels of the characteristics under consideration. For example, if all pies within a specified range taste the same to a person, he will be unable to discriminate between the small differences that a researcher in home economics is studying; or if all the girls involved in a beauty contest seem equally beautiful to a person, this person is useless as a judge who has to pick a winner; a referee unable to observe fouls and violations is not suitable to referee contests; if all plant strains appear to be equally infected with a disease to the judge when in fact they were not, this person's scores are useless in differentiating between the strains.

The ability to discriminate can often be sharpened with adequate training. However, some individuals may never be able to attain a high level of discrimination with regard to a particular characteristic despite considerable training. One of the key characteristics of outstanding research personnel is their ability to observe and to discriminate among the various types of evidence encountered and then to organize and to sort out the pertinent facts. Successful researchers and judges are keen observers.

2.2.8 The Questionnaire Measurements

Another very common type of measuring instrument is the questionnaire. It has many and diverse forms; its common goal is to elicit information from or about people, their activities, and their attitudes. A widely known form is the ordinary examination, with as many forms as there are instructors or persons giving the test. There may be true-false, multiple-choice, completion, matching, or discussion types of questions and various combinations of these on any given examination.

Another form of the questionnaire is the income tax form which we fill out every year for federal, state, and some city governments. This questionnaire seeks the answer to only one question, "How much income tax do you owe for the past year?".

Another type of questionnaire associated with surveys and censuses seeks to obtain information on type of dwelling, occupancy and ownership of dwelling, income and expenditures of occupants, attitudes of occupants toward various items ranging from prejudice to choice of political opponents, preference of automobiles and appliances, use of foods and products, attitudes about children or pets, ownership of cars and boats, attitudes toward government policies, and a multitude of other characteristics. A problem with these questionnaires is that they are often constructed by people who forget one simple fact, that is if the person being interviewed does not understand the question but some answer is given, the answer is as meaningless as one generated by a random or chance process. For example, suppose that we ask people to state which of three medical policies they favor and they have no choice but to select one of the three. If they know little or nothing about the three policies, we might expect about 1/3 of the people to favor each policy. If we toss a common six-sided die and record the proportion of times the events (1 or 6), (2 or 5), and (3 or 4) (or any other pairing) occur, we would expect approximately 1/3 in each group. Unfortunately, many people conclude from such results that the policies are equally favored. This could be true or it could be that the people did not understand the question. This situation may be of frequent occurrence and should be borne in mind when the results from a survey are interpreted.

Application forms may attempt to obtain information about an individual for university admission, credit rating, security clearance, admission to graduate school, or job application. Often these forms are very brief, but occasionally the inventor of forms becomes a little too enthusiastic. Does the age of a person's wife have anything to do with his being a national security risk? Probably not, because security may be granted even when this space is left blank. Forms of questionnaires are varied, and we complete one form or another almost daily. Many of you will be involved with developing questionnaires and the questions you formulate should be precise, exact, and unambiguous.

2.2.9 Chemical Measurements

Another type of measuring device is the chemical determination. Large laboratories are constructed for the sole purpose of performing chemical determinations on human, plant, animal, and mineral samples. The results are utilized in many ways. For example, the Food and Drug Administration checks on the contents and quality of foods, cosmetics, sprays, and drugs. Medical examinations utilize chemical measurements extensively to measure levels of cholesterol, triglycerides, sugar, and metals. Limits of variation are set for individual items, and the manufacturer of products and laboratory analyses must conform to these standards. Since not all people are good guys, large state and federal organizations have been set up to ascertain that John Q. Public obtains what is stated on the contents of a purchase or of a laboratory report.

Other chemical laboratories check soil samples for fertility content, milk samples for butterfat content, food samples for pesticide residues, concentration and content of drugs, concentration and content of alcoholic beverages, concentration of tars, resins, and nicotine in cigars and cigarettes, contents of cosmetics, and concentration and identity of weed seeds in crop or lawn seed. An important statistical problem in connection with all these is the design of the sampling procedure and the establishment of limits of variation that will be tolerated in the samples. For many items the statistical standards have not yet been established; there are too many items and too few statisticians. In other cases, standards have been set arbitrarily; many of these may be shown to be relatively impossible to attain when studied statistically. For example, for certain certified seed requirements, the presence of one specified noxious seed in a sample makes the entire lot of seed unsuitable for sale. In order to find such a seed in a lot and to be certain that it will appear in the sample, it would be necessary to inspect the entire lot seed by seed. This would be too expensive and time consuming for commercial seed production. Some other means, such as field inspection, must be used to eliminate the specified noxious weed seeds from the seed lot.

2.3 A CLASS SURVEY IN MEASUREMENT

In order to obtain some experience with measuring and measuring devices, and in order to have a set of measurements for class use in various ways, it is suggested that a class survey be performed. The object of the survey is to measure some characteristics of members of a class. As a part of this survey, each member of the class should supply the measurements requested in Problem 2.4. Here we have defined eye color to be one of five categories (blue, green, grey, brown, or black) and hair color to be one of four categories (red, blond, brown, or black). Age is requested in units of years and months. Class standing is defined in terms of year in college such as freshman, sophomore, junior, senior, graduate, or other. If the student is unable to fit himself or herself into one of the above categories (for example, hair color might be purple, orange, or mouse-colored), the instructor should be so notified. Perhaps a category such as "other" or "miscellaneous" should be added, but items which make up this category need to be defined.

As a second part of the survey, take a series of measurements in class. First of all, suppose that we wish to obtain the following measurements on each student present:

Name_____

Height to nearest millimeter_____mm.

Height to nearest foot_____ft.

Height to nearest yard_____yd.

Weight to nearest pound_____lbs.

Eye color (circle one) blue green grey brown black

Hair color (circle one) red blond brown black

At each of two stations (A and B) in the class, select a team of three students. Members of the team at station A could be the three shortest class members, and at station B the three tallest. The reason for height differences will be explained later. Have one member of each team be the recorder of the measurements on one of two forms, A and B, on which a student has written her/his name. Form A is to be used for station A measurements and form B for station B. A second member of each team is to measure height measurements in millimeters, to determine eye and hair color, and to read weights on a bathroom scale. The third member of each team is to measure the height of a student to the nearest foot and nearest yard. Measurements on the six members of the two teams should be obtained first in order that they are not missed; then the remaining students are measured as they file past at each of the two stations. Students should complete Problem 2.5 during the time that measurements are being obtained on the other students. The measuring devices required are two straight eight-foot dowels one inch in diameter and calibrated in yards, two eight-foot dowels calibrated in feet, two eight-foot one inch by two inch boards on each of which two metersticks have been mounted end-to-end, and one bathroom scale.

Before proceeding with the measurements, it should be decided how height and weight measurements are to be taken. Shall the students be measured with or without shoes? If without shoes, it is suggested that the student stand on sheets of paper. With regard to the height measurement in yards, all persons in most classes should have been recorded as two yards tall unless a recording or judgement error was made or unless very short or very tall people are in the class. People under four feet in height should be classified as one yard and people over seven feet should be classified as three yards tall. No one in the class measurements given in Chapter 9 was in either of these categories. There should be no variation in the heights of most classes measured to the nearest yard.

For most classes, which usually do not contain a dwarf or a giant, there should be two groups of measurements recorded in feet; these should be five and six feet. If any other measurement appears, we would suspect a recording error or an incorrect reading of the measuring device. Thus, measured in feet, all members of a class are either five or six feet tall. There is variation in heights when measured in feet but there is none when measured in yards. Also, we should not expect the different people performing the measurements to agree on all heights, as individuals near five feet six inches tall may be classified as five or as six feet tall.

In Problem 2.4 the student is to record height in inches; it is not suggested that heights be obtained in inches and a check be made on the student's version. Instead, consider the heights obtained in millimeters. Without summarizing the data obtained, one would guess that all heights should fall between 1450 and 1950 millimeters.. There will be many categories of height. Very few individuals now are recorded as having the same stature, and the variation in units of height is considerable. Furthermore, if we compared two heights recorded by the same individual and also two heights by two different individuals, we note that they would not always agree. In fact, they probably will agree less frequently than they disagree. This is due to the fine calibration and to the fact that the measured heights of an individual will vary as the subject's posture varies from measurement to measurement; also, the individual performing the measurement will not always read the measuring instrument in the same manner. There is a degree of subjectiveness in reading the meterstick when measuring the height of an individual. We could do better and have less variation in duplicate measurements if a flat surface such as the length of a table were being measured. With a little care, there should be little variation in duplicate measurements or between people performing the measurement. However, if the measurements were taken to the nearest one-half, one-fourth, or one-tenth of a millimeter variation would again occur. There is a limit of repeatability for any measurement; if the unit is fine enough there will be variation in measurements by the same and by different individuals.

The same problem will be encountered with the measurement of weights. The weights given by individual students are obtained from a variety of scales and a variety of conditions. The weights recorded in class are obtained on a single bathroom scale, read by a single individual, and can be checked by a second individual to minimize reading and recording errors. The variation or measurement error in the weight of an individual is influenced by such things as time of day in relation to eating, amount of clothes worn, method and correctness of reading a scale, the particular scale used, and the correctness and legibility in recording the weight. An individual's weight may vary by five pounds or more from the time just before breakfast until just after the last meal of the day even though the same clothes are worn and the same scale used.

Probably the most subjective measurement recorded will be eye and hair color; despite the few categories involved, the variation between the human measuring devices involved in the classification will be noticeable. When the student's age and weight are guessed, there would be even more variation and subjectiveness in these recorded characteristics. We shall utilize the data obtained in various ways as we proceed through the book.

2.4 STANDARDIZATION OF MEASURING DEVICES

The need for standardization is illustrated by an example. Four metersticks were purchased and arrived in a box which originally contained 12 metersticks. Two of the metersticks differed from the other two by more than one centimeter in the calibration marks. The metersticks carried the same brand name and lot number. This points up the fact that whenever a new measuring device is utilized it should be checked against a standard; the standard should have known accuracy. A measuring device with unknown accuracy may be useless for the purposes at hand. If a standardized meterstick were available, the newly purchased metersticks could be checked against it.

As was pointed out earlier, an inch was not always an inch. The United

States Congress in 1866 pronounced the international prototype meter bar in Paris as equal to 39.37 inches. A few years later, a platinum-iridium meter bar became the official standard for all United States measures of length. By 1940, the National Bureau of Standards was able to define an inch accurately to one-millionth of its length, but even this did not prove accurate enough; there was a difference in the British and the United States inch in that the former was equal to 2.539995 centimeters and the latter to 2.540005 centimeters. This difference of 0.00001 centimeter greatly affected the manufacture and use of precision instruments during World War II. In 1960, the official inch for both Britain and the United States was set equal to 2.54 centimeters exactly, and an international agreement made the meter "equal to 1,650,763.73 wave lengths in vacuum of the radiation corresponding to the transition between the energy levels of 2_{p10} and 5_{d5} of the atom Krypton 86". Thus, much of the confusion in linear measurements appears to be taken care of for the time being.

Standards have also been developed for gravity, time, temperature, voltage, weight, angular, and frequency measurements, and prototypes of almost all standards are kept in France, Britain, or the United States; it is therefore possible to obtain measuring devices of the desired calibration and accuracy. Copies of the prototypes have been widely dispersed for use in making measuring devices.

Measuring devices should be checked against a standard when first used and occasionally thereafter, since it is difficult or impossible to do accurate work with an incorrect device. Scales should be checked for accuracy throughout their usable range. Human judges should be checked for discriminatory power and for level of discrimination. Questionnaires should be pre-tested in order to eliminate ambiguities and lack of clarity. Unfortunately, instructors do not find it convenient to pre-test quizzes and examinations prior to use; occasionally some poorly designed questions appear on an examination. Chemical and physical procedures should be checked when first initiated and occasionally thereafter, in order to ascertain that the process remains accurate within the prescribed levels; procedures that are usefully accurate for one type of material may be inaccurate for a second kind of material.

Whenever possible, measuring devices should be calibrated against known calibrated standards and they should be recalibrated at intervals. Other types or devices can be checked by including duplicate samples and samples of known content along with unknown samples. However, one must be careful not to let this checking take an inordinate amount of time. Too much time spent in checking can be a gross waste of effort and can result in inefficient use of resources.

2.5 VARIABILITY

We have noted in Section 2.3 that variability is present in our measurements. When the unit of measurement is fine enough, we find variation between two duplicate measurements of height of a single individual even though the measurements might be made only minutes apart and by the same person. Variation occurs between the height measurements made by two different individuals. Even more variation is found between heights of the members of this class, as there is considerable variation due to causes other than measurement errors in our recorded heights. In fact, we need only look around us to note that two individuals who are "alike as two peas in a pod" are very difficult to find, especially since there are no identical twins in most classes. Variation is universal in characteristics of all populations. We live in a variable world. This is good because it has allowed the human race and individuals of other populations to survive. Since variability is universal, we must learn to live with it and to design experimental and survey investigations in such a way as to overcome its effects. There are several ways of accomplishing this, as will be described in Chapters 5, 6, 7, and 13. Some types of variability occur in an organized fashion (Chapter 12); as a result it is possible to devise statistical procedures for summarizing the sample facts and for making inferences about the population facts. The procedures are so many, varied, and detailed that numerous statistical books have been and are being written on the subject. The more one knows about the type and nature of variability, the more useful are the sample facts to make inferences about population parameters.

2.6 BIAS IN MEASUREMENT

An unstated tenet in the collection of numbers obtained from utilizing a measuring device is that the sum of errors in a positive direction is about equal to the sum of the negative errors of measurement; over a large number of trials it would be expected that the errors sum to zero or nearly so. Suppose that this is not the case, and that the magnitude of inaccuracies in one direction exceeded those in the other. The nature of this type of discrepancy is termed a systematic error or more commonly a **bias**. The reason for selecting a short person and a tall person for making the measurements described in Section 2.3 was to introduce an inaccuracy of this nature. Because the short person has to look up it might be assumed that he or she would record heights higher than would a taller person. Then, if on the average, short investigator A always reads the measuring device one unit higher than does tall investigator B, the bias of A compared to B is +1, and the bias of B compared to A is -1. Note we did not state which, if either, of the two investigators takes correct measurements in the sense that if they measured all individuals in the population they would obtain the population parameter for the characteristic measured.

2.7 ERROR IN MEASUREMENT

As we have seen, the causes of variation in measurement are many and varied. Barry (1964), Chapter 2, and Wilson (1952), Sections 9.1, 9.5, 9.8, describe some of these. There are systematic errors or biases, personal errors, mistakes, and errors due to assignable causes. In addition, variation in measurements may be caused by unassignable causes due to the cumulative effect of a number of uncontrolled and often unknown minor variables. If the magnitude and sequence of these variations are completely unpredictable and nonsystematic, that is, they form a random sequence, we denote them as random variations or random errors; the sum of the random errors over all individuals in a population should be zero.

A single measurement excluding mistakes or blunders may be written as

total variation = assignable causes + bias + random error.

The **error of measurement** is often defined as

error of measurement = bias + random error.

An **assignable cause** is a known factor or characteristic that causes differences in measurements. Consider height as an example. Differences in height measurements are due to age, sex, race, and nutrition level. These are all assignable causes of variation in height measurements. A **random error** is variation in measurements for which there are no assignable causes. In taking height measurements to the nearest one-tenth of a millimeter on a single individual and with a single measuring device, all measurements will not be the same; that is, variation is occurring. No assignable cause explains this variation. A bias is an assignable cause. Quite often the bias term is ignored when in fact it may be the larger factor in the error of measurement.

If differences in measurements between individuals in the population rather than the individual measurements are utilized, a constant bias term will disappear, that is, add to zero. Thus, the method of using measurements may affect the effect of the bias term; the investigator must be aware of this in choosing a method. In order for a measuring device to be useful, some measure of its reliability and of its accuracy should be available.

2.8 ACCURACY VERSUS PRECISION

In statistical terminology, the term **precision** refers to the repeatability of a measurement. Low precision means that there is wide variation in repeated measurements on the same object, whereas high precision means that there is little variation between repeated measurements. **Accuracy**, on the other hand, refers to the size of the bias term plus the precision. Thus, low accuracy could result from a large bias term with either low or high precision, or from a zero bias term coupled with low precision, as illustrated in Figure 2.1. High accuracy results when the bias term is small or zero and precision is high.

Perhaps the best illustration is to consider the drawings in Figure 2.1 as targets for darts, arrows, air pellets, or bullets. A rifle producing a pattern like the one on the upper right hand target when fired from a rigid fixed support would be considered highly accurate. If a similar pattern were obtained by a person holding and firing the rifle, we would say that this person is an excellent (highly accurate) shot and that the rifle is also highly accurate, provided, of course, that the person was not correcting for a bias in the rifle. A dart thrower who could produce the same pattern would also be considered to highly accurate. These ideas of accuracy and precision will be used in connection with estimates of parameters. We would like estimates that concentrate closely around the parameter value.

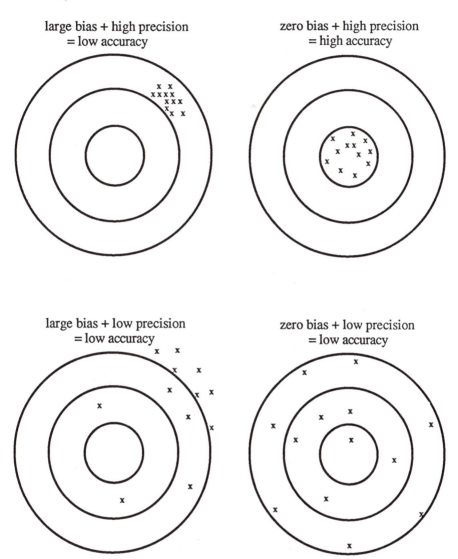

Figure 2.1. Graphical representation of precision and accuracy.

2.9 REFERENCES AND SUGGESTED READING

The three books listed below are easy to read and elementary in orientation; they are recommended reading:

Barry, B. A. (1964). *Engineering Measurements.* John Wiley and Sons, Inc., New York, London, and Sydney, x + 136.pp.

Wilson, E. B. (1962). *An Introduction to Scientific Research.* McGraw-Hill Book Company, Inc., New York, Toronto, and London. (Sections 4.3, 4.5, 4.6, 6.4, 7.1, 7.2, 7.7, 9.1, 9.5 and Chapter 5.), x + 373 pp.

Youden W. J. (1962). *Experimentation and Measurement.* National Science Teachers Association, Washington, D. C., 127.pp.

The following two references are for those who wish to pursue further the philosophy of measurement:

Campbell, N. R. (1928). *An Account of the Principles of Measurement and Calculations.* Longmans, Green, London.

Ellis, B. (1966). *Basic Concepts of Measurement.* Cambridge University Press, London, 220.pp.

An intriguing article on psychological testing and measurement is:

Anastasi, A. (1967). Psychology, psychologists, and psychological testing. *American Psychologist* 22:297-306.

2.10 PROBLEMS

Problem 2.1. Select a measuring device used in one of your laboratory classes and a set of measurements made on it; describe the types of errors of measurement and methods of eliminating or controlling these errors.

Problem 2.2. Construct a measuring device and calibrate it. Determine the limits of error. (See Youden, 1962, pp. 89, 114.)

Problem 2.3. Using an appropriate measuring device such as a micrometer or vernier caliper, measure the thickness of pp. 1-40, 41-80, 81-120, 121-

160, and 161-200 of this book, each 10 times. How would one estimate the thickness of a single page? Would the above method be more precise for estimating average page thickness than to measure one page 2000 times? Why or why not?

Problem 2.4. Supply the following data for yourself:

Name_____
Height to nearest inch_____(ins)
Weight to nearest pound_____(lbs)
Age to nearest month (as of Feb. 1 of the current
year)_____(yrs) _____(mos)
 Eye color (circle one) blue green grey brown black
 Natural hair color (circle one) red blond brown black
 Class standing (circle one) Freshman Sophomore Junior Senior
                           Graduate Other

Problem 2.5. Comment on the class survey with respect to the efficiency and conduct of the investigation; comment on the accuracy of taking the measurements and on ways in which the accuracy might have been affected.

Problem 2.6. What were your concepts of accuracy, precision, variability, and measurement error before reading Chapter 2?

Problem 2.7. Prepare a list of key words for this chapter and discuss relationships among them.

Problem 2.8. Discuss the various measurements obtained in Sub-sections 2.2.1 to 2.2.9 with respect to scales of measurement and to continuity.

Problem 2.9. Give five examples of the four scales of measurement as found in newspapers or magazines. Are the measurements discrete or continuous?

Problem 2.10. Describe the measurements of Problem 2.9 with respect to assignable and non-assignable causes of variation and with respect to precision and accuracy.

CHAPTER 3. DATA COLLECTION

3.1 INTRODUCTION

A collection of numbers, no matter how large, may contain no useful information. Suppose that one were to collect all the phone numbers of all Cornell University students for the given year. This would represent a lot of numbers. If no names were attached, if the phone numbers were from ten years ago, and if the sole purpose was only to obtain a large array of numbers, this collection of numbers would contain no useful information. If the number-collector was interested in height and weight measurement profiles for these students and had collected them, these numbers would contain the desired information. Under such circumstances, the numbers contain factual information and are designated as **data**. A **datum** (the singular form of data) is defined to be a fact from which a conclusion can legitimately be drawn. The distinction between data and just numbers is frequently ignored, and investigators make unwarranted assumptions about and conclusions from the numbers. This points up the necessity for appropriate planning for data collection. Voluminous sets of numbers can easily be collected; for example a computer may be used to generate numbers by the "side-foot" of computer paper, where a side-foot is defined to be a stack of computer paper one foot high; it refers to the thickness and not to the length or width of the computer paper.

As an illustration of obtaining numbers by the side-foot of computer paper, operations are performed on animals and humans and the progress of a patient is monitored via a computer. Many characteristics such as temperature, pulse rate, blood pressure, respiratory rate, and several others may be recorded each second or minute over a period of several days or weeks. This generates a vast amount of numbers and computer paper by the side-foot. The medical personnel most likely only require measurements about once an hour or even once a day. Also, the time interval for measurements could be short just after the operation and then lengthened as the patient improves.

There are several aspects to be considered in data collection. These are:

(a) Why should data be collected?
(b) What data should be collected?
(c) How are the data to be collected?
(d) Where and when are the data to be collected?
(e) Who will collect the data?
(f) Who will write a complete description of the data collected?
(g) Will the data be stored or discarded at the end of the study?

The above items are discussed in this chapter and are put in a more formal context under the Principles of Scientific Investigation in Chapter 4. A discussion of phony statistics with examples concludes the present chapter. This discussion is presented to illustrate a frequently occurring and undesirable situation. In general, we attempt to emphasize the positive side for collecting numbers containing useful information. The discussion of phony statistics represents a negative approach to data collection as they demonstrate what not to do. Except for describing phony graphical representations in Chapter 9, we do not dwell on phony statistics and procedures. Readers interested in additional examples and discussion of phony statistics are referred to Bross (1953), Huff (1954), *The Journal of Irreproducible Results*, Seilaff (1963), Reichmann (1964), Roll and Cantril (1972), Campbell (1974), Reichard (1974), and Wheeler (1976).

3.2 WHY COLLECT DATA?

As indicated in the previous section; there should be a valid reason for collecting data on a given characteristic. For example, in the class survey described in the previous chapter, data are to be collected on various characteristics of students. The reasons for collecting these data would be to obtain a data set with which the students are familiar, to introduce the idea of measuring devices and measurements, to illustrate errors, variability and bias in measurements, and to obtain a set of data for purposes of illustrating methodology.

Although it is imperative to collect the data required for drawing conclusions on the items of interest, it usually is not a good idea to collect data just because it is easy or because someone, somewhere, sometime may use them, even though data of this nature occasionally turn out to be exactly what is needed at the end of or in a later experimental investigation or survey. The latter possibility is of infrequent occurrence in well-thought-out investigations. Any investigation should have specific aims and objectives. These should be clearly defined and understood by all involved in the investigation.

3.3 WHAT DATA ARE TO BE COLLECTED?

Before starting an investigation, it is desirable to determine and describe the nature and characteristics of the data to be collected. The goals and objectives of the investigation should be clearly and precisely defined and understood. For the survey described in Section 2.3, it was decided to collect height, age, weight, eye color, hair color, and class standing measurements for each student. The classes or categories for eye and hair color and class standing were designated. Although the definition of these categories or classes was simple for the survey, this is not always the case. Categories must be made as distinct as possible and free from ambiguity. Wilson (1952), sections 7.1 and 7.2, discusses class definitions and gives examples.

In setting up weight classes, it is necessary to determine if weight is desired to the nearest pound, to the nearest kilogram, or to the nearest five pounds. Likewise, will the measurements be in the English or the metric system? The purposes of the study will indicate the fineness required of measurements. Some studies use too coarse a grouping while others use too fine a grouping. Careful consideration of the needs of a study will correct this. In medical studies, the recurrence or non-recurrence of a disease may be all that is recorded. The *time to recurrence* may be a much more informative measure of a procedure than simply whether or not a disease recurs. Although the actual disease titer of blood samples is measured, the laboratory analyst usually observes only whether or not the titer value exceeded a threshold level and then simply records diseased or not diseased or sick or not sick. It would be much more informative for many situations to have the actual value recorded.

3.4 HOW ARE DATA TO BE COLLECTED?

In determining the manner of collecting data, it is necessary to consider the conditions under which an investigation is performed. For example, if the class survey were to be performed out-of-doors during a snowstorm, it would have be difficult to get students to take off their shoes. Since there should be no such problem in a lecture room, height measurements could be standardized to some extent by taking all measurements without shoes.

The unit of measurement (yards, feet, or millimeters) was predetermined. Measuring devices, which were eight foot dowels marked in yards or in feet or which were eight foot by two inch by one inch boards on which two metersticks were screwed end to end were selected, as were the persons doing the measuring and recording. The form on which the data were to be recorded and the order of the data were determined. The results of one class survey are organized and presented in Chapter 9; the data are utilized for various illustrative purposes.

Wilson (1952, section 6.2) discusses and gives examples of the form and manner of recording data. Careless record-keeping can considerably decrease the value of data or even make them useless. Permanent strongly-bound notebooks are recommended for recording laboratory and other measurements. Data should be entered in the notebook at the time of observation, preferably in permanent ink. Recopying of data allows yet another chance for recording errors to appear. Photocopying the data sheets would circumvent the possibility of recopying errors. It is preferable to use ink or a ball point pen in the event that the notebook is used in legal cases as evidence. This is also necessary if the notebook is used extensively, as penciled figures tend to smear and rub off. The name of the observer and the period covered for the observations should be recorded in permanent ink. Each set of observations should be dated and described. If more than one person uses a book, each person should initial the material entered by him. The values for the observations should not be crowded on the page. Paper is cheap compared to having illegible data. Neat and legible recording is a desirable goal in data collection. A safe and permanent place should be selected for storing the notebook when not in use. If the data are extremely valuable and hard to replace, it may be desirable to store a photocopy or ozalid copy of the properly identified data sheet in a fireproof safe. Several instances have occurred when a fire destroyed all of a researcher's data.

As stated in the previous section, the observer must first determine what observations are to be made. Having made this sometimes difficult decision, he is in a position to determine the order of appearance of the characteristics in the notebook and the order of taking the observations, which may be very important in certain investigations. Time can be saved and accuracy of recording can be improved by consideration of the order of taking the observations and proper orientation of the form for recording. For example, let us suppose that a florist has 20 pots of flowers of variety A, and on these he is counting the number of saleable flowers produced each day. The 20 pots are arranged in two rows, one row on each side of a greenhouse bench. Suppose a notebook has been prepared with the numbers appearing in serial order from 1 to 20. Consider the following three numbering systems for the pots:

Numbering system 1

1	2	3	4	5	6	7	8	9	10
20	19	18	17	16	15	14	13	12	11

Numbering system 2

1	2	3	4	5	6	7	8	9	10
11	12	13	14	15	16	17	18	19	20

Numbering system 3

16	17	18	19	20	1	2	3	4	5
15	14	13	12	11	10	9	8	7	6

With numbering system 2, it is necessary to walk back to the beginning of the bench each time the observation for pot 11 is recorded. For small investigations this may not be important, but for investigations covering a relatively large area, the proper numbering system could become quite important as a time-saving device. Numbering system 3 could be more efficient in saving time than system 1 if the point of starting the observations were in the middle of the bench rather than at the end of the bench.

    The ordering of the questions on a questionnaire is important for continuity of thought by the person being questioned. It is highly desirable to order the questions and to categorize the answers in such a way that summarization of the data from a questionnaire is a relatively simple task. If the data are to be entered into a computer, it is wise to pre-code the questionnaire rather than to code each question individually after completion of the questionnaire. This saves an enormous amount of time and reduces the chances for error. As was pointed out earlier, the ordinary class examination is a form of a questionnaire. For example, true-false, completion, and matching questions are easy to grade on an examination. Discussion questions are difficult and time-consuming to grade, but they may be the only way to obtain the information desired.

Significant digits represent the number of figures retained in a measurement or observation. For example, the numbers 137, 137000, .137, .0000137, etc. have the same three significant digits, i.e., 137. The number of significant figures to be retained for observations should be determined prior to conducting the investigation, as difficulties in summarization arise when different numbers of significant digits have been retained. Usually, sufficient significant digits should be retained to have at least a range of 30 or more units in the sample in order that rounding errors in measurements have small effects on the statistics computed from the data.

Ordinarily the problem of how to measure or scale a characteristic poses little or no difficulty. There are situations in which this is not the case. For example, how does one go about measuring devotion, fidelity, racism, lawfulness, prejudice, belief, and so forth? In another situation, an investigator was testing chemicals to determine their efficiency in breaking dormancy of alfalfa seeds. He applied six different chemicals to dormant seeds and then tested the seeds in an experiment. He determined the number of treated seed lots that were not dormant, 10% dormant, 20% dormant, 30% dormant, 40% dormant, 50 % dormant, 60% dormant, 70% dormant, 80% dormant, 90% dormant, and 100% dormant. In addition, he determined the number of seeds that were killed by the chemical. There is no problem with percentages of dormant seeds. The problem arises in determining the distance between 100% dormant seeds and dead seeds. One could, as the investigator first did, rate dead seeds as 110%, but it is also logical to consider this distance as infinite since the dead seeds will never grow but the dormant ones will grow after the dormant period is over.

3.5 WHERE ARE DATA TO BE COLLECTED?

In some investigations there is only one possible place where the investigation can be conducted; consequently, there is no problem involved in making the choice between places. However, there will still be a necessity for determining the time for the investigation to take place; other groups or other investigators may need to utilize the space for other purposes. For our class survey, we determined that the survey would take place on a given day of a given month and year, at a specified time, and in a specified classroom. At any other hour, it would be difficult to assemble as many people as there are in a class in one place.

In other surveys, there may often be vast numbers of places where the survey could take place. For example, suppose that we wish to take a survey of 500 striking transit workers in the city of New York; workers could be interviewed at home, on picket lines, in meeting halls, or a variety of other places. If the decision were made to interview them in their homes between the hours of 10:00 a.m. and 8:00 p.m., it would further need to be determined whether other members of their family or other occupants of the building would be allowed to be present during the interview and what is to be done about refusals, about not-at-homes, about another person such as a union official who insists on being present during the interview, etc..

In sampling foods for determining pesticide residue, we need to determine the time, place, and material to be sampled and the laboratory or laboratories making the chemical determinations. In judging food products, as, for instance, cherry pies, the time and place, as well as the conditions, need to be explicitly stated in order to avoid difficulties in scheduling the investigation. The judges need to be selected and tested for their discriminating ability.

A record of the time and place of the investigation and a complete description of the items involved in the investigation and how they were collected are necessary for future use of the data. These should be recorded immediately and in an acceptable form. It is fatal to trust to memory because there are lapses for all humans.

3.6 WHO COLLECTS THE DATA?

If a number of people are involved in the collection of data, it is necessary to determine precisely which person is to record specific observations. This is required so that observations are not inadvertently omitted or duplicated. If possible, all measurements or observations of one type should be made by one person and/or measuring device, in order to standardize the results. Note that this was suggested for the class survey. There are differences sometimes small and sometimes large, between measuring devices and individuals doing the measuring; it is important to eliminate this source of variation from our data if at all possible as the more variation that can be eliminated due to assignable causes the less will be the variation among observations.

Clarity and preciseness of directions to technicians and associates are of utmost importance in minimizing confusion and in ascertaining that the desired data will be collected. It is best to have written copies of directions prepared well in advance of the investigation. The launching of a missile is an illustration of teamwork resulting from carefully prepared and rehearsed directions. In order for the launch to be successful, all members of the team must perform their work. If a technician fails to remove a plastic cap for the fuel lines, the missile may not be launched successfully. Likewise, if the designated person fails to water the plants in a greenhouse or to feed laboratory animals over a long week-end, the plants and animals may be dead or in very poor condition. Even when a reliable person has been designated to perform an important task, it is well to have a means of checking on him in most investigations.

An example of differences between individuals is given to illustrate this point. An investigator needed to go to a scientific meeting for four days but he was in the middle of an important experiment. Since he needed to go, he decided to have his laboratory technician take the necessary observations from the experiment. When the data were being summarized, the statistical analyst noticed that the observations for these four days were not at all like those for the rest of the experiment. He brought this to the attention of the investigator and they noted that these were the four days when the technician recorded the observations. The technician was not making the observations in the same way as the investigator.

3.7 COMPLETE DESCRIPTION OF COLLECTED DATA

As we have indicated throughout the preceding sections, it is absolutely necessary to have a complete description of the data available in a permanent and available form. If ten-year-old telephone numbers from 100 students from a given class together with associated height and weight measurements were available and if these numbers were unlabeled, they would be nothing but an aggregate of numbers. Any set or data can be rendered useless if they are not completely and correctly labeled and described. The best time to do this is immediately after the numbers have been obtained.

Record sheets must be sufficiently well identified to reorder them properly should they be placed in the wrong order. If the entire set of data is labeled and described adequately, but if individual record sheets are not, this could cause considerable difficulty. For example if an investigator were conducting two surveys simultaneously, unlabeled pages of individual questionnaires could become mixed, and it may be impossible to separate them if they are alike in all other respects; this could render the results of both surveys useless for the purpose at hand.

If the names of students were removed or omitted from examinations and problems, the class average would stay the same; however, if we allotted grades in an arbitrary manner to the student names, the grade associated with an individual would most likely change. This procedure would not lead to happy students because some students deserving high grades would receive low grades and *vice versa.*

Examinations and problems should be identified with one's name on every sheet; if the sheets are stapled together, a name on the front or back page would suffice.

3.8 DISPOSAL AND STORAGE OF DATA

After data have been collected, it is necessary to plan for their storage until the next process in handling them is started. Also, after the data have been utilized and the study is complete, it has to be decided what is to be done with the set of data. If all information has been gleaned from the study, the data should be destroyed. Data which have no further use should not be allowed to occupy valuable storage space.

If the data are to be saved for future use, one should be certain that they are stored in a safe, dry, fireproof place. Data sheets and graphs should be tightly packed to avoid curling and crinkling. Notebooks should be stored correctly to keep them from warping. Exposure to moisture may ruin the data set.

3.9 PHONY STATISTICS

Numbers and conclusions drawn from numbers are commonplace in the printed media of our Society. In order to evaluate the soundness of the conclusions, it is necessary for the reader to develop a constructively critical and questioning attitude toward the manner in which the numbers are obtained and toward the manner in which they are used. If valid and meaningful conclusions are to be drawn from numbers, the numbers should be obtained from a carefully planned and executed investigation. The reader should know when to consider a media report seriously and when not to. Readers should have little difficulty in making that decision about the conclusions cited in the following article (Reprinted with the permission of G. H. Scheer, Publisher, from *The Journal of Irreproducible Results*):

PICKLES AND HUMBUG
(A bit of comparative logic)

Pickles will kill you! Every pickle you eat brings you nearer to death. Amazingly, "the thinking man" has failed to grasp the terrifying significance of the term "in a pickle". Although leading horticulturalists have long known that *Cucumis sativa* possesses indehiscent pepo, the pickle industry continues to expand.

Pickles are associated with all major diseases of the body. Eating them breeds wars and Communism. They can be related to most airline tragedies. Auto accidents are caused by pickles. There exists a positive relationship between crime waves and consumption of this fruit of the cucurbit family. For example:

1. Nearly all sick people have eaten pickles. The effects are cumulative.
2. 99.9% of all people who die from cancer have eaten pickles.
3. 100% of all soldiers have eaten pickles.
4. 96.8% of all Red sympathizers have eaten pickles.
5. 99.7% of the people involved in air and auto accidents ate pickles within 14 days preceding the accident.
6. 93.1% of juvenile delinquents come from homes where pickles are served frequently.

Evidence points to the long-term effects of pickle-eating.

Of the people born in 1839 who later dined on pickles, there has been a 100% mortality.

All pickle eaters born between 1849 and 1859 have wrinkled skin, have lost most of their teeth, have brittle bones and failing eye-sight,--if the ills of eating pickles have not already caused their death.

Even more convincing is the report of a noted team of medical specialists: rats force-fed with 20 pounds of pickles per day for 30 days developed bulging abdomens. Their appetites for <u>wholesome food</u> were destroyed.

The only way to avoid the deleterious effects of pickle-eating is to change the eating habits. Eat orchid petal soup. Practically no one has any problems from eating orchid petal soup.

(Anonymous)

Such a report should be for amusement purposes only. Although many of the individual statements are true, they are useless. There are, however, situations for which the value and credibility of the report cannot be assessed easily, and it is here that the aids described in this section are helpful.

In the 1960 presidential election, John F. Kennedy stated that 25 million people in the United States go to bed hungry. In commenting on his statement, *Time* magazine stated that in their opinion 44 million people went to bed hungry every night because they were all dieting! This facetious remark illustrates the fact that such statistics have no basis of fact and cannot be substantiated. Our world is full of phony statistics. For example, when the little boy told of seeing a truck 40 times as large as a house, he was remonstrated for exaggeration by his teacher who said "Johnny, if I told you once, I told you a million times not to exaggerate". No data were involved in obtaining these phony statistics. A **phony statistic** is defined to be a statistic which has no basis of fact and/or whose meaning is ambiguous or unclear.

The first paper in Sielaff's (1963) book is one written by Daniel Seligman and is entitled "We're drowning in phony statistics", and it contains a wealth of phony statistics examples; it first appeared in the November, 1961 issue of *Fortune*. Mr. Seligman divides phony statistics into two categories, one which he calls Meaningless Statistics and the second Unknowable Statistics. He defines a **Meaningless Statistic** as one wherein a number or phrase is associated with an undefined term or ambiguous thought in such a way that it is unclear what is being added together to arrive at the figure. For example (from Seligman), New York's Mayor Robert F. Wagner announced in his annual report to the City Council that "over-all cleanliness of the streets had risen to 85 percent in 1960", that it was up from 56 percent in 1955, and that this figure was based on "personal inspections and evaluation of 4,493 city blocks". Mr. Seligman tried to ascertain what objective criteria were used to determine whether a street was clean or not. He met with no success, and one wonders whether the "personal inspection" of 4,493 city blocks would not have kept the good mayor so busy that he would have found no time to do anything else! Note the undefined terms such as over-all cleanliness, personal evaluation, and personal inspection. Also the base for the 85% figure is undefined.

An **Unknowable Statistic** is one whose meaning may be perfectly clear, but is derived from an alleged fact that no one could possibly know. Mr. Seligman gives the following as an example of this type. In the October, 1958 issue of *This Week* magazine, in a contribution by Dr. Joyce Brothers, it was stated that "the American girl kisses an average of seventy-nine men before getting married". Again Mr. Seligman attempted to obtain the source of this information, but his letter was never answered. Although the above statistic is of the Unknowable variety, it could also be of the Meaningless type in that it is not quite clear what is meant by men kissed. Are fathers, grandfathers, uncles, and other relatives included? Presumably she kisses men and not "an average".

Mr. Seligman, in giving examples of various loss statistics, finds that one quarter of the Gross National Product (G.N.P.) is being "frittered away". He states that enough loss statistics could easily be accumulated "to show" that losses in the United States exceed the G.N.P., but that anyone compiling such a list would run a considerable risk, since "there are a lot of people around who would take it seriously"!

The example in Table 3.1 is in the same vein as Mr. Seligman's interesting examples. It was found by a student answering Problem 3.1; the original author is unknown.

Over the years, several books have been written to discuss the abuse and misuse of statistics and numbers. Some of these are Campbell (1974), Haack (1979), Huff (1954), Moore (1979), Moroney (1951), Reichard (1974), Reichmann (1964), Roll and Cantril (1972), Seilaff (1963), and Wheeler (1976). These books cite many and diverse examples of phony statistics and procedures. Following Seilaff (1963) and Campbell (1974), phony statistics may be categorized as follows:

a. Meaningless statistic
b. Unknowable statistic
c. Eccentric theory

Table 3.1. 1967 POPULATION BALANCE SHEET
OR
WHO'S TO DO THE WORK???

Population of United States 198,000,000
People 65 years or over 55,000,000

Balance left to do the work 143,000,000
People 21 years or under 58,000,000

Balance left to do the work 85,000,000
People working for the government 35,000,000

Balance left to do the work 50,000,000
People on relief and Appalachian Program 24,000,000

Balance left to do the work 26,000,000
People in the Armed Forces 11,000,000

Balance left to do the work 15,000,000
People in City or State Government 12,800,000

Balance left to do the work 2,200,000
Bums and others who never work 2,000,000

Balance left to do the work 200,000
People in hospitals or asylums   126,000

Balance left to do the work 74,000
People in jail 73,998

* Balance left to do the work 2

* Two? Why, that's you and me! Say!! Then you'd better get a wiggle
on 'cause I'm getting awfully tired of running this country alone!!

-Anonymous

    d.   Preposterous statistic
    e.   Uncritical trend statistic
    f.   Dubious cluster statistic

Consider the following examples of meaningless statistics where the undefined or ambiguous words or phrase appear in italics:

President J. F. Kennedy in a State of the Union message stated, "Twenty-five million Americans are living in *substandard* homes".

Attorney General R. F. Kennedy stated, "ninety percent of the *major racketeers* would be *out of business* by the end of the year if the *ordinary* citizen, the businessman, the union official, and the *public authority stood up to be counted and refused to be corrupted*".

In analyzing the food intake of *more than* 50,000 men and women, 83% were found to be *overweight* while *undereating* and 17% were found to be *overweight* because they *overate*. This was for *one group of 4,500 cases*.

Ninety-five percent of *key government officials* read the most *popular* newspaper in Washington, D. C..

In the Soviet Union, physicists have a *status coefficient* of 7.64, pilots and radio mechanics have 7.62, mathematicians have 7.34, and geologists have 7.22!

The Bureau of Labor Statistics reported that it costs $9,191 for a *family* to buy a *moderate living standard.*

A number that is *very frequently used by experts.*

A *government official* who refused to be named.

Statistics may be unknowable because a physical barrier prevents accurate data collection, because people surveyed either do not know or will not tell, or only a subset of cases get reported. Some examples, which may also be meaningless, follow:

Number of *stray* cats or number of rats in New York City.

Traffic jams cost the nation $500 million.

Value of *kickbacks, payoffs, and bribes* is running at an annual rate of $5 billion.

American people *dribble away* $500 million on *home repair frauds*, $500 million on *worthless health and nutrition programs*, $100 million on *mail order robbery*, $100 million on *fake reducing preparations*, $50 to $200 million on other *fraudulent or half-fraudulent medications*.

Society loses $5 to $10 billion to *income tax chiselers*.

Society *throws away* $500 million a year to the *organized obscenity racket*.

There are at least five to seven million couples in the United States *who swap wives in one way or another* and this is *common knowledge* among sociologists.

University newspaper poll shows that there are 1,568 single, 16 married, and 11 *undecided* students in the incoming freshman class!

Each year moths *destroy* $400 million worth of *fine clothing and material*.

An **eccentric theory statistic** is one based on an unfounded and sometimes preposterous theory. Many of us use or know about one form or another of eccentric theory or superstitions. Some examples for forecasting the economy follow:

The Jupiter-Saturn Cycle. Favorable conditions for the economy exist when the planets form a sixty degree sextile with the Sun on a one hundred and twenty degree trine.

Levels of Lakes Michigan and Huron. Cost of steel rises as levels decrease which adversely affects the economy.

Tips at Sardi's Restaurant. As business falls off, buyers stop coming to Sardi's and the sellers get hit and then the economy.

Purchases of Gourmet Foods.   Economy rises and falls with sales of gourmet foods.

Ups and Downs of Women's Hemlines.   The economy rises and falls with the shortness of women's skirts.

A **preposterous statistic** is one that could not realistically be true. The example of the teacher telling a pupil that "if she told him once she told him a million times" is an illustration of this type.   Some other examples are:

Twenty million Alaskans voted in the 1980 Presidential Election.

New York City heroin addicts steal two to five billion dollars per year to pay for their habit. This figure was arrived at by assuming a cost of $30 per fix and that there are 100,000 addicts. Then, $30 times 365 days times 100,000 addicts equals $1.1 billion. The stolen property is sold to fences for one-fourth of its value. Thus, $1.1 billion times four equals $4.4 billion.

An **uncritical trend statistic** is obtained by forecasting or projecting beyond the data. Some examples are:

Longevity of life has been increasing and projecting to the year 5,000 we find that people will live to be 1,000 years old.

The Mississippi River is getting shorter every year and using the same distance decrease per year, in several hundred years it will be only two miles long. Conversely, a million years ago, it was 1,300,000 miles long!

Sprinters have been running the 100 meter dash faster and faster every year. By projecting into the future, we find that this race will be run in one second!

The number of passengers in autos has been decreasing over the last 80 years. By projecting into the future, we find that by the year 2,100, there will be 0.5 passengers per auto!

A dubious cluster statistic is one that uses a small, isolated, and/or unrepresentative subgroup or cluster of a population to obtain the data and then these data are used to form an estimate for the total population. Again we as individuals occasionally revert to using or producing such a statistic. Some examples follow:

The cost of tuition at Cornell University (Endowed) is, say, $12,000 per year. Since there are X million college students in the United States, the annual cost for tuition is $12,000 times X.

In one area, two out of 32 managers of supermarkets of chain A were "taken for a ride" by the Mafia. Since there are X = 80,000, say, supermarkets in the United States, 5,000 managers have been "taken for a ride".

My friends Mr. and Mrs. Average spend 10% of the family income for food. Therefore, 10% of the total income in the United States is spent for food.

If the source of the data is not known and trusted, any associated statistics should be carefully scrutinized. There is no infallible procedure guaranteeing that a phony statistic will be revealed, but the following questions should be asked about any statistic. Phony statistics usually will not meet the test posed by these questions.

a. What is the reputation of the supplier of a statistic? Does this person usually supply reliable statistics?
b. Does the supplier of the data and associated statistics have some pet cause or product to promote or advertise and are these affecting the data?
c. What supportive evidence is offered? Does the supplier clearly explain how the data were obtained and what is the population from which the sample was drawn?
d. What assumptions and methodology did the supplier use? Are they valid?
e. Are the statistics plausible? Do they make sense?

The watchword toward all statistics, projections, and forecasts should be BEWARE. Remember, to be a *Doubting Thomas* is much better than being a *Mr. or Ms. I. M. Gullible*!

3.10 REFERENCES AND SUGGESTED READING

Bross, I. D. J. (1953). *Design for Decision,* Macmillan Co., New York, viii + 276 pp..

Campbell, S. K. (1974). *Flaws and Fallacies in Statistical Thinking,* Prentice Hall, Inc., Englewood Cliffs, New Jersey, viii + 200 pp..

Haack, D. G. (1979). *Statistical Literacy. A Guide to Interpretation,* Duxbury Press, North Scituate, Massachusetts, xii + 323 pp..

Huff, D. (1954). *How to Lie With Statistics,* W. W. Norton & Co., Inc., New York, 142 pp..

Moore, D. S. (1979). *Statistics. Concepts and Controversies,* W. H. Freeman and Co., San Francisco, xviii + 313 pp..

Moroney, M. J. (1951). *Facts From Figures,* Penguin Books, Inc., Baltimore, Maryland, viii + 472 pp..

Reichard, R. S. (1974). *The Figure Finaglers,* McGraw-Hill Book Co., Inc., New York.

Reichmann, W. J. (1964). *Use and Abuse of Statistics,* Penguin Books, Inc., Baltimore, Maryland, 345 pp..

Roll, C. W., Jr. and A. H. Cantril (1972). *Polls: Their Use and Misuse in Politics,* Basic, New York.

Sielaff, T. J., Editor (1963). *Statistics in Action. Readings in Business and Economic Statistics,* The Lansford Press, San Jose, California. (First paper), vii + 251 pp..

Wheeler, M. (1976). *Lies, Damn Lies, and Statistics,* Liveright, New York.

Wilson, E. B. (1952). *An Introduction to Scientific Research,* McGraw-Hill Book Co., Inc., New York, London, and Sydney, (Chapters 1-7, 9), x +

373 pp.

3.11 PROBLEMS

Problem 3.1. Give two examples each of an Unknowable Statistic and of a Meaningless Statistic. Obtain examples from printed material you have read recently. Describe why they fit the category given.

Problem 3.2. An experimenter used 1000 ripe peaches (from Georgia) in an experiment. He wished to determine which quarter of a peach was the most tender. (We shall refer to this as the Georgia peach squeezing experiment.) He designated the peach fruit by quarters starting with the suture, or indentation, on the fruit, as follows: (1) left front, (2) left back, (3) right back, and (4) right front. He did not cut the peach. A device was utilized which measured the amount of pressure required to penetrate the peach skin. On each peach, the experimenter measured the quarters in the order numbered above. He reached the conclusion that the least amount of pressure was required to penetrate the right front quarter, the next lowest was for the right back quarter, the next in amount of pressure was for the left back, and the left front quarter required the greatest pressure for penetration. Comment on various aspects of this experimental investigation. How should the investigation have been designed to determine which quarter of the peach was most tender?

Problem 3.3. A television program showed a Golf Pro hitting ten balls each of four different brands. The experiment was performed on the show as follows. The Golf Pro hit ten balls of Brand A, then ten balls of Brand B, then ten balls of Brand C, and finally ten balls of the Condor brand. It was concluded that Condor balls travelled nine yards further when hit with a driver than any of the other brands. Comment on the design of the experiment and on the conclusion reached.

Problem 3.4. The New York State Lottery advertisements indicated that participants now had a better chance of selecting the correct six numbers when the possible numbers were increased from 48 to 54. Is this a valid claim or not? Explain your answer.

Problem 3.5. The following item appeared in a Cornell University Newsletter, January 22, 1968:

DIVISION HAS ADDED OUTSTANDING SCIENTISTS FROM DIVERSIFIED AREAS; STAFF HAS INCREASED BY NEARLY 30 PER-CENT TO TOTAL OF 71

In all there has been a net gain of 23 new faculty members. The Division's current faculty numbers 71 as compared to 48 in October 1964.

Describe how you think they arrived at the figure of 30% given in the headline.

Problem 3.6. Make a list of the key words in this chapter and show relationships among them.

Problem 3.7. The following by-line appeared in *The Ithaca Journal*, May 23, 1990, page 10B: "The cost of laughing has gone up 9.4 percent". They state that the cost of rubber chickens has gone up 33%, the cost of dancing chicken singing telegrams has gone up 18%, and the minimum fee for writing TV sitcoms has gone up 9.5%. What would your reaction to such a report be in light of the material presented in this Chapter?

CHAPTER 4. PRINCIPLES OF SCIENTIFIC INVESTIGATION

4.1 INTRODUCTION

The class survey described in Chapter 2 was carried out in an inefficient manner the first year it was taken; it was also incomplete, because not all individuals were present in the class when the measurements were taken. Consequently, we did not obtain all the data desired during the specified period, and, unless a part of another class period were relegated to securing the missing data, the investigation would necessarily remain incomplete. However, the data collection had served several purposes, and we had sufficient data for our needs. Hence, additional time was not taken to complete the first-year survey. This is an illustration of the fact that careful, precise planning is necessary for all investigations. For example, if the team taking the measurements goes from seat to seat as they did in that first year, we would note that finding the name and recording the data would take the most time. If all students had been lined up and if they had come to the recorder and measurer to be measured in the order in which their names appeared on the sheets, the measurements would have been obtained more quickly and efficiently. Alternatively, the plan described in Section 2.3 could have been followed, and all measurements could have been obtained in one class period. Better organization allowed the data to be taken during one class period the second and third years that the class survey was conducted.

There are many steps involved in planning scientific investigations, and a number of these will be discussed. In general, the material in the references cited is followed in presenting the principles of scientific investigation. First, however, a number of definitions and concepts will be required.

To illustrate the necessity of following basic principles for all investigations, several examples are cited. During a visit to the offices of a national survey and market research organization, the author was invited to participate in a discussion of a proposed survey on medical needs of the elderly. The survey leaders gave the investigator a thorough cross-examination on all aspects of the survey from beginning to end. In particular, they questioned him on the basis of the principles of scientific investigation as listed below in section headings. They asked about the purpose, usefulness, possible uses, and interpretation of the survey data.

62

After several hours of intense cross-examination, the investigator said, "Gentlemen and lady (a lady psychologist was present), I have never been subjected to such treatment as this in my whole life! But, I can tell you that your procedure truly impresses me as an excellent way to proceed! Let's continue". Afterward, the head of the organization told me that they had learned from sad experience that every client must be subjected to such a thorough cross-examination in order to prevent legal proceedings later because of a misunderstanding about the information desired and the segment of the population to be involved. Every investigator should likewise subject himself to such an examination to determine whether or not it is a worthwhile problem and what is the best way to obtain the desired information.

As a second example, the same organization was approached by a large publishing company and asked to conduct a national survey, costing over $200,000. Near the end of their "cross-examination period", the members of the survey organization found that the publishing company wanted only a small segment of the information that would have been obtained from the proposed survey. The desired information was made available to the publishing company for about 1/10 of the cost of the proposed survey. This example illustrates that careful planning and a thorough understanding of what is wanted can greatly lower the cost of an investigation, while in other areas, time saved can be the important item. Careful, precise, and rigorous planning of investigations often saves time and money and increases the value of the results. A person following the procedures of scientific experimentation will make more rapid progress than one who does not, and he or she will have more time for recreational or other activities will be more productive than his or her poorly organized colleagues.

4.2 PRELIMINARY CONCEPTS AND DEFINITIONS

Science is the term applied to a body of systematized knowledge. It is the observation, identification, description, experimental investigation, and theoretical explanation of natural phenomenon. Systematization of knowledge begins with the organization of everyday observations; these are then classified, their sequences become known, and their importance is evaluated through systematic investigation. **Scientific method or scientific inquiry** is the procedure whereby knowledge is acquired and is an attempt to extend our range of knowledge. The evaluated knowledge is then used to formulate a **hypothesis** which is a tentative or a postulated explanation of a phenomenon and a statement about a population characteristic. The hypothesis is tested by sampling or experimental investigations to ascertain its plausibility. A hypothesis that is relatively well verified and possesses some degree of generality is a **theory**. A theory that has been verified beyond all reasonable doubt at the moment is designated as a **law**. We say "at the moment" because new knowledge may later indicate that a law of the past was only a theory; for example, Newton's laws of motion were replaced by the laws of quantum mechanics; his laws are limiting forms of the new laws for heavy bodies (see Wilson, 1952, pages 29 and 30 for this and other examples). Also, Mendel's laws of inheritance, the exponential growth law, the growth and decay law, rate of increase law, etc., are laws used in biology today. Frequently we find it expedient to use a mathematical form of the responses for the phenomenon to express a law. Thus, the exponential growth law may be written as $W = Ae^{bt}$ = weight at time t; the rate of increase of growth at a specified time, t_0 say, is the derivative of W with respect to t, i.e., $dW/dt = Abe^{bt}$ evaluated at $t = t_0$. The letter e is the base for natural logarithms.

A hypothesis, a theory, and a law differ in degree of plausibility and generality. For example, suppose that we were to obtain the following sequence of sample observations:

0,   3,   6,   9,   12,   15,   18,   21 .

A reasonable hypothesis would be that these sample observations fit the sequence 3n for n = 0, 1, 2, 3, ... But, just because these data fit the 3n sequence is not proof that the next number in the sequence will be 24; there are several sequences which these numbers could fit. One such sequence would be

$$3n + n(n - 1)(n - 2)(n - 3)(n - 4)(n - 5)(n - 6)(n - 7).$$

The use of the latter sequence to predict the next number in the sequence would give a vastly different number (5,064) from 24. This illustrates the fact that sample data often support a number of hypotheses. While some of these may be ruled out because they conflict with known laws or theories, a series of investigations will be necessary to differentiate among the remaining hypotheses (See Problem 4.1.). Thus, support of one hypothesis by the data does not necessarily establish the truth of the hypothesis. This is a fact that is often overlooked.

The object of a scientific investigation is not to prove the scientist correct, but to establish the truth. In science, the investigator is seeking the truth and the true facts about a phenomenon. It ultimately does not matter what his dogmas, beliefs, or superstitions are because someone, somewhere, sometime will find the truth. The scientific world is completely impersonal and in the long run, is impervious to personal biases inserted in works. A great scientist said he believed that 95% (a phony statistic, of course) of the people were working on the wrong things (an undefined term) at any one point in time. Even if this statement resembles some of those in Seligman's article (*Fortune*, November, 1961) discussed in Chapter 3, there is some truth in it. Has the reader ever encountered articles on statistics written by a famous heart specialist or articles on heart surgery written by a famous statistician? When heart surgeons follow the heart surgeon's statistics and when the statisticians follow the statistician's heart surgery, both groups could be using erroneous procedures. Misplaced self-confidence introduces misconceptions which may lead investigators astray. Egotism often goes too far in science as it does in everyday life. Don't we, without any political experience, feel that we know how to run our government, or anybody else's government, better than the political experts who have made this their life's work? Aren't we then like the statistician writing articles on heart surgery? Do we attempt to force our point of view on others, or do we attempt to set up a logical, reasonable basis for our point of view and let the listener believe what he will? Although personal bias is often difficult to overcome, the investigation must be conducted in such a manner as to minimize or to eliminate all possible biases. The investigator must have self-esteem, but he should not indulge in self-worship; it may be necessary to point out to ourselves or to our colleagues that self-esteem is not a necessary and sufficient condition for a proof of a theorem in statistics or in mathematics or of a result in any other field of investigation. Personal opinions are irrelevant; the only item of relevance is whether or not the stated result is true. Often when a problem becomes difficult, it is human nature to resort to "appeal to intuition or good-will or accept me as an authority". Unfortunately, this type of reasoning crops up continuously in everyday life as well as in the social and political and other sciences. When hypotheses are difficult to state and much more difficult to prove or disprove, people often skip all steps of scientific experimentation and

a. accept the word of authority, friend, or idol as "law",

b. say something often enough so that they have propagandized
themselves, and perhaps others into believing that a "law" exists, or

c. camouflage the whole topic with half-truths, phony statistics, and
vague meaningless words and then derive a "law".

For example, we often hear such statements such as "hippies (all) have a message", "hippies (all) have brilliant minds", "artists (all) tell the story of social progress", "Negroes (all) are excellent athletes", "he is an odd-ball and therefore is intelligent", "a generation gap exists between (all) parents and their children", "a pietist-secularist gap exists", "change (all) represents progress for society and therefore is good", "he is a brilliant mathematician and therefore he is brilliant in political science", and so on. (The word "all" has been inserted in common statements to illustrate the falsity of these sweeping statements.) We definitely know that some of the hippies are stupid, that there are non-athletic Negroes, that no generation gap exists in many families, that a brilliant mathematician may be a complete ignoramus when it comes to politics. We know that some of these problems are not new; they have always existed in one form or another. We know also that the human being is very susceptible to propaganda, especially if it allows him to do what he wants to do and he will often go to great lengths to justify his position. Knowing this, we should always question statements involving sweeping generalities and currently popular adjectives and phrases. Do you really know what "doing his thing" means? Do you know precisely what your neighbor means when he uses the terms such as "liberal" and "conservative"? What is a conservative revolutionary radical? What is a liberal arch-right-winger? What do the words "modern" and "relevant" mean? When people speak of "civil rights", do they mean rights for themselves or for everybody? We must examine carefully and critically what we hear, see, and read, if we are to follow the principles of scientific investigation in everyday life as well as in scientific inquiry.

Furthermore, we should note that the scientific method builds on previous knowledge. Ignoring or destroying the knowledge and "know-how" of a previous "establishment" has been demonstrated time and time again to be a foolhardy method for making progress. The destruction of an organized society has in the past set civilizations back several hundred years in a social and technological sense. Radical changes in society often have disruptive effects, such as those which occurred in Iran in the 1980s. Therefore, the scientific method, an orderly, continuous quest for knowledge, making use of all previous knowledge, leads to the most rapid advance.

When a truth has been established (not assumed), we have a law. The law is often phrased in the form of a mathematical equation. As soon as the law is established, investigation ceases in the direction that produced the law, and proceeds in another direction toward establishing another law. Several laws later, the investigation may turn attention to a generalization of the current laws, if this is possible, as was the case with a generalization of Newton's law of motion.

The orderly quest or pursuit of new knowledge is know as **research**. Webster defines research as "(1) Careful search; a close searching. (2) Studious inquiry; usually critical and exhaustive investigation or experimentation having for its aim the revisions of accepted conclusions in light of newly discovered facts." Research should yield new knowledge. The term research is not to be confused with **re-search** which means a search or re-finding of already established and known facts, or merely follow-the-leader investigation. Scientific research results in an advancement of the state of knowledge and, therefore, in an advance of science.

It is useful to differentiate between **empirical research** and **analytical research**. The former deals with investigations involving measurements; the latter deals with laws, axioms, postulates, and definitions in the field of inquiry. In mathematics, philosophy, and theoretical physics research is generally analytic in nature, whereas in experimental physics, biology, social sciences, and business, much of the research is empirical in that it involves measurements and observations on various characteristics. The mathematician states a theorem within a framework of definitions and axioms. He then sets out to prove the theorem in a mathematical sense, utilizing the axioms and definitions. He does not collect data from observations to prove the theorem. The biologist, on the other hand, states a hypothesis and then conducts an investigation to collect data on the plausibility of the hypothesis. He may or may not accept the hypothesis on the basis of the facts obtained, but seldom, if ever, does this prove that the hypothesis is true or false. The empirical facts only substantiate the claim for the hypothesis; they do not prove it.

A necessary part of research is **inference**, which is a process of reasoning whereby the mind begins with one or more suppositions or propositions and proceeds to another proposition. It is a psychological process. Inference may be **deductive** or **inductive**. **Deductive inference** is the process of determining the implications inherent in a set of propositions. For example, in plane geometry consider the statement: "*If the three sides on one triangle are equal, respectively, to the three sides of a second triangle, then the triangles are congruent*", where the term "congruent" means that they have the same size and shape. The process of reasoning and the conclusion "the triangles are congruent" is the deduction and the remaining part of the statement is the proposition. The conclusion is reached through the process of deductive inference. This type of inference is associated with analytic research; it is also used in empirical research in various ways, such as in the construction of hypotheses, theories, and laws.

Aristotle was one of the first to stress the systematic nature of science and to teach the use of reasoning in the development of science. **Syllogism** is one form of logical deduction that was used to a large extent; this method of deductive inference begins with two premises, usually a major premise and a minor premise, or with propositions so related in thought that a person is able to infer a third proposition from them. The following examples illustrate this method of deduction.

Major Premise: All living plants absorb water (*induction*).
Minor Premise: This tree is a living plant (*observation*).
Conclusion: Therefore, this tree absorbs water (*deduction*).

Major Premise:	Human beings are men and women (*induction*).
Minor Premise:	This person is a man (*observation*).
Conclusion:	Therefore, this person is a human being (*deduction*).

The following excerpts and examples were taken from Professor Keewhan Choi's (formerly at Cornell University) introductory lecture in a course on model building on the subject of deductive inference:

Major Premise:	Every M is P.
Minor Premise:	S is M.
Conclusion:	Therefore, S is P.

Or, to cite a another example
Every virtue is laudable,
Kindness is a virtue,
Therefore, kindness is laudable.

It is really remarkable that from this kind of start, which seems so innocently simple, there has been developed a grand array of powerful procedures by means of which reasoning can be kept tidy and dependable.

Lewis Carroll, the author of *Alice in Wonderland* and a mathematician at Christ Church, Oxford, wrote several books and pamphlets explaining the application of the rules of logic. He showed that one could, from the three premises,

a. babies are illogical,
b. nobody is despised who can manage a crocodile, and
c. illogical persons are despised,

derive the conclusion that
babies cannot manage crocodiles.

This example of logical reasoning indicates, incidentally, that a logical conclusion is no more valuable than the premises on which it is based. The above deduction is not difficult. But take a look at another of Lewis Carroll's examples. What is the conclusion from the following nine interlocked premises?

a. All who neither dance on tightropes nor eat penny buns are old.
b. Pigs that are liable to giddiness are treated with respect.
c. A wise balloonist takes an umbrella with him.
d. No one ought to lunch in public who looks ridiculous and eats penny buns.
e. Young creatures who go up in balloons are liable to dizziness.
f. Fat creatures who look ridiculous may lunch in public provided they

do not dance on tightropes if liable to giddiness.
g. No wise creatures dance on tightropes if liable to giddiness
h. A pig looks ridiculous carrying an umbrella.
i. All who do not dance on tightropes and who are treated with respect are fat.

This one takes a bit of thinking, but it is straightforward (using the procedure of logic) to discover that these nine premises yield the conclusion:
    No wise young pigs go up in balloons.

One may not be overwhelmed by the significance of these examples, but it is rather impressive when it is realized that involved and complicated premises do in fact lead unambiguously to certain definite conclusions and that clearly formulated procedures of logical thinking can unravel a curious mass of interlocked statements of this kind. Knowledge of the past furnishes the major premise; a particular problem or situation supplies the minor premise. The deduction obtained by the psychological process of reasoning constitutes the conclusion.

Inductive inference forms a large share of the third part of the definition of the subject of Statistics. This type of inference is characterized by the fact that from the sample data, conclusions are drawn concerning population facts. For example, from tasting a small piece of cake, we draw conclusions concerning the entire cake; by sticking our toe in the water at one spot, we draw conclusions about the relative temperature of all the water in a swimming pool; a polling organization obtains answers from a small representative sample relative to preference of two candidates, A and B, for election to an office and from the sample proportion draws conclusions about the proportion of registered voters preferring candidates A and B; from a sample selected entirely by chance from two lots of light bulbs the experimenter sets up an experimental investigation to determine the average length of life of light bulbs and from the investigation draws conclusions about the average length of life of all bulbs in each of the two lots; from a sample of the weather early in the day, we draw inferences, dangerously of course, about the weather for the entire day, and based on these observations we decide the type of clothes to be worn for the day; and so forth for many other situations requiring decisions or conclusions in investigations and in everyday life experiences.

It may be noted that the scientific method or inquiry involves the following:
a. use of hypotheses, theories, and laws in the investigation,
b. use of a scientific attitude which involves critical and searching observational ability,

c. use of deductive and inductive inference, and
d. use of an orderly and organized quest for knowledge.

It may be noted that the above are followed very rigorously by detectives and successful prosecuting attorneys in solving fictional cases in novels and on television as well as in the real world. Such people as Sherlock Holmes, Maigret, and television show personalities such as Angela Lansbury in "Murder, She Wrote", Perry Mason, Hunter, and Spenser in "Spenser: For Hire", are exceptionally keen observers detecting the tiniest bits of evidence and organizing them into an impregnable case for or against a character.

In research, a scientific attitude is required. The scientist must have a passion for facts and truths, he must be cautious in his statements, he must have clarity of vision, he must be discerning, he must be persevering, he must be thorough, and he must have a good sense of organization in bringing together related facts. The person who disregards pertinent information in an interpretation of a scientific endeavor does not have the proper scientific attitude. The person who knows what he believes without wanting to be bothered by facts cannot fit into the scientific community without discord. In science one is seeking the truth, and someone will find it sometime. A scientist must be discerning and must work on important and general problems if he is to make significant advances in science. As Wilson (1952) states in the opening sentence of his book,

"Many scientists owe their greatness not to their skill in solving problems but to their wisdom in choosing them".

The scientist must be persevering, but he should also know when he is defeated and is unable to solve the problem. At the same time he must be absolutely thorough in his investigations. Adequate controls, observations, and measuring instruments are a necessity. In discussing the need for adequate controls, Wilson (1952), page 41, illustrates this by citing the following example: "It has been conclusively demonstrated by hundreds of experiments that the beating of tom-toms will restore the sun after an eclipse". On the question of adequate numbers of observations, two examples from Wilson (1952), pages 34 and 46, are pertinent. A nutritionist published an article on an experiment with a surprising result. When a visitor to his laboratory requested to see more of his evidence he replied, "Sure, there's the rat". The use of one rat in a nutritional study is obviously insufficient for drawing conclusions. In the second example, an experimenter in medical research reported that of the animals used in the study, 1/3 recovered, 1/3 died, and no conclusion could be drawn about the remaining subject because that one ran away!

By its very nature, research investigation requires inference. Inductive inference, as we have seen, proceeds from the results obtained from a sample, a part, or an experiment to characteristics of the entire population or of all the members of a class. This type of induction is based on partial evidence, and as such, it is characterized by a degree of uncertainty. The evidence for inductive inference may often be stated on a probability basis. A large part of the science of Statistics is devoted to procedures relating to inductive inference.

In model building, which is a process for finding a mathematical formula explaining a phenomenon, the process of deduction and induction is utilized over and over again, as illustrated in Figure 4.1. We postulate a model, construct an experiment to test the adequacy of the model, conduct the experiment, compare the data with the postulated model, and then start the cycle over again if the model is found to be inadequate. This process of investigation continues until a mathematical formulation is found which explains the phenomenon under a study. We then have Jane Doe's law.

In all of scientific investigation there are certain principles that must be observed if the investigation is to be a success rather than just an experience. Before listing these principles we define some additional terms. Two additional terms requiring definition are **sample survey** and **experiment**. A **sample survey** is an investigation of what is present in the population. When the sample is a 100% sample, it is called a **census**. Any observation that appears in the population could appear in the sample; any condition not represented in the population will not be observed in a sample or a census.

In many investigations, however, it is desired to investigate conditions which do not appear in a population, but could appear in an **experimental investigation or experiment**, where an **experiment** is defined as the planning and collection of measurements or observations according to a pre-arranged plan for the purpose of obtaining factual evidence supporting or not supporting a stated theory or hypothesis. The experimenter may, and often does, introduce conditions which do not exist in any naturally occurring population. In the experiment, the investigator controls the conditions; in a survey, the conditions are those that prevail in the population. For example, in a survey of the types of floodlights used for night-lighting slopes at ski centers in the Northeast in the 1964-65 season, we would find that mostly mercury vapor bulbs were used and that General Electric Lucalox sodium vapor light bulbs were not. In order to test the effectiveness of the latter bulbs for lighting, an experiment would need to be designed as was done at the Grosstal Ski Area (*Rochester Democrat and Chronicle*, 2/18/66, page 10D). This sodium vapor bulb would not have been found in a survey of bulbs used to light ski slopes but it was used in the experiment being conducted at Grosstal.

Some additional terms are defined below for use in the following sections:

A **treatment** is a single entity or phenomenon under study in an experiment. (E.g. a mercury vapor bulb or a sodium vapor bulb could be a treatment.)

An **absolute or single phenomenon experiment** is one which contains a single treatment.

A **comparative experiment** is one designed specifically to compare two or more treatments.

A **treatment design** represents the selected set of treatments for comparative purpose or for ascertaining responses to several levels of a number of variables. It could and often does, include controls and standards for comparison.

An **experiment design** is the arrangement of treatments in an experiment; it is used in *all* types of experimental investigations. In some literature, the term experimental design is used for experiment design.

4.3   FIRST PRINCIPLE: CLEAR STATEMENT OF PROBLEMS REQUIRING SOLUTION

In order to make progress in a research investigation, it is essential that the problem to be solved and the question to be answered are *explicitly and clearly stated in written form.* Explicitness and clarity are necessary in order to make certain that the resources of the survey or experiment are expended in a fruitful and efficient manner. The collection of unneeded data or the omission of essential data because of vagueness in stating a problem leads to inefficient investigations. Incompleteness of statement of the problem and of the questions to which answers are sought is wasteful of progress and of resources. Because of the fallibility of human memory, it is necessary to state the problem precisely in written form; for example, many items are forgotten during a term, and there is evidence of this in every course that a student takes. Since memory is fallible, one should not rely on memory for the important items associated with a research investigation. It is a relatively simple matter to follow the first principle of scientific inquiry, but it is surprising how often it is disregarded. As one outstanding biologist, A. C. Hildreth, once said, "One can't imagine any more of a fool's paradise than to experiment or investigate in whatever direction the mind may wander, without any purpose or objective. Such meandering investigations lead to naught but a wastage of funds and resources".

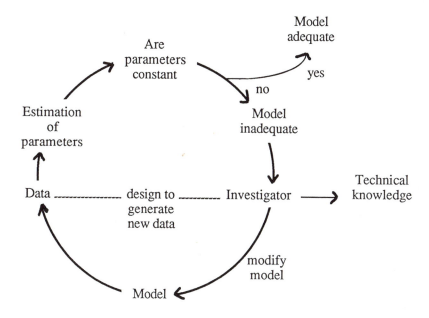

Figure 4.1. Model-building diagram (adapted from Box and Hunter, 1962).

In analytic research, the framework of definitions and axioms as well as the conclusions to be drawn must be explicitly and rigorously stated if there is to be any hope of effecting a solution and drawing a conclusion.

4.4 SECOND PRINCIPLE: FORMULATION OF TRIAL HYPOTHESES

As defined previously, a hypothesis is a statement of fact about some characteristic of the population; the statement may or may not be true. From the statement of the problem, a statement of the hypothesis or hypotheses is formulated and should be recorded in written form. Some of the essential conditions for the adequacy of a hypothesis as it relates to the problem under consideration are:
  a. It must possess sufficient clarity to permit a decision relative to sample or experimental facts.
  b. It must lend itself to testing by investigation.
  c. It must be adequate to explain the phenomena under consideration.
  d. It must allow a reliable means of predicting unknown facts.
  e. It should be as simple as possible.

In a proposed research project, a team of psychiatrists and psychologists was requested to submit hypotheses in written form in order that others could understand the goals and objectives of the study. The result was a 35 page jumble of words full of ambiguities and violating every one of the above essentials of an adequate hypothesis! After some time and after sufficient prodding, it was encouraging to note that these same individuals were stating hypotheses in one, two, or three line lucid and simple statements.

One type of hypothesis often used in statistics is the **null hypothesis** or the hypothesis of no difference. For example, suppose that two samples of light bulbs were drawn by a random process from the same lot, and the average length of life was determined for each of the samples. The null hypothesis, which may be stated as "the average length of life, say μ_1, of light bulbs for the population from which the first sample was drawn is equal to the average length of life, say μ_2, of those in the population from which the second sample was drawn, or alternatively as $\mu_1 - \mu_2 = 0$, would be true since $\mu_1 = \mu_2$ since both samples are drawn from the same population of light bulbs. In other cases, it is not known if the null hypothesis is true or not, as for example, in testing of the effectiveness of a new vaccine as compared to the effectiveness of a standard known vaccine. We postulate that there is no difference and then set up an investigation to ascertain whether or not the sample or experimental facts are in agreement with the null hypothesis. Statistical procedures have been devised to determine whether or not the sample facts agree with the null or any other specified hypothesis, where the term "agree" has been defined. "Agree" does not mean "proved".

4.5 THIRD PRINCIPLE: A CAREFUL, LOGICAL, AND CRITICAL EVALUATION OF THE FORMULATED HYPOTHESES

Once the hypothesis has been critically evaluated for adequacy, the next step in scientific inquiry is to search the literature to determine whether or not the proposed problem and formulated hypotheses will add new knowledge, whether or not they are plausible, and whether or not the problem is worthwhile. Wilson's (1952) statement that "Many scientists owe their greatness not to their skill in solving problems but to their wisdom in choosing them" may be applied to individuals in all walks of life including business and politics as well as to research investigators.

The problem of searching the literature can be systematized and made efficient. Wilson (1952) discusses this subject in Chapter 2 of his book, and he gives reasons for a literature search, the structure of scientific literature, methods of searching, and methods of recording the desired information. After completing the literature search, it may be decided that the problem is not worthwhile or that it has no solution; in this case, the first two principles listed above would have to be reformulated and one would proceed in another direction.

The first three principles are very important to the success of any investigation, scientific or otherwise. It is here that the investigator has a "go/no-go" situation and must decide whether to carry out an investigation or to wait until further information and/or resources are available.

4.6 FOURTH PRINCIPLE: DESIGN OF THE INVESTIGATION

Any empirical investigation involves a procedure for collecting the data. The observations may be collected in a haphazard manner to yield little information on the formulated hypotheses, or they may be obtained from a well-planned and well-designed procedure to yield the desired information. In planning the investigation, items such as those discussed in Sections 3.3 to 3.7 must be considered. The data collected must be such that evidence for or against the hypotheses is obtained. The lack of coordination of this step with the first three can, and often does, lead to serious deficiencies in the investigation.

In research investigations, consideration of the following items is required:

a. The characteristics to be observed on the different entities or treatments in the experiment or survey.
b. The design of the investigation, that is, the design of the sample survey or of the experiment (Chapters 5 and 7).
c. The treatment design for the experiment and selection of adequate controls and standards (Chapter 8).
d. The number and kind of observations to be made (Chapter 13).
e. The selection of procedures and measuring instruments resulting in meaningful data and controlling extraneous influences.
f. A detailed outline of costs, equipment, and personnel required to carry out the experiment or survey.
g. An outline of the statistical and computing procedures to be used in the summarization of the data.
h. An outline of the summary tables, graphs, charts, pictures, and figures desired for the investigation.
i. Safeguards against accidents and carelessness during the conduct of the investigation.

After a careful and critical review of all the above items, the investigator may decide that more facilities are required than are available or that the problem and the hypotheses will need to be reformulated in order to conduct an investigation within the available resources. For example, if the problem requires the purchase of an instrument costing $50,000, if the entire budget for the investigation is $25,000, and if it is not possible to rent the instrument or to use another measuring device, the problem will need to be reformulated and the investigation carried out within the available budgetary limitations. As a second example, suppose that the investigator wishes to conduct a sample survey using personal interviews from the population of individuals over 18 years of age in a preselected geographical area, say within the city limits of Syracuse, New York. Suppose that the average cost of a personal interview is $5.00, and that 500 interviews are desired. If the total budget is $500, other procedures such as mailed questionnaires, telephone interviews, a combination of the above, or a reduction in the number to 100 interviews will need to be utilized. It may be decided that none of these alternatives is satisfactory and that the survey should be delayed until additional funds of $2,000 are made available for the investigation.

After preparing the pertinent summary tables, graphs, and charts and the statistical analyses in outline form, the investigator should compare these with the stated problem and formulated hypotheses. If they do not conform, the procedure should be changed to obtain the information related to the objectives of the investigation. In preparing for an extension survey a number of years ago, the investigators stated the problem, the objectives, and the questions to be answered in precise form. When the questionnaire had been prepared, a comparison of the questionnaire with the objectives of the survey indicated that there were no questions on the questionnaire for five of the desired objectives. Luckily, the questionnaire had not been distributed, and it could be reformulated to conform with the goals of the survey.

The entities or treatments in the experiment must be selected with care. The success of the experiment, to a large extent, depends upon the selection of the treatments in an experiment. The importance of treatment selection varies with the investigation, but in general it is a very crucial item. We devote an entire chapter (Chapter 8) to the subject of treatment design.

4.7 FIFTH PRINCIPLE: SELECTION OF APPROPRIATE MEASURING INSTRUMENTS, REDUCTION OF PERSONAL AND OTHER BIASES TO A MINIMUM, AND ESTABLISHMENT OF RIGOROUS AND EXACT PROCEDURES FOR CONDUCTING THE INVESTIGATION

As was evident from the classroom survey, the selection of an appropriate measuring device was important for a differentiation between the heights of individual students. A measuring device calibrated in yards or meters gives units too coarse to differentiate and one calibrated in feet only separated the class into two groups, that is, five feet tall and six feet tall. The measuring device calibrated in millimeters was certainly calibrated finely enough for our purposes. Presumably, calibration in centimeters, or inches, would have been sufficient.

For all types of investigation, the selection of the measuring device and of the unit of measurement is very important. In order to accelerate progress in medical, economic, sociological, psychological, educational, and certain biological investigations, it will be necessary to devise appropriate and precise measuring instruments for the study of various phenomena. For example, how many people living in the world in 1958 could have predicted the interim state of affairs in Cuba, Czechoslovakia, Poland, East Germany, or Russia? Could knowledgeable people in social and political sciences devise a scheme which would eliminate deadly physical conflict between countries or even between juvenile and/or adult groups in the large cities of the world? The author believes that the greatest advances to date have been in the physical, medical, and biological sciences, but the *by far* the greatest need for progress is in the social and political sciences. We may learn the exact chemical and physical nature of the animal *Homo sapiens*, but unless we also learn how to control him, the law of the jungle may soon prevail. The sociological and psychological problems of welfare recipients, of children from broken homes or temporary common law marriages, of abused women and children, of migrating workers, and of a very fluid and mobile labor force considered necessary for a low percentage of unemployed, as well as the problems of teaching more material in the same amount of time, of increased pressures on students, of mass transportation, of food production for the future, of pollution, of the status of the ozone layer, of garbage disposal, of storage of toxic material, and so on, are with us now. Do we have measuring devices to obtain factual and exact information leading to hypotheses and theories which will allow appropriate action to be taken by politicians, by private citizens, and by government officials? As a professor of Sociology once said, "Our State Department asks the sociologist what will happen in country X if we institute policy A, policy B, or policy C. We sociologists only shrug our shoulders and profess ignorance. But the State Department has to have a policy, so the politicians have taken over the job of the sociologist by default". Perhaps, and the author emphasizes "perhaps" because of his incompetence in this area, the social scientist should spend more time devising measuring devices and less time conducting investigations, in order to accelerate progress in this important area. Perhaps more laboratories such as the Primate Laboratory in Madison, Wisconsin, are needed. Perhaps startling new techniques which as yet have not been envisaged are necessary. The problems are known in this area and hypotheses can be stated. But can an investigation be designed to obtain the relevant facts? If not, then the time should be spent on developing and calibrating measuring devices and in designing procedures to obtain the desired information rather than on obtaining irrelevant data.

Even though finely calibrated measuring devices are available, precautions are necessary to ascertain that the instruments will be used correctly and the measurements will be free from bias. For example, one could take three readings and discard the most discrepant one. Discrepant from what? One could use an "intelligent placement" of treatment in an experiment which could consistently allow one treatment to appear in the most favorable position. A "favorite" treatment or method could be scored high whenever the investigator knew its identity. In grading examinations, the grader should be unaware of the student's identity to avoid possible favoritism. The instructor may be biased toward neatness and clarity of exposition, and he may have to be constantly on guard against this bias in grading examination questions.

In taste-testing experiments, it is wise to keep each taster's score secret, since a known score of one taster may affect the score of the second. In judging diving events, the judges are required to show their scores simultaneously in an attempt to make their scores independent. Judges should be trained in a standard manner and should know the procedures of judging the material being considered. It benefits no one to allot the ribbons in a haphazard manner. Measuring devices and judges should not be changed in the middle of an investigation and they should not be affected by environmental conditions unrelated to the goals of the investigation.

Rigorous and exact procedures have to be devised for discarding or for utilizing "far out" observations. For example, suppose that the following data were observed:

Treatment

Control	9.8	17.8	10.1	18.2	10.2	10.3	9.7	9.9
New tr.	12.0	12.1	11.8	12.1	11.8	12.0	12.3	11.9

What should be done with data of this type? Are the numbers 17.8 and 18.2 discrepant observations ("wild things" or far-out observations) or are they representative of this type of data? It should be noted that excessively large or excessively small observations have considerable effect on the arithmetic average. Here the arithmetic averages for the two treatments control and new are identical, i.e., 12.0. An experimenter may have no criterion for discarding "wild things" but may wish to use statistical procedures for reducing their effect, e.g., using medians as averages rather than arithmetic averages. The medians are 10.15 and 12.00, respectively, for the two treatments.

4.8 SIXTH PRINCIPLE: RIGOROUS CONDUCT OF THE INVESTIGATION

Having proceeded according to the five principles above, the experimenter has again to decide whether or not the investigation should be conducted. If it is decided to proceed, then the experiment or survey must be conducted in an exact and rigorous manner, making certain that the data obtained from the investigation are free of all but random measurement errors and ambiguities. Prior to the execution of the investigation, it may be well for the investigator to reconsider the method for conducting the investigation. During the conduct of the investigation, he must be keenly aware of possible difficulties and be able to spot unexpected results that may come up. The investigator should be certain that extraneous variation is not brought into the investigation by the method of conducting the investigation. For example, the technician or method of measurement should not be changed in the middle of an investigation unless there is complete agreement between the technicians and the methods.

*4.9   SEVENTH   PRINCIPLE: A   COMPLETE   STATISTICAL
ANALYSIS AND INTERPRETATION OF THE DATA IN LIGHT OF
THE CONDITIONS OF THE INVESTIGATION AND FORMULATED
HYPOTHESES*

While statistical procedures are valuable aids for reducing data to
summary form, the object of an investigation should not be the application
of statistical analyses procedures. Instead, it should be the relationship and
interpretation of the results in light of the stated problem and hypotheses.
The results of statistical investigations should provide evidence for or
against the stated hypotheses. In addition to providing factual information
concerning the various hypotheses, the results of the investigation and the
statistical analysis should produce facts for the design of future sample
surveys or experiments. This is the second purpose of any investigation, as
listed in Chapter 1.

All statistical computations and computer software should be checked
and re-checked for correctness. The arithmetic and the algebra should be
free from computational, algebraic, and copying errors. The statistical
procedure should be checked for appropriateness; use of an incorrect
procedure can be as misleading as an arithmetical error. Unfortunately,
articles published in scientific journals contain all too frequently one or
more of the above types of errors. For example, one investigator
professed to support view A over view B with a small number of
observations, 15; unfortunately, he made a mistake in arithmetic which,
when corrected, supported view B. The fact that *eight* pages of a well-
known scientific journal had been used to publish this mistake is appalling!

In another instance, a famous psychologist presented the results of a
research study in a presidential address to a psychological society. The
results appeared to be in reverse order and contradictory in order to make
the interpretation he considered appropriate, and a re-check was requested
from the computing center. He was assured that the results were correct
and interpreted the results as best as could be done. Some time later an
additional table was needed for inclusion in the printed address. The
computations had to be performed on a desk calculator, as the high speed
computer was being repaired. It was found that the items had been coded
1, 2,..., 20 when they should have been coded 20, 19,..., 2, 1. He was now
faced with a complete reversal in interpretation. The psychologist did this
by emphasizing the need for complete checking on all phases of statistical
analyses and of data processing.

4.10 EIGHTH PRINCIPLE: PREPARATION OF A COMPLETE, ACCURATE, AND READABLE REPORT OF THE INVESTIGATION

No investigation is complete until a comprehensive, correct, and readable written report of the investigation has been made available in the required form. In some instances, the written report only has to be made available in handwritten or typewritten form to one or more individuals such as a teacher, administrative head, a graduate committee, fellow investigators in one or more laboratories, or personnel of a company; in other cases, it is printed and published as an article in a professional journal, newspaper, or book, and it is made available to the general public. The conclusions drawn should be substantiated and in agreement with the facts obtained from the investigation. In the written report, the description of the experiment or survey and of the statistical procedures utilized should be complete enough to allow the reader to decide whether or not the conclusions reached are in agreement with the facts of the investigation; the description of the data and of the method of collecting the data should be detailed enough to allow the reader to decide whether or not they bear on the hypothesis under investigation or are so mixed up with other effects that the data do not or may not allow testing of the hypotheses. The mixing up of effects sometimes occurs in investigations, and the resulting data may support several hypotheses in addition to the one of interest.

A possible misconception among investigators is that the experimental results should indicate differences, that is, the null hypothesis should be rejected and that this represents "positive results" from an investigation. However, correct and informative data *always represent positive results whether or not the null hypothesis is rejected.* There are *no* "negative results" if the principles of scientific investigation are followed. It is just as important to know that the null hypothesis is probably true as to know that it is probably false. Improper procedures in an investigation leading to meaningless results are negative, but reliable data on a phenomenon are never negative. Whether or not differences are found, a written report should be prepared and submitted to the appropriate public.

A principle could appear to be established were investigators only to report the differences found and to ignore those cases when differences were not found. For example, in an examination of the effect of drug A on patients, suppose that only cases of recovery were reported while cases of non-recovery were not. These reports would erroneously indicate 100% effectiveness for drug A when in fact it could be that drug A had a detrimental effect or was ineffective on certain groups of individuals. If these results were never reported by investigators, one would only find this out by conducting a crucial experiment of one's own. It has happened in

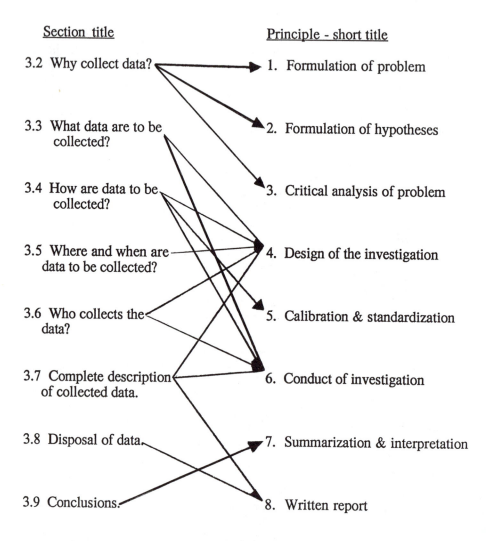

Figure 4.2. Interrelations between Sections 3.2 to 3.9 and the eight
        principles of scientific investigation.

some fields of inquiry that the reporting of "positive results" and the omission of "negative results" in scientific publications has led to misconceptions.

In this connection, the failure to find a solution to a problem should not be confused with the fact that the results of an investigation indicate no differences among the treatments or variables. The confusion of these two ideas could be one reason for the attitude developed toward "negative results" in certain areas of investigation.

4.11 DISCUSSION

Close adherence to the principles of scientific inquiry will lead to progress and efficient utilization of resources. The investigator who follows these principles will make considerably faster progress than his counterpart who just investigates or who plans poorly. The planning stage of an investigation is probably the most important step in an investigation, but all steps are important. The old adage that "a chain is no stronger than its weakest link" is definitely true of any investigation. Failure to observe any one of the principles of scientific experimentation may doom, or at least weaken, the entire investigation by leading to difficulties in the design, in the analysis, and in the interpretation of the data.

Many of the steps in scientific investigation do not involve statistics, but failure to observe any one often leads to difficulties in statistical analyses. The fourth and seventh principles listed above are highly statistical in nature, the fifth is partly statistical, and the remaining principles are mostly nonstatistical in nature. Figure 4.2 shows the relationship between the topic headings in Sections 3.2 to 3.9 and the eight principles of scientific investigation. There are other interrelations, but the major ones are indicated in the diagram.

4.12 REFERENCES AND SUGGESTED READING

Box, G. E. P. and W. G. Hunter (1962). A useful method for model building. *Technometrics* 4:301-318.

Carroll, Lewis (1865). *Alice's Adventures in Wonderland.*

Federer, W. T. (1955). *Experimental Design, Theory and Application*, The Macmillan Co., New York, xix + 544 + 47 pp.. (Chapter 1) (Republished by Oxford & IBH Publishing Co., New Delhi, India, 1967 and 1974)

LeClerg, E.L., W. H. Leonard, and A. G. Clark (1962). *Field Plot Technique.* Burgess Publishing Co., Minneapolis, Minn. (Chapter 2).

Sielaff, T. J., Editor (1963). *Statistics In Action. Readings. In Business and Economic Statistics.* The Lansford Press, San Jose, California, (First paper by D. Seligman), vii + 251 pp.

Wilson, E. B., Jr. (1952). *An Introduction to Scientific Research,* McGraw-Hill Book Co., New York, ix + 373 pp.. (The reader may omit Chapters 10, 11, and 12, if there is not time to read all of the book.)

In addition to many books on scientific experimentation, the reader will find it valuable to read:
Bennett, J. H. (Editor) (1965). *Experiments in Plant Hybridization, Gregor Mendel (with introduction and comments by R. A. Fisher).* Oliver and Boyd, Edinburgh and London, ix + 95 pp..

4.13 PROBLEMS

Problem 4.1.   This problem is designed to illustrate the selection of treatments or kinds of observations bearing on the hypotheses being considered. In all cases, select the *minimum* number of observations (readings on the hour) differentiating among the five hypotheses listed below.   Our clock has only an hour hand and we can obtain only observations on the hour. One of the observations is to be at 12:00 noon. The five hypotheses are:
  a. The clock has stopped.
  b. The clock keeps perfect time.
  c. The clock revolves forward three complete revolutions every 12
     hours.
  d. The clock is running backward at the rate of one revolution every 12
     hours.
  e. The clock is running backward at the rate of three revolutions every
     12 hours.

Given that readings can be observed only at 3:00, 6:00, 9:00, and 12:00 o'clock, what observations are necessary to differentiate among the above hypotheses? Suppose we hypothesize that the clock is running either forward or backward 1, 2, 3, or 4 revolutions per 12 hours, or that it could have stopped. If readings are available only on the hour, what readings are necessary to differentiate among these nine hypotheses? Would it be helpful if an observation at 12:00 noon on 1000 successive days were available? Why, or why not?

Problem 4.2. Given 12 objects, 11 of which weigh the same and the 12th is either lighter or heavier and a balance scale which will accommodate up to 6 objects on each side, design a procedure of four or fewer weighings such that the odd object can be identified, and that it can be determined whether it is lighter or heavier than the others. (Note: obviously seven weighings would determine the relative weight of the odd object if the 12 objects were first divided into six pairs and each pair weighed to give six weighings; one of the six pairs would not balance. By noting the direction of the balance and by weighing one of the objects from the pair not balancing against one object from a pair that balanced, one can determine the odd object and whether it is lighter or heavier on the seventh weighing. It is possible to find a solution in three weighings but a four weighing design will suffice for this class.)

Problem 4.3. The dictator of a "People's Republic" has ten workers making gold bars weighing ten kilograms each. Each worker is to make gold bars of identical size and place ten of them in a box each day as the day's quota. The dictator, being an efficient person and knowing that some or all of the workers are materialistic and would filch gold whenever possible, wants to check on the workers as simply as possible and wants to detect any worker who does not make gold bars weighing ten kilograms. Furthermore, the process is such that if any amount is taken from a bar it would be precisely x units. How would the dictator do this with one weighing?

Problem 4.4. A man buys a newspaper and a bag of peanuts. He becomes engrossed in reading the newspaper while absent-mindedly munching on the peanuts from the bag. He has eaten all but two peanuts from the bag. In observing the remaining two, he notes that they are full of worms. What inferences might he make from the sample of peanuts consisting of the two remaining peanuts? What is the type of inference used? What elements would tend to make the sample of two peanuts representative of the population of peanuts in the sack with respect to worminess?

Problem 4.5. Prepare an outline for an investigation either to be or which has been conducted by you. Follow the steps in scientific investigation in preparing your outline.

Problem 4.6. Prepare a list of key words in this chapter and show relationships among them.

CHAPTER 5. SURVEYS, SURVEY DESIGN, AND INTERVIEWING

5.1 INTRODUCTION

A survey is an investigation of what is present in a particular population. If every member of the population is investigated or measured relative to the characteristic being considered, this is called a **census** or a 100% sample survey. If only a fraction of the population is investigated, this is called a **sample survey**. Whenever data are collected in a survey, a population and a characteristic of the population are automatically implied. Therefore, whenever a survey is contemplated, it is absolutely essential that the population and the characteristic to be measured are precisely defined. Failure to do this can lead to a set of meaningless numbers and endless difficulty in interpretation.

In the class survey described in Chapter 2, our universe or population was defined to be those students registered in the course for credit. Visitors were not a part of the population. Furthermore, we decided to measure the heights of class members in yards, in feet, and in millimeters, to measure weights of individuals in pounds, and to ascertain hair and eye colors. But was this a sample survey or a census? We had attempted to obtain a census, but some students were absent and some refused to have their weight taken. Hence this survey ended up as a sample survey of those attending class on the given day and of those who cooperated in the survey. All surveys should have a goal. Ours was to acquaint students with measurements and with a survey and to provide a set of data for future class use.

There are many ways of conducting a survey. Some methods are better than others, some lead to erroneous conclusions, some lead to naught, and others to providing the desired information. The sampling procedure should not be selected "to prove one's ideas or prejudices" but rather to seek the truth as a true scientist should. Mother Science is an extremely hard taskmaster; if one does not find the truth, someone, sometime, somewhere will. Some investigators may fool others, and sometimes even themselves, with biased information, but in time the truth will come out.

Although very few of us have ever thought of ourselves as samplers, we are continually taking sample surveys. One of the first surveys that is taken every day is of the weather. From the sample of weather facts observed or heard on the radio, we decide on wearing apparel and perhaps on activities for the day. Then some of us sit down to breakfast to continue our sampling. If the first taste of hot cereal is satisfactory, we continue eating. If the cereal is burned or is too heavily salted, we may discontinue eating after the first bite (sample). This process of sampling continues all day long, in one form or another, from sipping coffee to answering questions on an examination. Each of us has our own peculiar method of sampling and in drawing inferences from the sample facts about the population parameters. We shall discuss various methods of taking surveys, that is, sample designs that are useful in survey work.

Another question that arises is whether a sample or a census should be taken. Since we know that samples may vary considerably, why not always take a census in order not to have sampling errors affecting our results? In certain situations, it may be impossible to take a census. For example, would it be possible for a girl to date *all* boys in the city in which she lives in order to decide on a husband? The answer is obviously "no"; she samples until she finds the "right one" and obtains a proposal. Likewise, in sampling firecracker stocks for exploding, ammunition for firing, wines for tasting, light bulbs for length of life, pies for suitability for human consumption, and so forth, the sampling is destructive; in order to have some material left to sell or to use, a census is out of the question. In order to ascertain the temperature of a cup of coffee, a small sip is recommended over the entire cupful simply because of the possible consequences. Likewise, one should not dive into a pool of water until one knows the depth (from a census) and the temperature of the water (from a sample), and it is fatal to let the doctor take too large a sample of your blood. In heavily populated countries such as India and China, a census of people and some of their characteristics is not feasible because of the logistics problem of training and supervising so many census reporters. Also, it would be almost impossible to take a census of a species of fish in a large lake such as Cayuga Lake. Even if the logistics problem could be solved, the cost of taking a census could become astronomical. Thus, we see that not only are we all samplers but that some situations require that we take a sample of a very small fraction of the total population.

The accuracy of a result may even be inversely related to the amount of sampling. For example, it has been found in taking inventory of liquor stocks that the reliability of the results and legibility of writing decreases and the "shrinkage" in stocks increased as the size of the sample was increased! Also, suppose that we wished to know the weight and length of 1000 fish caught on a given day by 200 fishermen. We could ask the fisherman, but we all know about "fish stories"! Alternatively, we would obtain a more reliable estimate by actually weighing and measuring a randomly selected sample of 1000 of the fish caught by the fishermen. As another example of the necessity of taking a sample, the Iowa State Soil Conservation Service needed to know how much erosion of soil had been caused by the 13-inch rainfall in the month of June. After the rains subsided, they had only a short period of time before the farmers started working their fields and covering up the evidences of erosion. The survey was planned and executed during a period of four days (and nights). Obviously, a census would have been impossible in this situation. (These last examples bring up questions on sample size; this will be considered in a later chapter.)

So far we have considered only situations in which sampling is possible; there are cases where sampling is impossible. For example, students are not allowed to take only a sample of tests, and a teacher is not allowed to grade only a sample of the students' papers. The Internal Revenue Service will not let us estimate deductions from only a sample of deductions. The banks cannot determine whether or not we are overdrawn by considering only a sample of the checks that have been written. Employers are not allowed to pay workers only a sample of their earnings. Students should not be allowed to attend only a sample of the classes given even though this may be a common practice for some students. Thus, many situations require a census or complete enumeration rather than a sampling of the entities under consideration.

For additional comments, the reader is referred to Slonim's (1962) book entitled *Sampling.* The book is almost devoid of mathematical results and contains a wealth of examples from all walks of life. Several examples used herein were taken from this book. Other examples may be found in the selected readings listed in Section 5.7.

Sample survey design is only one aspect to be considered in conducting a survey. In Section 5.2, the various steps to be followed in conducting a survey are presented. Types of sample survey designs are presented in Sections 5.3 and 5.4, with biased designs being discussed in the former and unbiased designs in the latter. Randomization or random allotment in the selection of sampling units is discussed in Section 5.4.1 in relation to fairness and equal opportunity. Questionnaire construction for interviews and checks on interviewers are discussed in Section 5.5. There are many situations wherein anonymity of responses is not required. This situation is considered in Section 5.5. In other situations, anonymity of respondents answers may be required because of the sensitive, embarrassing, or incriminating nature of the questions. Methods for obtaining anonymous responses are given in Section 5.6.

5.2 STEPS IN A SURVEY

The steps listed under principles of scientific investigation as given in Chapter 4 could be used for any survey, or the steps could be listed in a variety of different manners. An alternate form specifically related to a survey is given below.

The *first* step in planning a survey is to define the population or universe to be surveyed for the desired characteristics as related to the objectives or goals of the survey. For example, if we decide to study selected characteristics of dental patients who patronize Clinic HURT staffed by Drs. Filmore, Payne, and Toothacher, we must decide what is meant by "patronize" and by "dental patients" and whether or not we wish to consider the patients of the three dentists separately, i.e., stratify them into three groups. If we select those individuals who enter the clinic, we might find that the girl who is very interested in young Dr. Filmore may appear quite frequently in our survey, but that no dental work was performed on her teeth. She might be "patronizing" but still not fit our definition!

After determining precisely what population is to be studied, the *second* step in planning a survey is to develop a list or a description of all individuals in our population. This is called the **sampling frame**. From the patients' records, we could obtain a complete list, the sampling frame, of names of all patients (the sampling units) of Clinic HURT during specified time period. From registration records, we could obtain a listing of all members of this class; from voter registration lists, we could obtain a listing of voters in a precinct, and so on to obtain sampling frames for a variety of populations. In another case, we could describe our population to be all goldfish in a pond which did not slip through a net with holes one-half inch in diameter. It is imperative to describe the population and the individuals in a population for all surveys.

The *third* step in planning a survey is to describe precisely the information being sought. In a medical study, the information desired may be average time to recurrence of tuberculosis or cancer rather than the incidence of recurrence of the disease. In a food preference study, it may be important to know what food a person would eat if all types were available; an individual may be eating what he does simply because the local supermarket does not stock his favorite foods. In a study on clothing sizes in an army, it is essential to define whether we wish information on sizes of clothing issued, on actual sizes that an individual would wear if he or she had a choice, or on actual sizes that an individual should wear as determined by a panel of "experts". The type and amount of information desired should be carefully and thoughtfully considered.

The *fourth* step in planning a survey is to determine whether or not the information is already available. It is much cheaper and less time-consuming not to take a survey than to take one. Nothing is to be gained by obtaining a second, third, or even a fourth survey on the same item. In fact, in surveying people there is much to lose in that people quickly become tired, and sometimes angry, if they are interviewed too often and quite often this means more than once. When this happens, the non-response problem increases in later surveys. Someone who has gone through three long interviews in one week is quite likely to become a non-respondent!

The *fifth* step pertains to solicitation of only essential information. For example, many of you would become quite perturbed, and rightly so, if on each quiz you were required to write, in addition to your name, your parents' names and ages and their educational background, and your social habits during the week. This information, with the exception of your name, is useless for purposes of obtaining information on your comprehension of subject matter in a given course. Anyone will occasionally, or even frequently, be guilty of obtaining useless information or information that is readily available elsewhere, and hence everyone needs to be constantly on guard against this. As Slonim (1962) says, "With the exception, perhaps, of stilts for a serpent, there is nothing more useless or ridiculous than a mass of figures collected at great travail, added, multiplied, divided by the cube root of π, and converted (by a \$350-an-hour electronic computer) to homogenized index numbers - that have no bearing on the problem".

The *sixth* step concerns sample size which in turn is related to precision desired for the sample survey, to the available funds, and to the method for obtaining the information (for example, personal interview versus mailed interview). After the population, sample size, and other information about the universe are precisely and clearly defined and determined, the observational unit and the sampling unit must be clearly described and understood. The **observation unit** is the *smallest unit* on which information is obtained, while the **sampling unit** is the *smallest unit* that defines the population under consideration. For example, suppose that a list of all N houses in a city is available. Furthermore suppose that numbers from one to N are placed on tags, the tags are placed in a large container and thoroughly mixed; then a sample of n tags is drawn blindly from the container to obtain a random selection of houses. The characteristic of interest is the type of roof on the house. Suppose that two enumerators independently determine the type of roof. The house is the sampling unit and the determination of roof type by an enumerator is the observational unit. As a second example, suppose that a list of all N classes in high schools in Tompkins County for 1989-90 is the sampling frame. By the process described above, a random sample of size n < N classes is selected. Suppose the characteristic of interest is the teaching method employed in the class and responses related to teaching method are solicited from each student in the class. The response from an individual student is the observational unit and the class is the sampling unit. If the characteristic of interest were type of breakfast eaten that morning by a student, then the student would be the sampling unit, and the response for type of breakfast eaten on a particular morning would be the observational unit.

The question concerning the desired number of sampling units for a survey is taken up in Chapter 13. However, in order to obtain an idea of the effect of increasing sample size on the variability of a statistical estimate, let us consider the following example. Suppose that a sample survey is designed as a simple random sample as described in Subsection 5.4.1, suppose that n = 100 individuals are polled in a city as to their preference for one of two mayoral candidates A and B, and suppose that 60 favor A and 40 favor B. The proportion favoring A is $p = 60/100 = 0.6$; the proportion favoring b is $(1 - p) = 40/100 = 0.4$. One measure of the variability in p is called the estimated standard deviation of p, or of (1 - p), and is defined as $[p(1 - p)/n]^{1/2} = [0.6(0.4)/100 = 0.0024]^{1/2} = 0.049$. (This statistic is described in Chapter 10.) We may note that as n increases, the estimated standard deviation of a specified proportion p decreases; for example, for n = 1000, $[0.6(0.4)/1000 = 0.00024]^{1/2} = 0.0155$. The value for n = 1000 is about one-third of the estimated standard deviation for n = 100 and p = 0.6. If the sample size had been 50, the estimated standard deviation would be $[0.6(0.4)/50 = 0.012]^{1/2} = 0.11$. Writing the estimated standard deviation in the form $[p(1 - p)]^{1/2}/[n]^{1/2}$, we note that this ratio decreases as the square root of the sample size n increases. As the sample size n approaches infinity, the estimated standard deviation approaches zero for any p not equal to zero.

In the above example, A appears to be leading, but how does one know whether or not the proportion of people favoring A in the sample is the same as in the entire city? One cannot know this without taking a census, but one can obtain an interval estimate on the true proportion preferring A (or B). To do this, we may (as explained in Chapter 12) construct an interval estimate of the true proportion as $p - 2[p(1 - p)/n]^{1/2}$ to $p + 2[p(1 - p)/n]^{1/2}$ which for the above example with n = 1,000 is from 56.9% to 63.1% who prefer A. If such were the situation, A would have more confidence in his victory than previously. However, for a sample size of n = 100, the interval estimate of the true proportion of people favoring A would be from 50.2% to 69.9%, and A would be less confident of victory. If the sample size had been only 50, then the interval estimate would be from 38.0% to 82.0%, and A would have little confidence in obtaining a victory.

Sample survey design is a *seventh* step in the planning of a survey. Although we shall discuss several sample survey designs and several additional ones are discussed in Cochran (1963), Hansen, Hurwitz, and Madow (1953), Kish (1965), McCarthy (1957, 1970), Slonim (1960), Sukhatme (1954), and Yates (1960), it is always a good idea to consult with a statistician on a survey design. The statistician should be conversant with the theory and application of theory for survey design procedures. The advantages and disadvantages of a number of sample survey designs are discussed in the following sections.

As an *eighth* step in planning a survey, the reporting form or questionnaire needs to be constructed. Simplicity, brevity, and clarity are three very important ingredients for any questionnaire. The form of the question is also very important. Beware of ambiguous questions such as the "wife-beater", implicating, or biasing question such as:

"Do you still beat your wife?"

"Do you still cheat on examinations?"

"Are you pregnant again?"

"Have you ever been unemployed before?"

"In light of the high cost of no-fault insurance, would you favor eliminating it?"

"Considering that one-man one-vote has destroyed the economies of countries A, B, and C, would you favor one-man one-vote for country D?"

If the answer is "no", we don't know whether the person has stopped beating his wife or has never beaten her. The same would be true for the question on cheating. The person may not be unemployed now as implied in the question. Since the cost of no-fault insurance is "high", many people would oppose it simply because it was stated that it was high and not on the merits of no-fault. A better question here would be, "Do you favor or oppose no-fault insurance?" with a possible "don't know" category. The conditions in country D may or may not be the same as in countries A, B, and C but this would need to be determined. Some steps to consider in questionnaire construction are:

1. comparing the list of objectives of the survey with the list of questions to determine whether they coincide,
2. making certain that only relevant questions are included,
3. using as short a questionnaire as possible to achieve the objectives,

4. critically examining each question for ambiguity, clarity, and pertinence,

5. pre-testing the questionnaire on a small sample and revising, if necessary, and

6. consulting an expert on questionnaire construction.

If all of the above are carefully considered, useful information should be obtainable. Also, it is necessary to decide whether the interviews are to be conducted by mail, by telephone, from a panel, by personal interview, or by some other means. It is essential to realize that the method of obtaining an answer to a question may, and often does, influence a respondent's answer. An answer to a question of age may be quite different over the telephone than in a personal interview. Answers by telephone may not be considered as binding as answers written in response to a mailed inquiry. The non-response problem is often quite high with mail questionnaires. What are we to do if only 5% of our mailed questionnaires are returned? Such questions as these must be carefully considered and decisions reached before proceeding with a survey.

As a *ninth* step in planning a survey, it is necessary to train the individuals who obtain the interviews and record the results. It is essential that interviewers do not influence the answers given by interviewees. The method of asking a question can influence the answer, and the interviewer must (unless the respondent does) write down the answer correctly. A method of checking on the interviewers is necessary, since they might be tempted to fill out questionnaires without even contacting the interviewees. One method of training individuals to obtain and to record answers correctly is to have them interview each other during the training sessions and to illustrate how to obtain unbiased answers. It has been demonstrated that certain questions may be influenced by the age, sex, race, and dress of the interviewer.

The *tenth* step in planning a survey is to determine what will be done with "not-at-homes" or "refuse-to-answer" interviewees, how the returns are to be audited, what information is desired from the survey, and what safeguards are to be utilized to guard against reporting and processing errors. It is a well-known fact that individuals often report less income than they actually receive and that the age or weight of men and women is often given as less than it actually is.

To stress further the importance of adequate training of enumerators and of adequate checks, consider two incidents that happened to a fellow statistician, Mr. F, when he was enumerated in two different ways. In the first instance, a young lady came to Mr. F's home and told him that she was taking a survey for a laxative company. After the proper introductions, she asked a number of questions. One question was, "How often do you use laxatives?" (A loaded question.) Mr. F answered that he never had. Then, to his horror, he observed that she wrote down that he used laxatives four to five times a week. Mr. F asked why she had deliberately written down a falsehood. She replied that the company would not like it if she were to write down his answer. Mrs. F answered in the same manner as her husband, but the enumerator indicated on the schedule that Mrs. F used laxatives every day of the week. In a second survey in which Mr. F happened to be one of the "unlucky" individuals drawn, he was asked why he did not shop in Ithaca. (Another loaded question). Mr. F, being a very serious individual, explained in detail why he did not. Again the enumerator did not write down his answers but something else. When asked by the indignant Mr. F why his correct answer was not written down, the enumerator replied that it really did not matter because everything would average out anyway!

After a careful consideration of each of the above ten steps in planning a survey, the surveyor will need to do the following steps in order to complete the investigation:

Step 11. conduct the survey,
Step 12. summarize and interpret the results from the survey, and
Step 13. report on the results of the survey in a careful, precise, and scientific manner.

5.3 TYPES OF BIASED SAMPLE SURVEY DESIGNS

Although we cannot hope to cover all possible types of sample survey designs, we shall present some of the simpler and more commonly used ones. Both non-probability and probability sample survey designs will be discussed in the following sections. When the chance of selecting any one of the possible samples of size n is known, the sample survey is denoted as a **probability sample survey design**. When the probability of selecting a sampling unit is not known, this is known as a **non-probability sample survey design**. The latter are the ones discussed in this section. Again it should be emphasized that it is wise to consult with a person well-versed in sample design before conducting the survey.

5.3.1 The Purposely Biased Sample

In certain instances, an individual wants to obtain some numbers which he hopes to pass off as data. When the investigator obtains responses in a survey by purposely biasing the results, this is defined to be a **purposely biased sample survey design**. Extremists, politicians, organizations, advertisers, labor unions, and so forth, often obtain information only from people who respond in the desired manner. They may take a sample of those people who attend one of their meetings and then use the results as if they represented the entire population of people. Statements such as "two to one prefer Morning Cough Cigarettes", "nine out of ten dentists prefer Evening Cough Cigarettes", and "every person interviewed stated that All Day Cough Cigarettes were milder than those of their own brands", are examples of purposely biased investigations. This is especially evident when it is reported that the two-to-oners did their survey in "several large cities" as follows. Four large cities were selected. Two interviewers went to a busy area of the city. One member of the team gave away free samples of Morning Cough Cigarettes. He walked along with an individual who accepted a free cigarette and lit up. The other member of the team was stationed a short distance down the block; he interviewed two of those his colleague had walked with to one of those that he did not walk with but who was smoking a cigarette. Thus, for every three individuals interviewed, two were found to be smoking Morning Cough to one who was not. For the "9-out-of-10" group, it is reported that free samples of Evening Cough Cigarettes were distributed in a cafeteria line and that "9-out-of-10" had smoked the free cigarette. This was reported as a preference for Evening Cough by all people in the population of smokers. With regard to the statement that All Day Cough Cigarettes were *always* milder than a smoker's own brand, if a cigarette is lit and if the smoker takes a puff and after a short time takes a second puff, the first puff always appears milder! This was the procedure used to compare the cigarettes, that is, All Day Cough was always smoked first.

5.3.2 Convenience Sample

A convenience sample is defined to be one which is convenient to take without any regard to bias or representativeness of a universe. For example, it might be decided to interview 100 people on some selected subject; the 100 interviews might be taken at Sloe's Bar and Grill because the interviewer frequents the place, at a National League Football game because the interviewer wished to see that game, at the 100 households nearest his home, or the 100 interviews may be limited to the interviewer's boy friends and their girl friends. As examples in other areas, the interviewer may take 100 lumps of coal from the top of a carload, the interviewer may select the sample solely from the members of his class, a comrade may interview fellow comrades attending a given meeting, or the doctor may select the interviews from patients treated at the hospital where she works.

It is easy to continue giving many such examples of convenience samples, since man is naturally a lazy creature and often sacrifices quality to do less work. It is amazing how many investigators use the excuse, "Oh, I couldn't do it that way because it is too difficult", when this is the correct way to obtain unbiased results. Ease or difficulty should not be the criterion used in scientific investigations. Rather the criterion should be to obtain unbiased and interpretable results. The big question about convenience samples is that we have little or no idea about their representativeness of *any* population and no idea about their reliability.

5.3.3 Judgement Sample

The **judgement sample** is one obtained by a self-styled expert in a subject matter area wherein the "expert" reasons that his intimate knowledge of all sampling units in an area allows him to select a sample which represents specified characteristics of the population. For example, suppose a School Board member, Mr. I. M. Smart, uses Mr. and Mrs. I. Don'tknow, Mr. and Mrs. I. M. Rich, Mr. and Mrs. I. M. Fashionplate, and Mr. and Mrs. U. R. Average to represent the opinion of the members of a school district. Mr. Smart may firmly believe that these four couples represent the opinion of all members, but his confidence in his great discerning abilities by no means makes the sample representative. Individuals do exist who are able to take the pulse of a population with regard to various issues, but the ordinary person using judgement samples is usually only fooling himself (and perhaps others). Even the true experts can go astray when something unusual happens which they do not take account. Judgement samples are to be used with extreme caution but even when so used, they provide no measure of reliability as do the sample designs discussed in Section 5.4.

Politicians have a great desire (and need) to know how they are faring with the voters who will vote on Election Day. As a result there are numerous polls conducted during election years. The poll might be taken by a roving reporter such as Sam Lubell who travels throughout the United States "taking the pulse of the average voter". There is little doubt that the survey is a judgement one but it is also a conveniece type of survey. Other polls consist of surveying responses of alleged voters on Election Day from "key counties". The key counties are judged to be representative of the voting responses of the entire population simply because of the past voting-for-the-winner record of the county. Many things could happen to the population of one of the selected key counties to make them vote with a loser at the next election.

5.3.4 Quota Sample

The population may be sub-divided into several sub-populations or strata representing the various categories. In this sense the sample is representative. Suppose then that the interviewer or surveyor is allowed to select individuals or units within each stratum or sub-population in any manner he desires, until the specified number or quota of individuals is obtained. This type of sampling is called **quota sampling**. To illustrate this, suppose that our population is stratified into men over and under forty, women over and under forty, and persons with or without a college degree to give eight strata or groups. The surveyor then selects 50 individuals in each category to obtain the desired information. There are many ways in which the sample could be selected. Many quota samplers simply obtain a convenience sample or a judgment sample in each stratum. More than likely, any difficult-to-obtain individuals are omitted. The representativeness of such a sample within each stratum is highly suspect, since there is no way of ascertaining the reliability of samples taken in this manner, as there is with the probability samples discussed in the next section.

5.3.5 Haphazard Sample Survey Design

A **haphazard sample survey design** is defined to be one where the investigator selects individual sampling units whenever his fancy strikes. He simply goes and without any plan selects units for response at his inclination. There may be a mixture of judgement and convenience in the selection of the sampling units. Responses obtained in this fashion are often stated to be taken at "random". In this text we differentiate between haphazard and random sampling. The latter involves using specific randomization procedures as described in the following sections. As an example, consider the following. Sam Blum in an article entitled, "What makes a perfect husband" (*McCalls,* August 1967, page 61) wrote, "I interviewed, at random and with relish, secretaries, college girls, school-teachers, a young woman standing in line to see a movie, two dancers, a salesgirl, several women temporarily between marriages, the wives of friends--a whole array of husband watchers". This type sampling may have been done with relish but it was haphazardly and not randomly done.

5.3.6 Some Comments on Biased Sampling

As we have noted, most people are surveyors and some even make their living taking surveys. Most of us have heard of Gallup, Roper, Nielsen, National Analysts, and so forth, and there are many survey firms of which we have never heard. We could rightly presume that these firms use a variety of procedures to obtain results. This writer was rather surprised by an article appearing in *The Ithaca Journal* dated March 16, 1967 and entitled "Survey Firm Apologizes." The survey firm was called the "Voice of the Voter." In the article, the survey procedure was described as follows: "Some 1500 questionnaires were distributed in the county the weekend of Feb. 4-5, but only 35 were completed and returned.... 'Our organization considered that at least 100 replies back would be necessary before we could effectively go ahead with further action'." This firm appeared to imply that 100 responses out of 1500 would have been satisfactory; however, the firm did use only 35 returns out of 1500. What about the opinions of the 1465 out of 1500 who did not respond? Even if 100 returns had been available, what could one really conclude about the opinions of the 1400 non-respondents? Better survey procedures would have eliminated the need for an apology from the survey firm.

Another mistake often made by surveyors is that of taking a prejudiced sample of size one and then drawing conclusions from said survey. The proud husband often talks about his wife's shopping habits, cooking procedures, and so on. Likewise, the biased parent believes their child to be unusually bright and thinks of all children's reactions in light of their child's reaction. This idea of taking a sample of size one goes further than within the family. Slonim (1962), pages 86-7, describes a situation where the owner of a large hosiery factory believed that 100,000 dozen pairs of nylon hose worth about one megabuck were disappearing from his plant every year. They used all means of ferreting out the culprit who might be appropriating them. As everyone and everything was searched, it developed that the company estimated the annual output of nylon based on the performance of *one* operator. It turned out that this one operator was the best in the plant and she used less yarn than any other operator; the 1,200,000 pairs of nylons that were presumed to have been filched had never been made!

In another situation, a hybrid seed corn company in Iowa estimated the amount of seed corn for sale in the spring from the corn that had been harvested in the fall. The latter often contained 40% or more of moisture before the moisture content was reduced by drying to 15% moisture for storage. A chart was used to determine the amount of corn at 15% moisture content that would be obtained from X pounds of corn at 45%, say, of moisture. Every spring, the company appeared to have lost several thousand pounds of valuable seed corn. They suspected piracy, but from all appearances they had reliable and trustworthy employees. After considerable investigation, attention was turned to the chart itself. It turned out that the chart had been developed for moisture contents of less than 35% and that a straight line prediction was used. After additional investigation, it was found that both the chart and the machine used for determining moisture content were biased and inaccurate for samples of corn with more than 35% moisture content. No one had taken any valuable seed corn. It had simply never been there in the first place.

The above are examples one would not like to experience personally. With proper survey procedures one can avoid most, if not all, of the pitfalls of inadequate or biased sampling. Close adherence to the ten steps for survey work as listed in Section 5.2 will result in representativeness of samples and reliability of results.

5.4 TYPES OF UNBIASED SAMPLE SURVEY DESIGNS

Since most investigators desire information that could be used to make inferences about some specified population, they desire sample survey designs which accomplish this objective. The class of designs known as **probability sample survey designs**, i.e., the probability or chance of selecting each individual sampling unit is known and used in the selection of the sample, represent designs which yield results applicable for the entire population. The chance of selecting specific sampling units in the designs discussed in Section 5.3 is unknown before the units are selected. Hence, these designs are all classified as non-probability sample survey designs. There are many types of probability survey designs. Some of these are discussed in the following sub-sections, but first we discuss randomization and randomizing devices and procedures.

5.4.1 Random Selection, Chance Allotment, and Equal Opportunity

We hear a great deal about equal opportunity these days, and it is a concept that we shall utilize in order to obtain unbiased estimates of parameters. To assure unbiasedness we are forced to use probability samples, which give an equal opportunity or chance of selection to all possible samples of size n. In order to attain the "lofty goal" of equal opportunity, we need some device for doing this. In particular, a device is required which gives all numbers 1, 2,..., N, where N = the number of possible samples, an equal chance of being selected. One way of doing this is to obtain N ping-pong balls and number them consecutively 1, 2,..., N. The ping-pong balls are then put into a hypothetical "hat", for example a wastebasket. The balls in the "hat" are thoroughly mixed, for example by throwing all those in the "hat" against an uneven wall and then sweeping them up and putting them back into the "hat". The idea of thoroughly mixing the balls is crucial, as this produces the independence of selection between two balls. (In a proposed draft-lottery system, Jim, Bob, Bill and Sam could each have an equal chance of selection, but the selection of one should not affect the chances of any of the others being drafted.) Then we close our eyes, reach into the "hat", and blindly draw out a ball. The ball could have any number from 1 to N on it; any number from 1 to N has an equal chance, or equal opportunity ,of being selected, that is one out of N or 1 / N.

As is evident from watching lottery numbers being drawn on television, lottery officials make frequent use of ping-pong balls. In fact they go to great lengths to ascertain that the balls are all of the same weight and shape. In an article appearing in *The Cornell Daily Sun*, 2/29/80, it is stated that the balls cost $70 apiece and are replaced every two to four months. Several sets of ten balls are available for use. There are many regulations pertaining to the balls and their use in drawing lottery numbers. Considerable precautions are taken to assure that no individual knows which set of balls will be used in the drawing. The device used to draw balls with the winning numbers involves forced air which has all balls in a set suspended and bouncing. Balls are inspected frequently to assure that none are chipped, broken, or damaged in any manner. These elaborate precautions are taken so that each player knows that they have the same chances of winning as any other player. Participation in lotteries would decline drastically if the players felt that others had a greater chance of winning than they do. Players do not knowingly play in "fixed games".

Now if we return the ball to the "hat, and after thoroughly mixing the balls, again draw one in the manner described previously, any number 1, 2,..., N has an equal chance of selection, even the one we returned to the "hat". Thus the same ball could be drawn the second time by mere chance alone. When such a scheme is utilized repeatedly, a series of random numbers can be generated. Also, when equal opportunity of selection by a chance process is possible for all elements of the universe, this is defined to be **random or chance allotment** and the process is called a **random process**. By returning the ball to the "hat" after every draw and repeating the sampling we have been **sampling with replacement** of the ball.

If the process described above is used to produce a series of random numbers but the ball is *not* returned to the "hat" after each draw, this is described as **sampling without replacement**. Thus no item is repeated, and we have n different numbers resulting from n drawings. In most surveys we do not wish to have any one person or item appearing or being interviewed in the sample more than once; thus, sampling without replacement would be utilized for most surveys. For the random number generator described above, we would thoroughly mix the ping-pong balls and blindly draw one; then the remaining N - 1 balls would be thoroughly mixed and a second ball drawn; and so on, until n balls have been drawn. The numbers on the n balls would be the random sample of n numbers from N numbers. The elements in the population, which had been previously numbered from 1 to N, then correspond to the numbers on the N balls; this represents the sampling frame. The selected random sample would correspond to those numbers on the n randomly selected balls.

In the examples considered in this chapter, we have been talking mostly about sampling without replacement. Hence, in drawing a random sample of three different letters from the six letters a, b, c, d, e, and f, we would put six ping-pong balls numbered a, b, c, d, e, and f in the "hat". After thoroughly mixing the balls, we blindly draw one from the "hat"; the remaining five balls are again thoroughly mixed and again we blindly draw one ball from the "hat", and then the remaining four balls are thoroughly mixed again, and again a ball is blindly drawn from the "hat". This results in a random sample of three letters from the universe of six letters, this being one of the 20 possible samples listed in the next sub-section. Since the thorough mixing allows an equal and independent opportunity of any ball being selected, we have a random sample of size n = 3. Alternatively, we could have numbered the 20 possible samples from 1 to 20, and we could have obtained 20 ping-pong balls numbered consecutively from 1 to 20. The 20 balls could have been put into a "hat" and thoroughly mixed; then one of the balls is blindly drawn, resulting in a random sample of n = 3 letters corresponding to the number on the ball. The process would result in a random sample of n = 3 letters. The randomization procedure here selects from the same 20 possible samples and with the same opportunities as the procedure described in the preceding paragraph.

Another device for obtaining a set of random numbers is through the use of a telephone directory. Suppose we number the digits starting from the right of a seven-digit telephone number as the 1st, 2nd, 3rd, 4th, etc. digits. If we drop the 1st digit and use only the 2nd, 3rd, and 4th numbers of a telephone directory for a large city, it has been demonstrated that these three digits may be used as random numbers of 000 to 999 or for N = 1000 numbers. Using any two of these three columns of digits would result in random numbers of 00 to 99 or for N = 100 numbers. Likewise, use of any one of the three columns may be used for random numbers from 0 to 9 or for N = 10 numbers. If N is smaller any unused numbers are ignored. Numbers in the directory which are in bold face type should be omitted, as they usually contain consecutive or repeating digits, and an arbitrary start (randomly selected if possible) in the directory should be made. One method of starting is to pick the next word you hear, find its alphabetical place in the directory, and start from there, either proceeding forward or backward. To illustrate, let us select a random number from the set 01, 02, ..., 20. From the Madison, Wisconsin, telephone directory, let us start after the name U. R. Hoarse to obtain the following numbers (only the last 4 columns given):

7646
7<u>20</u>3
7422
2224
7688
4<u>08</u>3
3560
2537

.
.
.
.

The first number between 01 and 20 that occurs in the second and third columns is 20 and the second number appearing is 08. If one wished to use all numbers between 00 and 99, each number could be divided by 20 and the remainder used to obtain the number, with 00 being equated to the number 20. In the above list of numbers, the remainders would be (columns numbered from the right):

3rd and 4th columns	2nd and 4th columns	2nd and 3rd columns
16	14	04
12	10	00
14	12	02
02	02	02
16	18	08
00	08	08
15	16	16
05	03	13
.	.	.
.	.	.
.	.	.

Any number that repeats is ignored, as is a number not in the set of numbers used. The use of remainders allows many more numbers to be used without so much skipping.

There are several methods of checking a randomizing device to ascertain whether or not the numbers are appearing randomly. Usually no one method is infallible but there are some easy checks that are useful. One is to check whether or not each number is appearing equally often in a large set of random numbers. For example, draft eligible people were given numbers 1 to 366 and one of these numbers was drawn each day of the year in 1972. Let us group these numbers into the lower one-third of the numbers, i.e., 1-122, the middle one-third, i.e., 123-244, and the upper one-third. The following data by months were obtained:

Number drawn	Jan	Feb	Mar	Apr	May	June	July	Aug	Sep	Oct	Nov	Dec	Total
1-122	10	10	12	12	11	13	10	14	5	10	4	11	122
123-244	10	6	8	7	9	9	9	11	15	10	16	12	122
245-366	11	12	11	11	11	8	12	6	10	11	10	8	122
Total	31	28	31	30	31	30	31	31	30	31	30	31	366

The expected number for any third by month would be one-third of the days in that month which for a 30 day month would be 10. The 4 for November is rather low from expectation but this appears to be simply random sampling variations. As the number of data sets such as the above are accumulated, the results from each entry should be closer to its expected number.

There are many so-called random tables published in books and many programs for generating random numbers on micro-computers, mini-computers, and high-speed computers are available. One of the most extensive, if not the most extensive, publication contains a million random digits; it was put out by the Rand Corporation, and is entitled *"A Million Random Digits with 100,000 Normal Deviates"*.

5.4.2 The Simple Random Sample

A **simple random sample** of size n may be defined to be one in which all elements in the population have an *equal and independent* chance of being selected in the sample, where independent means that the selection of one element in the sample does not affect the selection of another element. Alternatively, if *all possible* samples of n elements have an *equal* chance of being selected, the sample is said to be a simple random sample of size n. To illustrate, suppose that the population consists of the elements a, b, c, d, e, and f; suppose that a sample of size n=3 is to be selected. The possible samples of size three are abc, abd, abe, abf, acd, ace, acf, ade, adf, aef, bcd, bce, bcf, bde, bdf, bef, cde, cdf, cef, and def, resulting in 20 samples of size n=3. If the samples are numbered, and if a number between 1 and 20 is randomly selected such that any number has an equal chance of being selected, then a simple random sample results. Alternatively, we could randomly draw one member from the population of six letters, then randomly draw one member from the remaining five letters (sampling without replacement), and then randomly draw one member from the remaining four letters. The latter procedure would give an equal and independent chance of selecting any element in the population, resulting in a random sample of size n = 3.

It is not enough to give an equal chance of selection to each sample member. The chance of selection must be *equal and independent.* Suppose that we decide to have two samples, abc and def, of size n=3 of the 6 letters, and suppose that we flip an unbiased coin to determine whether we take the sample abc or the sample def. Now every element in the population has an equal chance of 1 / 2 of being selected, but, every time a is selected so are b and c. Thus, the chances of selection are not independent, even though they are equal.

In a simple random sample, the observational equation of the ith observation from the population may be expressed as:

ith observation = true population mean + bias + deviation.

When an observation is for a randomly selected individual in the population, the term deviation is denoted as random error. Symbolically, the above equation may be written in the form:

$$Y_i = \mu + bias + \varepsilon_i$$

where the symbols and the words in the two equations are pairwise equivalent, and i = 1, 2,..., N = the number of observations in the population. In the sample of n observations, the ith observation may be expressed as:

$$Y_i = \text{sample arithmetic mean} + \text{deviation from mean}$$

$$= \bar{y} + (Y_i - \bar{y}) = \bar{y} + e_i.$$

The statistic $\bar{y}$ estimates the true population mean μ plus the bias. If the bias is zero, then $\bar{y}$ is an unbiased estimate of μ. The statistic $N\bar{y}$ is an estimate of the population total.

5.4.3 Stratified Simple Random Sample

If the population is divided into sub-populations and if a simple random sample is selected from each sub-population, the resulting sample design is

defined to be a **stratified simple random sample**. The selected
sampling units are grouped, blocked, or stratified according to sub-
population. A simple random sample of size n_i, n_i not necessarily equal, is
selected in the ith sub-population for all i (i.e., in every sub-population).
This type of grouping of the entire population into sub-populations is
known as **stratifying or stratification**. The sub-populations are known
as **strata**, and an individual sub-population as a **stratum**. Instead of
stratum we could use the terms groups or blocks. Suppose that our
population consists of the first nine letters of the alphabet and that we
block, group, or stratify the nine letters into two strata such that vowels are
in stratum 1 and consonants are in stratum 2, resulting in

Stratum 1	Stratum 2
a	b
e	c
i	d
	f
	g
	h

Now suppose that we take a simple random sample of one item from *each*
stratum. The 18 possible stratified random samples are:

	1	2	3	4	5	6	7	8	9	10	11	12	13	14	15	16	17	18
Stratum 1	a	a	a	a	a	a	e	e	e	e	e	e	i	i	i	i	i	i
Stratum 2	b	c	d	f	g	h	b	c	d	f	g	h	b	c	d	f	g	h

Taking the same number of sampling units from each stratum is denoted as
equal allocation.

Suppose on the other hand that we wish the sample size, n_i, in each
stratum to be proportional to the stratum size, say N_i. Since stratum 2 is
twice as large as stratum 1, we should select twice as many observations in
stratum 2 as in stratum 1. This type of allocation of sample sizes is denoted
as **proportional allocation**. The latter form has several desirable
characteristics and is used frequently in sample survey design. In this case,
the 45 possible stratified random samples of size three are:

	1	2	3	4	5	6	7	8	9	10	11	12	13	14	15	16	43	44	45
Stratum 1	a	a	a	a	a	a	a	a	a	a	a	a	a	a	a	.a	e	i	i	i
Stratum 2	bc	bd	bf	bg	bh	cd	cf	cg	ch	df	dg	dh	fg	fh	gh	bc		fg	fh	gh

Other types of allocations are possible. For example, the variability in stratum 2 might be twice as large as in stratum 1, and hence we might wish to take four times as many individuals in stratum 2 as in stratum 1, twice as many because of size and twice as many because of greater variability. Cost considerations and other factors may also be involved. In certain situations, one item or a few items may make up a relatively large proportion of the population; for example, one large farm may produce 10% of a commodity, a second farm 5%, and a third farm 4%. The surveyor may, and often does, decide to enumerate these three farms and to sample the remaining farms in the county. The three large farms would make up one stratum or group, large farms, in which a 100% sample would be taken and the remaining farms in the county would be the second stratum, small farms.

Suppose a population contains k subpopulations of size $N_1, N_2,...,N_k$, for $i = 1, 2,...,k$ and $N = (N_1 + N_2 + ... + N_k) =$ total population size. In the absence of bias, an observation Y_{ij} may be expressed as

Y_{ij} = true population mean + deviation of true stratum mean from true population mean + deviation of observation from true stratum mean.

Symbolically, this may be expressed as:

$$Y_{ij} = \mu + (\mu_{i.} - \mu) + (Y_{ij} - \mu_{i.})$$

$$= \mu + \delta_i + \epsilon_{ij}$$

where $\mu = (N_1\mu_1. + N_2\mu_2. + ... + N_k\mu_k.)/N$, $\delta_i = \mu_i - \mu_.$, and $\mu_i.$ is the true stratum mean. For a randomly selected observation, ϵ_{ij} is the random error deviation. In the sample of size $n = (n_1 + n_2 + ... + n_k)$ where n_i individuals are randomly selected from the ith stratum, the observational equation may be written as:

$$Y_{ij} = \bar{y}. + (\bar{y}_{i.} - \bar{y}.) + (Y_{ij} - \bar{y}_{i.})$$

$$= \bar{y}. + d_i + e_{ij}$$

where $y_i. = (Y_{i1} + Y_{i2} + ... + Y_{in_i})/n_i$, $\bar{y}. = (n_1\bar{y}_1. + n_2 \bar{y}_2. + ... + n_k \bar{y}_k.)/n$, $d_i = (\bar{y}_i. - \bar{y}.)$ is the estimated stratum effect for the ith stratum which is $(\mu_{i.} - \mu)$, $\bar{y}_{i.}$ is an estimate of the true stratum mean $\mu_i.$, and $\hat{e}_{ij} =$

$(Y_{ij} - \bar{y}_{i\cdot})$ is the estimated random error component for the ijth observation. If n_i / N_i is a constant for all i, then the allocation is proportional; $\bar{y}_{i\cdot}$ is an estimate of μ_i. and $(N_1 \bar{y}_{1\cdot} + N_2 \bar{y}_{2\cdot} + \ldots + N_k \bar{y}_{k\cdot})$ is an estimate of the population total $N\mu$.

5.4.4 Cluster and Area Samples

If the population is divided, naturally or not, into sub-groups or clusters and if a simple random sample of clusters is selected, this is denoted as a **cluster sample design**. The clusters could be sub-populations or strata. If the clusters form areas, then this is defined to be an **area sample design** or **area sampling**. A 100% sample of the clusters results in a stratified sample where the clusters are the strata.

In certain areas of the United States, the land is blocked into sections which are one mile square, and the cluster is the land which falls within the section lines. One of the early area samples in the United States was a random selection of sections (one square mile) of land in Wyoming. Apple trees are clustered together in an orchard; a random selection of orchards results in a cluster sample. The group of individuals within a household represent a cluster; a simple random sample of households in a given city results in a cluster sample of clusters of individuals in a household.

Nothing has been said so far about how many individuals are selected within each cluster. All sampling units or individuals within each cluster could be enumerated, or alternatively, a simple random sample of n_i individuals could be obtained within each of the selected clusters. Likewise, equal or proportional allocation in the clusters could be used. Proportional allocation is frequently used in cluster sampling. The statistical analyses are simpler for proportional sampling than they are for disproportional sampling. The observational equation for a cluster sample is of the same form as for a stratified simple random sample.

Because natural phenomena frequently lead to clusters of sampling units, a cluster sample design is frequently utilized in survey work.

5.4.5 Stratified-Cluster Sample Design

As might be suspected, the sample design could become more and more complicated as more and more levels of stratification are used. Moe (1952) conducted a survey on farmers' opinions of agricultural programs. He stratified New York by counties; then each county was divided into areas of approximately five farms each. A simple random selection of clusters or areas of five farms was made within each county; the number of clusters selected being proportional to the number of clusters in a county. All full-time farmers within an area or cluster were sampled. We would denote such a sample design as a **stratified-cluster (or area) sample with a 100% sampling in each cluster and with proportional sampling of the clusters.** This design proved effective in obtaining results and the statistical analysis remained simple. This survey of about 1500 farmers was planned, conducted, and summarized in a two-month period. The teamwork necessary to conduct such a survey is described by Moe (1952); the results are reported in an easily understandable manner since Mr. Moe uses ratios such as 6-out-of-10 or 9-out-of-10 to report the opinions of farmers on various topics.

5.4.6 Every kth Item With a Random Start Sample Design

Another type of probability sample design often used when a listing or an ordering of all sampling units is possible is to select an interval, say k, to select a random number between 1 and k, say c, (See Subsection 5.4.1.) and then to take the cth, k + cth, 2k + cth, etc. item on the list. Such a sample design is denoted as an **every kth item with a random start sample design.** For example, suppose that we have a list of 5000 houses in a city, and we wish our sample to be 100 houses. A random number between 1 and 50, say 13, is selected, and then take the houses numbered 13, 63, 113, 163, 213, ..., 4913, and 4963, as the sample. There are 50 such samples possible; hence, the chance of selecting any given sample is one in fifty.

We may wish to take several such samples in order to ascertain reliability. Suppose that we wish to take five such samples, and still have a total sample size of 100. We would then randomly select five numbers between 1 and 250, and for each sample the sampling interval would be 250. Thus, suppose our first randomly selected number were 90, the first sample then would consist of the 20 items numbered 90, 340, 590, 840, 1090, 1340, 1590, 1840, 2090, 2340, 2590, 2840, 3090, 3340, 3590, 3840, 3090, 3340, 3590, 3840, 4090, 4340, and 4590. The remaining four samples would be obtained similarly.

This type of sampling often results in a considerable saving of travel time in planning the sample and in taking the survey. However, caution must be exercised against cyclical variations with sample designs of this type. For example, a sample of every kth (say 50th) business from a list of businesses in New York City was obtained by a surveyor. The results appeared absurd until he noted that every business appearing in the sample was located on or near Fifth Avenue!

5.4.7 Sieve Sampling

The total amount of money M may be fixed for an account and the investigator may wish to consider only this amount. It may be desirable to sample the individual accounts rather doing a census of all accounts. Let n_o be the proposed sample size and let M/n_o equal the **sieve size**. For an individual account $Y_i \geq M/n_o$, select a random number between 1 and Y_i with probability $1/(M/n_o =$ sieve size). For all $Y_i > M/n_o$, sample at 100% rate, i.e., take all of these. A random number between 1 and M is drawn. If the random number is between 1 and Y_i, the account is included in the sample. If the random number is larger than Y_i, the account is not sampled. This procedure of sampling is known as **sieve sampling** (See Wurst *et al.*, 1989). The actual sample size will be n_1 rather than n_o but the averages of of these two values over a large number of such samples will be the same.

5.4.8 Panel Surveys

In some situations, the investigator wishes to try out a new procedure or product for general use by an entire population. He selects a group of individuals by a random process (Often the process is not random and then this type of investigation would be in the previous section.), and uses these individuals to determine their reactions to the product. Market surveys, readership surveys, retail manufacturing surveys, nutritional surveys, and many other types are often conducted in this manner. A recent book on panel surveys is the one by Kasprzyk *et al.* (1989).

5.4.9 Some Comments

If an investigator spends the time, money, personnel, and materials to conduct a survey, the information obtained should be useful and of value. The value could, of course, be for advertising, political, or for deceiving people purposes, and in such cases biased information may be desirable. On the other hand, if unbiased information about a population is desired, then a sample survey design yielding unbiased information is desired. The designs discussed in this section are of this type. Judgement sample survey designs as described in the previous section, may be useful in the hands of a real expert on population structure, but in general they should not be used. Usually individuals using judgement survey designs are not the experts they portray themselves as. Potential users of survey results should not be hood-winked by the self-styled experts who use judgement survey designs.

5.5 QUESTIONNAIRE CONSTRUCTION AND INTERVIEWING

Some aspects of questionnaire construction were discussed in Section 5.2. In addition to these, the questionnaire may need to be revised one or more times as the result of the pre-testing. The form and format of the questionnaire should be such that a small, preferably none, amount of editing is necessary before the results are entered into a computer for summarization purposes. The form of the questionnaire will be influenced by the type of interviewing being used. Types of interviewing that may be used in a survey are:

      a. Personal interview,
      b. Telephone interview,
      c. Mail interview,
      d. Newspaper, magazine, or other interview,
      e. Combination of above interviews,
      f. Direct question interview,
      g. Indirect question interview,
      h. Anonymous response interview, and
      i. Non-anonymous response interview.

The type and form of the question will vary considerably for each of the above. In a mail interview, the respondent will be unable to request clarification for a given question but they could on a telephone or personal interview. Certain information, e.g., prejudice, honesty, discrimination, etc., can only be obtained from indirect questioning about items related to the desired information. If asked, most people would say they do not discriminate against the opposite sex. Instead of asking such a question, the interviewer desiring information on sex discrimination may ask a series of indirect questions which would be a measure of prejudice. For example, questions like, "Do you believe in equal pay for the same position for both sexes and why or why not?", "Which sex would you prefer to have as your supervisor?", "What is your opinion on maternity leave for the mother and for the father?", etc.. Such indirect questions require careful consideration to ascertain that they are fulfilling their purpose.

For many questions, the answers will not be embarrassing, not incriminating, and/or not sensitive. In this case, the anonymity of the respondent is not required in order to obtain factual answers. The conductor of surveys should be aware of the following possibilities that can occur with regard to responses. An anonymous response is sought by the investigator but the following may be the true situation:

      a. Interviewer and respondent know the response is not anonymous.

b. Respondent believes response is anonymous but it is not.
c. Interviewer believes response is anonymous but it is not.

If an anonymous response is sought, the following situations may pertain:
a. Interviewee is convinced of anonymity of responses.
b. Interviewee is not convinced of anonymity of responses.

Anonymous response procedures such as discussed in the next section will fail to illicit the desired information if a above is not the case. It is imperative that the respondents be fully convinced of anonymity before they will give answers that could implicate them. This fact may be overlooked when using procedures for obtaining anonymous responses.

The best questionnaire in the world could be constructed but if the training of interviewers is inadequate, the results from a survey may be useless. Here are some steps to follow in the training of interviewers:
a. Select honest, intelligent, persevering, trustworthy, and conscientious individuals.
b. Acquaint interviewers with the true, not necessarily desired, facts to be obtained.
c. Have interviewers give interviews to each other and have them discuss ways of interviewing in order to obtain truthful and reliable answers from respondents.
d. Acquaint interviewers with survey design and reasons for using this particular design.
e. Install hidden checks so that interviewers can be checked for reliability.
f. Check to determine the effect of appearance, dress, race, sex, age, or other characteristic that could influence a response.
g. Train interviewers how to put the respondent at ease and to prevent them from giving answers they believe the interviewer might want rather than the true answer.
h. Be certain that interviewers thoroughly understand that they are not to bias the responses in any manner, shape, or form.

5.6 METHODS FOR OBTAINING ANONYMOUS RESPONSES

5.6.1 Introduction

Answers to certain questions may be sensitive, embarrassing, and/or incriminating for the individual responding. In such cases, it will be necessary to assure the respondent that the answers given cannot be traced back to him and that the true answers are necessary in order to attain the objectives of the survey. He must be convinced of the necessity for giving factual information. It may not be enough to assure anonymity, but it may also be necessary to assure respondents that a class action will not be instituted as a result of the information obtained from the survey. Provided that the respondent can be assured of the above, the following six sub-sections describe methods for obtaining information to embarrassing, sensitive, and/or incriminating questions.

5.6.2 "Black Box" Method

A survey of any type can be conducted in such a manner that the responses are anonymous. In a mail survey, the return envelopes could be unmarked in any manner and the only identification would be the postal mark as stamped on the envelope. This would provide anonymity. In a personal interview, the responses to questions could be secretly written and deposited in a sealed box with many other responses. The box would be thoroughly shaken to assure mixing of the questionnaires in the box. This **black box method** allows the respondent to deposit his unidentified responses in some sort of container and to retain anonymity. A. J. King, while at Iowa State University, conducted a survey on incomes and sources of income for Iowa farmers. The Internal Revenue Service felt that much of the income farmers received was not being reported for tax purposes and they wished to obtain information on the extent and amount of this under-reporting. Any sort of direct questioning or any indication that this information was being sought by the Internal Revenue Service would doom the survey for obtaining factual information. Each interviewer drove a pick-up truck with a large padlocked wooden box with a slit for inserting questionnaires.   The box could be shaken to assure mixing of questionnaires, assuring anonymity of response for individual farmers. The results of the survey showed that the incomes were under-reported. The Internal Revenue Service did institute a class action against Iowa farmers in that they instituted a system of auditing tax returns. If the farmers had anticipated this, they more than likely would have given their income as the amount reported on their tax returns.

5.6.3 Randomized Response Procedure

There are several forms that this procedure can take. The **randomized response procedure** described here is for the respondent to answer one of two questions by some randomizing procedure; one of the questions is sensitive and the other unrelated question is not sensitive. The interviewer does not know which question the interviewee answered, thus assuring anonymity of the response. The procedure is suitable for yes or no type of questions. What the investigator does know is the expected proportions of respondents answering the sensitive and the non-sensitive questions. He also knows the expected proportion of yeses for the non-sensitive question. With this information, he is able to estimate the proportion of yeses to the sensitive question even though he does not know which question the respondent answered.

The randomized response procedure (RRP) is illustrated with three examples. Let A be the sensitive question and B be the non-sensitive one. Also, let the proportion of time the sensitive question is answered be 0.7, or 70% of the time. (Note that the more sensitive the question, the lower this proportion will need to be in order to obtain factual information.) Some randomizing device such as tags in a jar or a spinning wheel with numbers where the interviewee can secretly obtain a number may be used. When the interviewee becomes comfortable with the device, proceed as in the following examples. Additional information on RRPs may be found in papers by Campbell and Joiner (1973), Folsom *et al.* (1973), Greenberg *et al.* (1969a, 1969b), and Warner (1965, 1971).

Example 1: Answer question A if the last digit of your Social Security Number is 0, 1, 2, 3, 4, 5, or 6; answer question B if the last digit is 7, 8, or 9.
Question A: Did you smoke marijuana during the past month?
Question B: Is the last digit of your Social Security Number 7 or 8?
    Answer: Yes_____   No_____

Example 2: Answer question A if your birthday falls between 1/1 and 9/12 and answer question B if your birthday falls between 9/13 and 12/31.
Question A: Have you stolen $5 or more from any person or organization in the past two years?
Question B: Does your birthday fall between 9/13 and 11/25?
    Answer: Yes_____   No_____

Example 3: You have a bottle with ten corks numbered from 0 through 9. The corks are of two colors, red and white. Shake the bottle thoroughly and secretly draw one cork from the bottle. Note your number and return the cork to the bottle. Answer question A if the number on your cork was 0, 1, 2, 3, 4, 5, or 6 and answer Question B if the number is 7, 8, or 9.
Question A: Have you given or received payment in any form for sex during the past year?
Question B: Is the color of your cork red?
    Answer: Yes_____   No_____

Let N be the number of individuals answering each set of questions in each example above. Let Y be the number of yeses obtained from the N individuals. Then the proportion answering yes is Y/N. Also, for each of the three examples (corks 7 and 8 were red), the expected proportion of individuals answering yes if they answered question B would be 2/3. Knowing this, the following formula may be used to estimate the proportion of individuals answering yes to question A:

prop(yes on A given A is answered) = [prop(yes answer) - prop(B is answered)| × prop(yes given B is answered)] / prop(A is selected)

which for the above examples is

prop(yes on A given A is answered) = [Y / N - 0.3(2 / 3)] / 0.7.

If Y / N was 0.4, say, then the proportion of yeses on A given A was answered is [0.4 - 0.3(2 / 3)] / 0.7 = 0.2 / 0.7 = 2 / 7, or about 29%.

5.6.4 BIBD Block Total Procedure

A balanced incomplete block design (BIBD) is one in which there are v questions to be answered but a respondent only answers k of them. The set of k questions is called a block and the respondent gives only the total score for the k questions but not the answers to the k individual questions, i.e., only a block total is obtained. The sets or blocks of k questions are constructed in such a way that *every pair of questions occurs together an equal number of times in the b blocks.* (More detail on a design of this type may be found in Chapter 7.) Such a procedure is denoted as the **BIBD block total procedure.** A more detailed description of the procedure and its properties may be found in Smith *et al.* (1974) and Raghavarao and Federer (1979).

The BIBDBT procedure will be illustrated with an example from Smith *et al.* (1974). A set of v = 7 questions as follows was given to 84 students:

1. Are you under 21 years of age?
                                            Yes (0)   No (1)
2. Did you cheat in any way on the Statistics 200 prelim you took last week?
                                            Yes (2)   No (3)
3. In general, are you happy with your decision to come to Cornell?

Yes (1)   No (0)
4. While at Cornell, have you ever stolen money or any other article worth over $5.00 from a friend, roommate, employer, or anyone else?

Yes (3)   No (2)
5. Does your parent earn more than $25,000 per year?

Yes (0)   No (1)
6. Have you smoked any marijuana during the past two weeks?

Yes (2)   No (3)
7. Are you enrolled in the College of Agriculture and Life Sciences?

Yes (1)   No (0)

The numbers in parentheses are the ones to be added to obtain the ith block total response for individual j, say Y_{ij}. Care has to be taken in setting up these numbers so that an individual's response cannot be identified. The blocks were as follows:

Questions in block	Block total	=	Responses in block total
1, 2, & 4	Y_{1j}	=	$X_{1j} + X_{2j} + X_{4j}$
2, 3, & 5	Y_{2j}	=	$X_{2j} + X_{3j} + X_{5j}$
3, 4, & 6	Y_{3j}	=	$X_{3j} + X_{4j} + X_{6j}$
4, 5, & 7	Y_{4j}	=	$X_{4j} + X_{5j} + X_{7j}$
5, 6, & 1	Y_{5j}	=	$X_{5j} + X_{6j} + X_{1j}$
6, 7, & 2	Y_{6j}	=	$X_{6j} + X_{7j} + X_{2j}$
7, 1, & 3	Y_{7j}	=	$X_{7j} + X_{1j} + X_{3j}$

For individual j who answered questions 1, 2, and 4, the sum of answers to these three questions is Y_{1j}. There were m = 12 individuals who answered this same block of questions. Denote the total for these 12 individuals as $Y_{1.}$, where the dot in the subscript means summed up over the subscript j = 1, 2, ..., 12. The block totals are observable but what is desired is the values for the X variates. The solutions are obtained as:

$$X_{1.} / (m = 12) = [Y_{1.} + Y_{5.} + Y_{7.} - (Y_{2.} + Y_{3.} + Y_{4.} + Y_{6.}) / 2] / 3m$$
$$X_{2.} / 12 = [Y_{1.} + Y_{2.} + Y_{6.} - (Y_{3.} + Y_{4.} + Y_{5.} + Y_{7.}) / 2] / 36$$
$$X_{3.} / 12 = [Y_{2.} + Y_{3.} + Y_{7.} - (Y_{1.} + Y_{4.} + Y_{5.} + Y_{6.}) / 2] / 36$$
$$X_{4.} / 12 = [Y_{1.} + Y_{3.} + Y_{4.} - (Y_{2.} + Y_{5.} + Y_{6.} + Y_{7.}) / 2] / 36$$
$$X_{5.} / 12 = [Y_{2.} + Y_{4.} + Y_{5.} - (Y_{1.} + Y_{3.} + Y_{6.} + Y_{7.}) / 2] / 36$$
$$X_{6.} / 12 = [Y_{3.} + Y_{5.} + Y_{6.} - (Y_{1.} + Y_{2.} + Y_{4.} + Y_{7.}) / 2] / 36$$
$$X_{7.} / 12 = [Y_{4.} + Y_{6.} + Y_{7.} - (Y_{1.} + Y_{2.} + Y_{3.} + Y_{5.}) / 2] / 36$$

Note that the solution is obtained by summing the block totals where a
question appears and subtracting the sum of the block totals where the
question does not appear divided by 2. Since each question appears three
times in the seven blocks, the divisor 3 appears in the above solutions.

5.6.5 Partially Balanced Block Total

There may be situations where a BIBD is not available to form blocks.
There is, however, a large class of partially balanced incomplete block
designs, PBIBD, which may be used for setting up blocks of questions. In
particular, when the number of blocks b is equal to the number of
questions v, it is possible to construct any size of block, say k. To
illustrate, suppose that there are v = 5 questions and that blocks of k = 3
questions will be used. The following is one such design:

Questions in block	Block total	=	Responses in block total
1, 2, & 4	$Y_{1.}$	=	$X_{1.} + X_{2.} + X_{4.}$
2, 3, & 5	$Y_{2.}$	=	$X_{2.} + X_{3.} + X_{5.}$
3, 4, & 1	$Y_{3.}$	=	$X_{3.} + X_{4.} + X_{1.}$
4, 5, & 2	$Y_{4.}$	=	$X_{4.} + X_{5.} + X_{2.}$
5, 1, & 3	$Y_{5.}$	=	$X_{5.} + X_{1.} + X_{3.}$

The solution for the X responses is not as simple as for the previous case
and is obtained as follows:

$$X_{1.}/m = (2Y_{1.} - Y_{2.} - Y_{3.} - Y_{4.} + 2Y_{5.})/3m$$
$$X_{2.}/m = (2Y_{1.} + 2Y_{2.} - Y_{3.} - Y_{4.} - Y_{5.})/3m$$
$$Z_{3.}/m = (-Y_{1.} + 2Y_{2.} + 2Y_{3.} - Y_{4.} - Y_{5.})/3m$$
$$X_{4.}/m = (-Y_{1.} - Y_{2.} + 2Y_{3.} + 2Y_{4.} - Y_{5.})/3m$$
$$X_{5.}/m = (-Y_{1.} - Y_{2.} - Y_{3.} + 2Y_{4.} + 2Y_{5.})/3m$$

In the above, there are m individuals answering any one block of questions
and each question appears three times in the five blocks. There is a pattern
in the solutions and this is due to the fact that there is partial balance in that
pairs of questions occur either once or twice in the five blocks. When v =
b, the block size can be 2, 3, or any size desired by the investigator.

5.6.6 Supplemented Block Total

A **supplemented block total response** procedure, SBTR, uses an incomplete block design where the sensitive question appears in every block but the non-sensitive question does not and respondents only give a total of the responses for the questions in their group (See Raghavarao and Federer, 1979). Thus, this procedure has the advantage over the above described procedures in that the sensitive question is answered by every member in the sample. Their answers still remain anonymous since they only give a total of their responses to the questions in the block. The SBT design will have a set of blocks where some of the non-sensitive questions are included with the sensitive question and a set of blocks where all the questions are in the block. The following two examples illustrate the procedure.

Example 1: For one sensitive question and four non-sensitive questions, use the following blocks and questions:

Block	Questions in block	Block total = responses to questions in block
1	1 and 2	$Y_{1.} = X_{1.} + X_{2.}$
2	1 and 3	$Y_{2.} = X_{1.} + X_{3.}$
3	1 and 4	$Y_{3.} = X_{1.} + X_{4.}$
4	1 and 5	$Y_{4.} = X_{1.} + X_{5.}$
5	1, 2, 3, 4, & 5	$Y_{5.} = X_{1.} + X_{2.} + X_{3.} + X_{4.} + X_{5.}$

The solutions for the responses to the individual questions are:

$$X_{1.} / m = (Y_{1.} + Y_{2.} + Y_{3.} + Y_{4.} - Y_{5.}) / 3m$$
$$X_{2.} / m = (Y_{1.} - X_{1.}) / m$$
$$X_{3.} / m = (Y_{2.} - X_{1.}) / m$$
$$X_{4.} / m = (Y_{3.} - X_{1.}) / m$$
$$X_{5.} / m = (Y_{4.} - X_{1.}) / m$$

There are m individuals answering any particular block of questions.

Example 2: Suppose that blocks of three questions are desired instead of blocks of two as in Example 1. The following is an SBTR design for this case.

Block	Questions in block	Block total = Question responses in block
1	1, 2, and 3	$Y_{1.} = X_{1.} + X_{2.} + X_{3.}$

2	1, 2, and 4	$Y_{2.} = X_{1.} + X_{2.} + X_{4.}$
3	1, 2, and 5	$Y_{3.} = X_{1.} + X_{2.} + X_{5.}$
4	1, 3, and 4	$Y_{4.} = X_{1.} + X_{3.} + X_{4.}$
5	1. 3. and 5	$Y_{5.} = X_{1.} + X_{3.} + X_{5.}$
6	1, 4, and 5	$Y_{6.} = X_{1.} + X_{4.} + X_{5.}$
7	1, 2, 3, 4, & 5	$Y_{7.} = X_{1.} + X_{2.} + X_{3.} + X_{4.} + X_{5.}$
8	1, 2, 3, 4, & 5	$Y_{8.} = X_{1.} + X_{2.} + X_{3.} + X_{4.} + X_{5.}$
9	1, 2, 3, 4, & 5	$Y_{9.} = X_{1.} + X_{2.} + X_{3.} + X_{4.} + X_{5.}$

The solution for the response to question 1 is:

$$X_{1.} / m = (Y_{1.} + Y_{2.} + Y_{3.} + Y_{4.} + Y_{5.} + Y_{6.} - Y_{7.} - Y_{8.} - Y_{9.}) / 3m$$

That is, the solution involves the sum of all incomplete blocks minus the sum of blocks where all questions appeared. The solutions for the other effects are:

$$X_{2.} / m = [Y_{1.} + Y_{2.} + Y_{3.} - (Y_{4.} + Y_{5.} + Y_{6.}) / 2] / 3m$$
$$X_{3.} / m = [Y_{1.} + Y_{4.} + Y_{5.} - (Y_{2.} + Y_{3.} + Y_{6.}) / 2] / 3m$$
$$X_{4.} / m = [Y_{2.} + Y_{4.} + Y_{6.} - (Y_{1.} + Y_{3.} + Y_{5.}) / 2] / 3m$$
$$X_{5.} / m = [Y_{3.} + Y_{5.} + Y_{6.} - (Y_{1.} + Y_{2.} + Y_{4.}) / 2] / 3m$$

Note that the design for 2, 3, 4, and 5 is a BIBD for $v = 4 = b$ and $k = 3$ and that the solutions are easy to obtain when this is true. They will be somewhat more difficult when a PBIBD is used. Raghavarao and Federer (1979) give the general method for constructing SBTR designs. This design will be more efficient than the previous ones, especially for the larger values of v.

5.6.7 Randomized Block Total Response

The degree of perceived anonymity by the respondent may not be sufficient to get them to respond truthfully to sensitive questions. In order to add another level of anonymity, let the respondent randomly and anonymously select the block of questions that is to be answered. The interviewer not only does not know the questions being answered but does not even know what block of questions is being answered. This procedure is known as the **randomized block total response** procedure. When the degree of anonymity is understood by the respondent, there should be little trouble to get them to answer truthfully.

In conducting the survey, the mb questionnaires could be placed in a large container and thoroughly mixed. The respondent would be asked to select a questionnaire from the container, answer the questions, obtain a block total, record the block total, and place the form in another large container containing the forms from other respondents. There should be several blank questionnaires placed in the answer container for the first people answering and several copies of the last questionnaire in the first container in order to assure the respondent. The interviewer would know the identity of the last block of questions by this method.

5.6.8 An Example of RR, BIBD, and RBIBD Methods

Smith *et al.* (1974) conducted an investigation using the set of questions in Subsection 5.6.4. Eighty four students were involved in the survey, i.e., $m = 12, v = 7 = b$, and $k = 3$. They used the RR, BIBDR, and RBTR procedures using the same seven questions. The results obtained for questions 2, 3, and 4 were:

Question	RR method	BIBDR method	RBIBDR method
2	0.03	0.01	0.08
3	0.83	0.64	0.83
4	0.01	0.01	0.12

A higher percentage of individuals answering yes to the sensitive questions, 2 and 4, was obtained using the RBIBD method. When students were queried about their feelings of anonymity, the following comment summarizes their feelings, "If you people are smart enough to devise procedures like the RR and BIBDR, you are smart enough to determine how I answered each question. For the RBIBDR method, there is no way you could find out how I answered each question". At least some of the students were not assured of their anonymity for the RR and BIBDR methods.

5.7 REFERENCES AND SUGGESTED READING

Folsom, R. E., B. G. Greenberg, D. G. Horvitz, and J. R. Abernathy (1973). Two alternative questions randomized response model for human surveys. *J. American Statistical Association* 68, 525-530.

Greenberg, B. G., J. R. Abernathy, and D. G. Horvitz (1969a). Application of the randomized response technique in obtaining quantitative data. *Proc., Social Statistics Section, American Statistical Association,* pp.40-43.

Greenberg, B. G., A. A.Abul-Ela, W.R. Simmons, and D. G. Horvitz (1969b). The unrelated question randomized response model. Theoretical framework. *J. American Statistical Association*, 64, 520-539.

Haack, D. G. (1979). *Statistical Literacy. A Guide to Interpretation.* Duxbury Press, North Scituate, Mass., xii + 323 pp.

Huff, D. (1954). *How to Lie With Statistics.* W. W. Norton and Company, Inc., New York, 142 pp..

McCain, G. and E. M. Segal (1973). *The Game of Science, 2nd edition.* Brooks/Cole Publishing Co., Monterey, Calif., 198 pp.

McCarthy, P.J. (1970). Sampling. Elementary principles, 3rd Printing. Bulletin No. 15, New York State School of Industrial and Labor Relations, Cornell University, Ithaca, New York, iii + 31 pp..
(This bulletin is more technical than Slonim. The presentation is elementary and is directed more toward the analysis aspects.)

McCarthy, P.J. (1957). *Introduction to Statistical Reasoning.* McGraw-Hill Book Company, Inc., New York, Toronto, and London, xiii + 402 pp..
(Chapters 6 and 10 are recommended reading. The basic ideas of sample design are presented in these two chapters.)

Moe, E.O. (1952). New York farmers' opinions on agricultural programs. *Cornell University Extension Bulletin 864*, Cornell University, Ithaca, New York.
(An example of the planning, conduct, and analysis of a survey.)

Moore, D. S. (1979). *Statistics. Concepts and Controversies.* W. H. Freeman and Co., San Francisco, Calif., ix + 313 pp.. (Chapter 1)

Slonim, M.J. (1962). *Sampling* (original title was *Sampling in a Nutshell*). Simon and Schuster, New York, xii + 144 pp..
(The entire book should be read. The presentation is elementary, and the ideas and concepts of sampling are easy to grasp.)

For those who wish more of the theory, application, and analysis of sampling designs, the following books are recommended. The reader will need to have a relatively good background in statistical theory and methodology in order to comprehend the material presented.

Cochran, W. G. (1963). *Sampling Techniques, 2nd edition.* John Wiley and Sons, Inc., New York and London, xvii + 413 pp..

Deming, W. E. (1950). *Some Theory of Sampling.* John Wiley and Sons, Inc., New York, xvii + 602 pp..

Hansen, M. H., Hurwitz, W. N., and Madow, M. G. (1953). *Sample Survey Methods and Theory. Volumes I and II.* John Wiley and Sons, Inc., New York.

Sukhatme, P. V. (1954). *Sampling Theory of Surveys with Applications.* The Iowa State College Press, Ames, Iowa, xix + 491 pp..

Yates, F. (1960). *Sampling Methods for Censuses and Surveys. 3rd edition, rev. and enl..* Charles Griffin and Company, Ltd., London, and Hafner Publishing Company, New York, 440 pp..

5.8 PROBLEMS

Problem 5.1. Figure 5.1 has been drawn (by D. S. Robson, Cornell University) to illustrate a population (e.g., fish). The parallelograms are of various lengths ranging from 1 / 8 inch to 2 inches. By visual or judgement selection alone, select one parallelogram to represent the mean of the population, and measure it in millimeters. As a second part of the problem, stratify the parallelograms into large, medium, and small, then select by judgment one parallelogram in each to represent the mean of the stratum; obtain the average length of the three sampling units selected. How does the mean of the three compare with the one selected previously? As a third part of the problem, number the 100 parallelograms from 00 to 99 and randomly select three sampling units. Use a local telephone directory to obtain a set of random numbers. Measure the three sampling units and compare the mean of the simple random sample with the judgment sample obtained previously. (Note: It is a good idea to summarize the sample results from all class members to illustrate variability among samples and possible biases in judgement samples.) As a fourth part of the problem, the class members are to select a simple random sample of size ten to illustrate the decrease in variability in samples

Figure 5.1. One hundred parallelograms.

of size ten as compared to samples of size three. Then compare all methods with the true mean of the population. All sampling should be without replacement.

Problem 5.2. This is to be an *independent* survey to be conducted by the student. The results are to be checked against all other surveys of the class for independence. The survey is given the weight of one class examination and therefore is considered to be an important problem. It is to be submitted prior to the end of the course in order to allow time for grading. In the write-up of the survey, indicate how each of the steps in planning and conducting a survey have been followed. The survey may be on any topic of interest to the student. Prior to conducting the survey, the student is to submit the proposed project and outline of procedure to the instructor for approval. The feasibility of obtaining results should be evaluated at this stage. (Note to Instructor: It may be desirable to have groups of three students on each project in order to obtain more substantial results.)

Problem 5.3. Devise a procedure for generating random numbers for the set 0, 1, 2, 3, 4, 5, 6, 7, 8, and 9. Use identical circular tags, construct a ten-sided die, or develop any other procedure. From 100 trials on the random number generator, record the number of times each of the ten digits occur. Does the procedure appear to be unbiased in that each number occurs equally frequently? Why or Why not?

Problem 5.4. Obtain a telephone book and write down the phone numbers for the names appearing after your name. Obtain a sufficient number so that using the 2nd and 3rd last digits you can obtain
    a. a random permutation of numbers 1 to 20 and
    b. a random permutation of numbers 1 to 35.

Problem 5.5. Describe an investigation where you believe an anonymous response would be desirable. Can you think of methods other than those described in this chapter for obtaining such responses? Which of the methods described in this chapter do you think would work for your example? Why?

Problem 5.6. Given the following set of four observations on the number of children in a family:

No. of Children
$1 = Y_1$
$4 = Y_2$ $\qquad\qquad\qquad\qquad$ $y = 16 / 4 = 4$

$7 = Y_3$
$4 = Y_4$
$16 = Y.$

Explain in words what the symbols $Y_.$ and y mean. Suppose that Y_i is the ith observation for the number of children in family i and that we have a sample of size n. What are the possible values of i and what are the possible observations? Use the above data as an illustration of the general formulation.

Problem 5.7. The following statement appeared in an alumni publication of Colorado State University in Spring 1978: "Several major suggestions for improving alumni services by the College of Agricultural Sciences are being considered by the College's Alumni Relations Committee following a recent survey of agriculture students. Of the approximately 5,000 survey forms mailed out, 1,101 were returned, which is considered an adequate sample size for such a survey". What is your reaction to this statement?

Problem 5.8. In a warehouse, there are 20 stacks with each stack consisting of 100 bags of potatoes. Each bag weighs 120 pounds. An investigator wishes to draw one pound random samples from a selected bag to determine the degree of blackening of potatoes when cooked after 3, 6, 12, and 18 months of storage. At each period, say at 3 months, she selects a random sample of two bags from each stack; then, she obtains a one pound random sample from each of the selected bags. She will end up with $2 \times 20 = 40$ one pound samples for each length of storage.
  a. What is the sample design called relative to bags?
  b. What is the sampling design relative to one pound samples?
  c. What is the sampling unit for a and for b?
  d. What are the assignable and non-assignable causes for a and b?

Problem 5.9. Suppose that the population consists of six individuals whose heights are: A-5 feet 2 inches, B-5 feet 6 inches, C-5 feet 8 inches, D-5 feet 10 inches, E-5 feet 10 inches, and F-5 feet 12 inches. Describe the sampling and observational units. What is the population size N equal to? What is the variable of interest? List all possible samples of size four.

Problem 5.10. In a poll, 2,500 adults were interviewed and asked: "In the discussion over auto insurance methods, have you heard or read about the 'no-fault' insurance?". The 53% who answered in the affirmative were then asked this question: "Do you approve or disapprove of the idea of voluntary 'no-fault' insurance?". Of the 1,325 adults who indicated awareness of the program, 63% approved, 24% disapproved, and 13% had no opinion. Briefly comment on the following headline used with the article: "'No-fault' insurance wins public approval". You may assume that the sampling procedure and sample size were adequate.

Problem 5.11. Suppose you have a population with N = 40 sampling units. and want to take a random sample of size n = 36. Is there any problem if you take a simple random sample of size four and exclude these units to obtain your sample of size 36? Justify your answer.

Problem 5.12. Suppose a population is made up of several sub-populations and a uniform sampling fraction is used to obtain the sample. In order to obtain an unbiased sample is
  a. it necessary to take a simple random sample from each selected sub-population?
  b. the sample size smaller for a than it would be if the sub-population structure were ignored, assuming the same precision is desired, and a simple random sample were obtained from the entire population?

Explain your answers and use assignable causes in answering b.

Problem 5.13. The Draft Lottery assigned numbers 1 to 366 to the individuals of draft age with one number being drawn each for the 366 days in 1972. The following represent the number of the lowest one-third, of the middle one-third, and the highest one-third of numbers drawn by months. A lottery carries with it the connotation of randomness and fairness. Often there is the assertion that low numbers or high numbers are appearing too frequently. Here are the numbers drawn for 1972:

No. drawn	Jan	Feb	Mar	Apr	May	June	July	Aug	Sep	Oct	Nov	Dec
1-122	9	7	5	8	9	11	12	13	10	9	12	17
123-244	12	12	10	8	7	7	7	7	15	15	12	10
245-366	10	10	16	14	15	12	12	11	5	7	6	4

Given the above, what are your comments on the assertion that high numbers appeared much less frequently than they should, especially in the last third of the year? Combine these data with those in the text. Does this change your opinion about randomness of this lottery system?

Problem 5.14. In a Gallup poll, 1,500 adults were interviewed and asked:

  In the discussion over birth control methods, have you heard or read about the voluntary male sterilization procedure?

The 63% who answered the question in the affirmative were then asked this question:

Do you approve or disapprove of the idea of voluntary sterilization?

Of the 945 adults who indicated an awareness of the operation, 53% approved, 34% disapproved, and 13% had no opinion. Briefly comment on the following headline used with the article:

Male sterilization wins public approval.

You may assume that the sample design was an unbiased one and that sample size is not questioned.

Problem 5.15. Suppose you have a population with 36 sampling units, i.e., N = 36. You want to take a simple random sample of size 20, i.e., n = 20. Is there any problem in taking a simple random sample of size 16, excluding these 16 units, and using the remaining 20 as your sample? Briefly support your answer.

Problem 5.16. Suppose a population of N sampling units is divided into several sub-populations based on some characteristic and a uniform sampling fraction is used to sample from the sub-populations.
   a. Is it necessary to take a simple random sample from each sub-population?
   b. Is it possible to argue that the sample size for the above sampling procedure could be smaller than the sample size if the population weren't divided into sub-populations before sampling?

Concisely explain your answers.

Problem 5.17. A reporter for the Cornell Daily Sun wanted to determine the popularity of candidate A among the students at Cornell University. He stated that he wished his results to be "statistically solid" and then selected the 100 students in his introductory statistics class as his sample. Each student in the class was asked the following question:

Would you vote for A in the upcoming Presidential Election?

The observations were recorded as "yes"(1) or "no"(0). The reporter calculated that 45% said they would vote for A in the Election.

    a. Comment on the validity of the sampling procedure.
    b. How would you calculate the 45% figure?

Problem 5.18. Define a population as those persons physically present in a building at a particular time. A sample of the population is needed to participate in an experiment to be done in a central location. Each person in the building was asked if they would be willing to participate in the experiment. In this way every person in the population had an equal opportunity to participate in the experiment. The sample was those people who came to the central location and voluntarily participated in the experiment. The sample size was small compared to the population size. Would you consider this to be a simple random sample? Why or why not?

Problem 5.19. In an issue of *Science,* Linus Pauling, Nobel Prize winner in Chemistry and champion of Vitamin C for decreasing the incidence of the common cold, questioned that *Science* had contacted most of the members of the National Academy of Sciences (NAS) concerning an article written in *Science.* This article contained the following statement:

> the overriding feeling was that Pauling had a right to express his views, in spite of the fact that most other NAS members took issue with their scientific validity

Suppose there were 40 NAS Members on the Cornell campus. You questioned these 40 and find that 19 had been contacted by Science. Would you support Pauling's questioning that most members were contacted? Why or why not?

Problem 5.20. A college newspaper conducted a mail questionnaire on dormitory housing. Five thousand questionnaires were mailed out and 411 were returned. What would you conclude about any inferences the reporters of this newspaper might make about the responses from this survey?

Problem 5.21 Suppose a listing of the 3,000 buyers of high resolution color television brand X. was available for sets purchased in the last two years. It is desired to do a sample survey of this population to estimate the number of service calls on the sets. All sets are covered by a service contract for one year and one-third are covered for an additional six months. The latter group make up the last 1,000 names on the list. How would you draw a sample of 150 names from this listing. Justify your selection of the design.

Problem 5.22. A College Concert Commission and a Faculty Committee on Music conducted a joint survey to determine the musical interests of the College community. They wished to use this information in planning future concerts. Their questionnaire read, "We would appreciate you completing this survey and filling out either the popular or classical music sections, or both. PLEASE RETURN THE SURVEY AS SOON AS POSSIBLE TO THE FOLLOWING PLACES: THE MAIN DESKS OF WSH, NOYES CENTER, OR THE NORTH CAMPUS UNION, OR MAIL IT TO THE CONCERT COMMISSION, ROOM 32, WSH." Describe how you would advise the Concert Commission in using the results of such a survey.

Problem 5.23. Quota sampling may be described as stratified sampling where potential respondents are contacted until all the strata are filled. What is the major disadvantage of quota sampling? Describe a hypothetical example of quota sampling using a college undergraduate population.

Problem 5.24. A student leaving a lecture in a statistics class was heard to say:

It seems to me that one would always take a stratified random sample if the only alternative was a simple random sample.

Comment on this quote, specifically including a couple of reasons why one might prefer a simple random sample.

Problem 5.25. Comment on the following statement in a student's report:

............randomly distribute a questionnaire.........

Give some possible interpretations of this partial sentence.

Problem 5.26. While shopping, you look at a Milton Bradley game entitled "Go to the head of the class". On the cover are the words "average playtime 90 minutes". Also, on the cover is the information "ages 8 - adult" and "2 - 9 players". How do you think the value of 90 was obtained? Is this sufficient information for deciding for a gift to a friend? Is there additional information that would be helpful in making your decision to purchase or not?

Problem 5.27. A university's long range planning committee distributed a questionnaire on a College Report to all undergraduates students at the time they registered for the Spring semester. The students were polled on their attitudes about enrollment, tuition, size of classes, use of facilities, length of undergraduate education, and other areas of particular interest to students. The completed questionnaires were returned to seven locations on campus. What advice would you give the Committee about such a survey? How would you have conducted such a survey? Be concise.

Problem 5.28. In an article concerning the banning of DES (diethylstilbestrol) in cattle feed, the author made the statement:

Since some 30 million cattle are slaughtered each year, 1023 is not too healthy a sample from which to draw statistically valid conclusions.

Comment on the quote in light of the contents of this chapter.

Problem 5.29. A headline in the local paper read, "Majority Sympathizes With Indians". The study was done by Louis Harris with a national cross section of 1,472 households during the period March 15-23. It is reasonable to believe that Harris did not take his sample of 1,472 from a Master List of all households in the U. S.. How do you think that a large survey organization such as Harris selects a sample? What are the important aspects of the presumed procedure?

Problem 5.30. A university senate student senator claims that 75% of the students favor legislation that she is proposing in the senate. You would like to obtain some data to support or refute the senator's claim. You plan to make 50 interviews. How would you do this to obtain unbiased results?

Problem 5.31. An RCA commercial indicates that in a recent survey, it was found that TV station engineers favor RCA color TV over other color TVs by a margin of two to one. Make two specific comments about this statement.

Problem 5.32. You wish a simple random sample of adults living in a large apartment complex which has efficiency units (limited to one adult) and two-bedroom units. You cannot obtain a listing of the adults but you can construct a listing of all apartments by looking at the floor plan of the complex. How could you obtain your simple random sample of adults?

Problem 5.33. A population is defined to be the homeowners in the city of Ithaca, New York. From the city directory, a list of all addresses is found. A simple random sample of addresses is obtained. If the person(s) at address is found to be a homeowner, the homeowner is interviewed. If the resident(s) is not a homeowner, the interviewer flips a coin to select another house on either the right or left side of the renter. If the person at the second residence is a homeowner, he/she is included in the sample. If not, then another randomly selected address is drawn. What are your comments about this sample design?

Problem 5.34. You wish a simple random sample of next year's resident advisors. You plan to conduct the survey during the Summer. You can obtain a listing of the present year advisors who plan to continue and a second listing of new advisors. The former listing also contains the names of all advisors who do not want to continue (an asterisk after their names). How would you obtain a simple random sample from the two lists?

Problem 5.35. Suppose you want to sample stores in Ithaca, New York, selling fresh, frozen, or packaged meats to determine the effect of Lent on the sales of meat. Outline a procedure for obtaining this information.

Problem 5.36. Calspan, a research organization, did work on the effect of stopping and taking short breaks for preventing drowsiness during interstate or toll road type driving. The hypothesis is that the more frequent the short breaks, the less will be the chance of falling asleep during driving. The New York Thruway between Buffalo and Syracuse was used. Drowsiness was measured by attaching sensory devices to the driver and monitoring with equipment in the back seat of the car. The sensors are easily attached to the driver's head and body. Briefly outline what is important to consider in this study and how to handle these considerations.

Problem 5.37. Suppose a college newspaper stated that "students are supporting a week-long boycott of meat products, as indicated by a random survey of students frequenting dining areas on campus". What would be an alternative to taking a survey to obtain this information?

Problem 5.38. Give a one-sentence reply to each of the following:

a. Difference between a random sample and a simple random sample.
b. As a method of selecting a person who is a good discriminator for a taste testing panel, present three samples, two of which are identical and the

third different. If the person picks the one sample that is different among the three, use that taster as part of the taste-testing panel.
c. Method of using 100 people (simple random sample from a well defined population) for testing the difference between two brands of hand cream.

Problem 5.39. During World War II, a special research group in England recorded data on holes in planes returning from missions over Europe. The objective was to determine the places on planes to put extra armor by examining the pattern of hits made on planes during combat missions. After weeks of data collection, the research group could not agree on how to interpret the data. As they neared the deadline for their report and the limit for their frustrations, they heard a voice, softly but firmly, saying, "Put the extra armor where no holes appear. You are looking at the planes that made it back from battle". What is your comment about this recommendation?

Problem 5.40. Prepare a list of key words for this chapter and show relationships among them.

CHAPTER 6. A SAMPLER IN AN ANALYTICAL LABORATORY AND OTHER AREAS

6.1 INTRODUCTION

An analytical laboratory performs biological, chemical, and/or physical analyses on samples obtained from investigations on a routine basis. In order to minimize costs and maximize use of laboratory facilities, a laboratory director needs to use efficient procedures and to employ statistical sampling procedures whenever appropriate. If the director and his personnel are not conversant with the necessary statistical procedures, the services of a statistician, *who is knowledgeable about the theory and application of sampling procedures*, should be employed. Sampling techniques and procedures can be extremely useful, practical, time-saving, and cost-effective in handling routine analyses in a laboratory. It is often possible to effect considerable savings in time and resources by utilizing appropriate sampling procedures. Several examples will be used to amply demonstrate this. It also will be shown how to use certain sample designs or sampling procedures to accomplish goals and objectives which would be difficult or impossible to attain otherwise. Examples will be used to demonstrate this.

In the next section, the nature and problems of an analytic laboratory are discussed; then, the usual procedures employed by laboratory directors to cope with or ignore the problems are presented. In Section 6.3, a discussion of the approaches that a sampler can take to resolve these problems is given. In the following sections, sub-sampling, sequential sampling, pooling samples before analyses, double sampling, and other procedures may be utilized to reduce the total number of analyses performed by a laboratory and still obtain the desired information. Usefulness of a procedure depends upon the particular situation. Several examples have been mentioned in previous chapters. These and others will be used to demonstrate the various sampling designs and procedures. Possible savings from using efficient procedures are demonstrated. Procedures for obtaining information on population parameters and still preserving the anonymity of the individual are presented.

There are a variety of areas where sampling, sampling plans, and sampling procedures are desirable. There are numerous analytical laboratories in the World but only a few who use any formalized procedure for ascertaining and assuring that the *quality* of analyses is high and remains so (See Section 6.11.). Too often the investigator *assumes* that everything is in order when it is not. Procedures and sampling plans for disasters and catastrophic events should be established at a national level (See Section 6.12.). Surveillance and control procedures need to be established and implemented for various diseases of humans and of animals (See Section 6.13). Sampling procedures are needed to assure the uniform distribution of a compound throughout a batch of material.

6.2 NATURE AND PROBLEMS OF ANALYTIC OR DIAGNOSTIC LABORATORIES

Many laboratories doing analytic or diagnostic analyses on samples are characterized by:
   a. many types of analyses,
   b. large numbers of analyses,
   c. a large back-log of samples to be analyzed, and
   d. shortage of facilities, space, and personnel to reduce back-log.

To illustrate, the New York State Diagnostic Laboratory at Cornell University performs about 750,000 sample analyses per year with an annual budget of around four million dollars. This works out to a cost of $5.33 per sample. If the same information could be obtained by using more efficient procedures for three million dollars, the average cost per sample would be reduced to $4.00. Whether or not this can be done would require a consideration and an institution of some of the procedures discussed in this chapter.

A laboratory director's solution to an overload and large back-log of samples to be analyzed may be to
   a. let the back-log increase,
   b. destroy all samples that are more than Y years old,
   c. discourage individuals from sending additional samples for analyses,
   d. hire more technicians,
   e. buy more equipment,
   f. use faster but less precise methods, and/or

g. acquire additional laboratory space.

All of the above do not contribute to more efficient and cost-effective procedures. Now, a statistician's solution for handling a large back-log of samples would be to conduct analyses on a sub-sample of all the samples sent in for analyses or to use other sampling procedures. Performing analyses on only a sample of the samples, pooling of samples, and/or using other procedures can cut cost and time considerably in many cases.

In summary, a laboratory director's solution for handling large volumes of laboratory analyses is to expand facilities to meet the demands whereas a statistician's solution is to sample or to pool samples for analyses, to work within present facilities, and to be cost-efficient.

For any analytic procedure used for laboratory analyses, there should be

a. a written description of the procedure,

b. a measure of the bias, precision, and variation of results,

c. a measure of the effect of dilution levels on the accuracy of an analysis,

d. a written description of the meaning of "zeros" and "traces", and

e. a measure of how accuracy decreases and variation increases with increasing dilution rates.

Without the above, the validity of an analytic procedure should be questioned.

6.3 REDUCING THE NUMBER OF ANALYSES

There are several methods of reducing the number of samples to be analyzed, several methods of pooling samples, and several reasons why pooling samples may be desirable. Samples may be pooled in order to

a. obtain sufficient material to run an analysis,

b. obtain sufficient material to run many analyses on the same sample,

c. to reduce cost of analyses,

d. to reduce time for obtaining results from a laboratory, and

e. to make the most efficient use of facilities.

Some statistical methods for reducing the number of analyses and for pooling samples are:

a. sub-sampling,

    b. sequential sampling,
    c. pool to attain objectives, and
    d. double sampling.

These and others are discussed in the following sections.

6.4 SUB-SAMPLING

One method of reducing the back-log of samples waiting to be analyzed is to do an analysis on a simple random sample of all the samples. This smaller sample may provide the desired information, thus making analyses on the remaining samples unnecessary. In other cases, the sub-sample could consist of those samples most likely to show an effect and if these samples showed no effect then the remaining samples need not be analyzed. To illustrate this, suppose that a radioactive disposal dump is located in a given area and it needs to be determined how far the radioactivity extends out from the dump. Obviously, samples most likely to show radioactivity would be those located nearest to the dump. Samples from concentric circles around the dump could be analyzed and if these showed no radioactivity, then there would be no need for taking samples further away from the dump.

Quite often investigators send in a large number of samples for analyses on one or more characteristics. If there is a back-log, this means that analyses will not be done. If the investigator were told that a simple random sample of all his samples could be done more quickly, many of them would be agreeable to this. It could be that from the results of the small sample, the investigator may decide not to have analyses performed on the remaining samples. Such a result would definitely reduce the back-log of samples to be analyzed. If there had been 10,000 samples in the original lot and if a simple random sample of 500 (or even 1,000) had been analyzed, there would be 9,500 (or 9,000) samples that may not have to be analyzed.

6.5 SEQUENTIAL SAMPLING

Sequential sampling involves taking samples and/or analyzing them one at a time (or even a batch at a time), and, based on the results, decisions are made to continue or discontinue sampling and/or analyses. Sometimes results not otherwise anticipated become apparent without performing the pre-determined number of analyses. For example, John H. Whitlock, Cornell University, had planned an experiment on 50 sheep to study a certain biological phenomenon. The sampling was destructive. After six sheep had been examined, the answer began to be apparent and after 12 sheep had been examined, the result was abundantly clear. At this point the investigator consulted the author who suggested stopping at 15, thus saving the remaining 35 sheep for other experiments. The additional three sheep were recommended to make the final results mostly independent of the decision to stop the experiment.

Sequential sampling should have been made for analyses of samples collected at the Love Canal chemical dump site. Considerable savings could have been achieved for this 4.5 million dollar study. The chemical dump was located as indicated in Figure 6.1. A drainage ditch was dug in the ground above the chemical dump. Because of community and political pressure, it was necessary to take all the samples as quickly as possible. Granted this, there may be no need to analyze all samples, which reportedly cost about $2,000 per sample per analysis. The samples marked x should have been analyzed first with perhaps sparse sampling of +, *, and o samples. If the chemicals were not detected in samples nearest the dump, then the analyses would stop. Actually, this is an approximation of the true situation since the chemicals were buried in a heavy clay soil which did not allow much water movement and hence little movement of chemicals. Also, samples should have been taken at various depths below the chemical dump to ascertain if there had been downward movement of the chemicals. Only horizontal movement of chemicals would be detected by the sampling scheme in Figure 6.1.

6.6 POOLING SAMPLES - INTRODUCTION

In order to reach the objectives of a study using analyses from pooled samples, several conditions must be met. Some of these are:

a. The characteristic is a presence or absence phenomenon (e.g. diseased or not diseased) or a measurable quantity (e.g., amount of drug or alcohol in a sample of blood or wine).

b. The percentage of positives (e.g., diseased) is relatively small, say less than 20-25%.

c. The composite sample mean is the same as the mean of the individual samples.

d. The compositing of samples does not alter the characteristic in the samples.

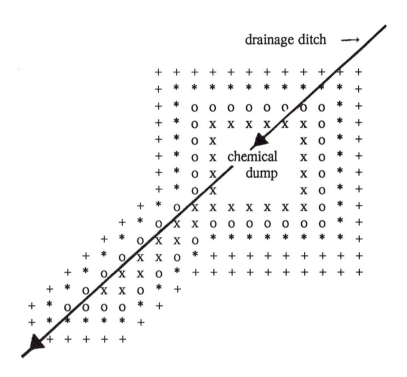

x - samples to be analyzed first
o - samples to be analyzed only if x samples were positive for the chemical
* - samples to be analyzed only if o samples were positive for that indicated
    that chemicals were diffusing out from dump
+ - samples analyzed only if * samples were positive for the chemical

Figure 6.1. Sampling scheme for chemical dump with a drainage ditch
through the dump. (Arrow indicates the direction of flow.)

e. The laboratory technique and analysis is accurate enough to detect the presence of the characteristic even if only one member in a pool of g samples has the characteristic.

Many characteristics are measured by presence or absence of the characteristic. Most types of diseases fall in this category. Pooling techniques are especially useful for detecting presence or absence of a disease caused by bacteria. The size of the pool g would have to become very large in order to miss finding at least one bacterium in a sample, and usually one or two bacteria are sufficient to start bacteria growing on a test plate.

In relation to other pooled samples for measuring the amount of alcohol or drugs in urine or blood samples, if the pool g becomes too large, the laboratory analyses may be unable to detect the presence of the characteristic. In Figure 6.2, various curves are drawn to illustrate that laboratory analyses for some characteristics may require a much smaller group or pool size than others. In curve I, the percentage of analyses that miss the characteristic, even when it is present (false negatives), starts increasing even with small group sizes and with a group size of 32, one positive in a pool of 32, it would go undetected 100% of the time.

On the other hand a curve like III would indicate that one sample with the characteristic in a pool of 64, would be detected 100% of the time. Such is the case with an ELISA test for *Bovine leucosis* where this test is sensitive enough to detect one diseased cow in a pool of blood from 75 cows (see Federer *et al.*, 1985). Thus in order to pool effectively, the investigator must know the percentage of false negatives encountered as the group size increases.

Tests can also give false positives but pooling can be a safeguard against this. Suppose that the analysis indicates that that characteristic is present in the pool but that this is a false positive. In order to determine which member(s) of the pool has (have) the characteristic, analyses could be made on the individual samples. Then unless another false positive occurred, it would be found that none of the members of the pool have the characteristic and that the test result from the pool was a false positive.

For some characteristics, e.g., hemoglobin, pooling samples may actually alter the amount and nature of the characteristic. For such characteristics, pooling of samples should not be used. The investigator needs to know the reactions taking place when samples are pooled.

6.7 POOLING SAMPLES BECAUSE OF INSUFFICIENT MATERIAL FOR AN ANALYSIS

There are situations wherein the amount of sample material obtained is less than is required for a laboratory analysis of the sample. To illustrate, L.C. Clark, Epidemiologist at the University of Arizona, was conducting a survey on the amount of arsenic absorbed by individuals exposed to arsenic sprays used on tobacco plants. The amount of arsenic can be measured from hair samples but 200 milligrams (mg) of hair are needed for an analysis. It would not be possible to get this much hair from one sampling for many of the individuals in a sample. Therefore, for the hair samples obtained, these were all ordered according to the amount of exposure an individual had to arsenic tobacco sprays. Then, the hair of adjacently ordered individuals was pooled for analysis in order to obtain samples of 200 mg or more. If all samples with less than 200 mg had been discarded, there would have been few individuals left in the survey with more than 200 mg of hair.

As was mentioned in previous chapters, in order to obtain a sufficient amount of material to conduct approximately 300 biochemical analyses in the Beijing-Cornell cancer mortality-dietary study, pools of 25 were used. Only a part of the individual's sample was used to make the pool, thus allowing analyses to be made on the individual samples should that be desired at a later date.

Another situation that often occurs in surveys is to have no sample values for many categories. For example, consider that the survey involves a summarization of number of individuals at 12 income levels and five levels of education of the following nature:

Education level	|	1	2	3	4	5	6	7	8	9	10	11	12
	|					Income level							
1	|	X	X	X	X	X	X	X	X	X			X
2	|	X	X	X	X	X	X	X	X	X			X
3	|	X	X	X	X	X	X	X	X	X		X	
4	|	X	X	X		X		X	X		X		
5	|	X			X		X			X			

where X denotes that one or more individuals fell in that category. For

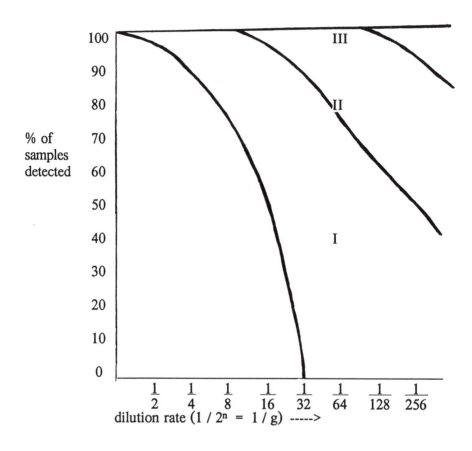

Figure 6.2. Effect of dilution rate on percentage of samples indicating presence of a characteristic.

presentation purposes, it may be desirable to pool or group income levels 10 to 12 together and to pool education levels 4 and 5 together. Such grouping would have some individuals in every category and is appropriate for certain situations.

In the Love Canal study described earlier, additional savings could have been made by pooling g adjacent samples marked x, i.e., those closest to the dump. Suppose groups of ten adjacent samples were made. Then, only 10% of the analyses on the x's would need to be made. If no chemicals were detected in these pools of ten, further analyses would not be needed. This would have resulted in considerable savings in time, facilities, and money.

6.8 GROUP TESTING

The original form of **group testing** was put forth by Dorfman in 1943. Although this was a creative and potentially useful idea for reducing the number of laboratory analyses, very little use has been made of the procedure. The Dorfman group testing procedure was supposed to have been developed for reducing the very large number of tests required for screening recruits for venereal diseases for the United States Defense Forces in World War II. It is stated in Feller (1950) that use of the procedure resulted in up to 80% savings in the number of laboratory analyses. The pooling procedure used for the cancer mortality-dietary study described previously resulted in a 96% savings in the number of analyses that needed to be run. Likewise, the Dorfman group testing procedure can result in considerable savings in number of samples, time, laboratory space, personnel, and materials. Effort needs to be made to acquaint laboratory directors of this potentially useful procedure.

The Dorfman group testing method proceeds as follows. For N samples to be screened for a characteristic, say a disease, and a group size of g, there will be $G_1 = N/g$ groups or pooled samples that require testing for the first phase. Suppose that the characteristic is binomially distributed.(See later chapters for a description of the binomial.) and that p percent of the population have the characteristic. Then $G_2 \le G_1$ of the groups will test positive for the characteristic. For the G_2 groups testing positive, the second phase of the procedure would be to test each sample in any positive group or pool, to determine which individuals test positive, i.e., have the characteristic. The expected total number of sample analyses that would be required would be

$$G_1 + gG_2 = N(1/g + 1 - q^g), \tag{6.1}$$

where $q = 1 - p$. When

$$1 / g = q^g \text{ or } q = (1 / g)^{1/g},$$ (6.2)

it will be a toss-up as to whether to do individual sample analyses or to use this group testing procedure. This is called the **break-even point** and occurs when p is approximately 29%. The smaller p is the greater would be the savings in number of samples analyzed. This is illustrated in Table 6.1. When individual sample analyses are performed, i.e., the group size is one, the fraction of N samples analyzed is one, or 100%. For the values of p in the table, a group size of two or more results in fewer samples being analyzed. For p = 10%, the lowest number of samples is done when the group size is four. For p=5%, the group with maximum saving is 5; it is 11 for p = 1%, and 32 for p = 0.1%. These points are called **optimum group sizes**. These can be calculated from the following formula for any percentage q = 1 - p:

$$g^2 = 1 / q^g \ln(1 / q),$$ (6.3)

where ln is the natural logarithm. For example, for p = 0.1% or q = 99.9%,

$$g^2 = 1/0.999^g \ \ln(1/0.999)$$

or

$$32^2 = 1/0.999^{32} \ (.0010005003) = (32.125)^2.$$

and 32 is the nearest integer satisfying the equality in (6.3). Using the optimum group size results in a savings of 41% for p = 10%, of 57.4% for p = 5%, of 80.4% for p = 1%, and 93.7% for p = 0.1%. As may be noted from Table 6.1, for small p, the group sizes around the optimum group size achieve almost the same savings. Using the optimum group size is more critical for larger p. For example, for p = 10%, using g = 2 or 8 instead of g = 4 results in a 10% loss in savings. For p = 5%, using g = 3 or 9 results in a 5% loss in savings. The lower the value of p in the population, the greater the savings in sample analyses will be. Thus, the smaller p is the more important it is to use a procedure like that of Dorfman (1943).

Even greater savings may be achieved with variations of the Dorfman procedure. This is illustrated with the following example. For a certain breed of horses, natural rather than artificial breeding methods are required for registration. It was suspected that one or more of 600 mares might have a newly found venereal disease. No owner of a stud would

Table 6.1 Effect of group size and percentage p on the number of samples analyzed.

| Group size | Expected fraction of total number of samples analyzed | | | |
	p=10%	p=5%	p=1%	p=0.1%
1	1.00	1.00	1.000	1.000
2	0.69	0.60	0.520	0.502
3	0.60	0.48	0.363	0.336
4	0.59	0.44	0.289	0.254
5	0.61	0.43	0.249	0.205
6	0.64	0.43	0.225	0.173
7	0.66	0.44	0.211	0.150
8	0.69	0.46	0.202	0.133
9	0.72	0.48	0.198	0.120
10	0.75	0.50	0.196	0.110
11	0.78	0.52	0.196	0.102
12	0.80	0.54	0.197	0.095
13	0.82	0.56	0.199	0.090
14	0.84	0.58	0.203	0.085
15	0.86	0.60	0.207	0.082
16	0.88	0.62	0.211	0.078
17	0.89	0.64	0.216	0.076
18	0.91	0.66	0.221	0.073
19	0.92	0.68	0.226	0.071
20	0.93	0.69	0.232	0.070
21	0.94	0.71	0.238	0.068
22	0.95	0.72	0.244	0.067
23	0.95	0.74	0.250	0.0662
24	0.96	0.75	0.256	0.0654
25	0.97	0.76	0.262	0.0647

allow his stud to be used unless the mares were certified free of the disease. A diseased stud would be useless for future breeding purposes. Hence, blood samples from the 600 mares were sent to the New York State Diagnostic Laboratory (NYSDL) for analysis. NYSDL charged $100 per sample for analysis and the job required the lab's facilities for about one week. Since the test for the disease was quite sensitive and since the prevalence of the disease was suspected of being low, pooling was definitely indicated. A suggested pooling procedure would be to obtain 20 pools of 30 samples each, assuming the test was accurate enough to detect one diseased mare in a pool of 30. Then for those pools testing positive, three pools of ten samples each would be made; from the pools testing positive, pools of five would have been made. For any pool of five testing positive, perform the analyses on each of the five samples in the positive pool. If no mares had been infected only 20 sample analyses, rather than 600, would have been made. This would have resulted in $60,000 / 20 = $3,000 per sample, considerably more profitable than $100 per sample. Also, very little of laboratory personnel, space, and time would have been used. If one mare was infected 20 + 3 + 2 + 5 = 30 analyses would have been made, resulting in $2,000 per analysis.

If the sensitivity of the test had been very high, a more efficient procedure would have been to:

    a. Obtain a pool of all 600 and test.
    b. If positive, split into pools of 300 and test.
    c. For positive pools in b, split into pools of 150 and test.
    d. For positive pools in c, split into groups of 75 and test.
    e. For positive pools in d, split into groups of 37 and 38 and test.
    f. For positive pools in e, split into pools of 18 or 19 and test.
    g. For positive pools in f, split into pools of 9 or 10, and test.
    h. For positive pools in g, split into pools of 4 or 5 and test.
    i. For positive pools in h, test individual samples.

Step (i) might be instituted after steps f or g if the number of positive groups was fairly high. This procedure follows one suggested by Sobel (1967) and Sobel and Groll (1959, 1966). If no mares had been infected, then only one sample analysis would have been made. If one mare was infected, then only 19 or 20 analyses would have been made. For two infected mares, at most 39 to 40 analyses would have been made. As should be obvious from the above, many other modifications are possible.

Raghavarao and Federer (1973) and Bush *et al.* (1984), considered the following scheme. If the test is sensitive enough, make a pool of all N samples and test. If the test is positive, make an r row by c column array of individual samples, or even pools of size g, such that r is as nearly equal to c as possible. If, e.g., N = 72, use r = 8 and c = 9 rather than 6 and 12, 3 and 24, or 2 and 36. If N does not factor into two integers r and c, then use the nearest N* = N + k which has factors r and c where k will be the number of empty cells. Suppose N = 61. If k is taken equal to three, an eight row by eight column array with three empty cells can be formed. Then, the row pools and column pools are analyzed. For the row pools and column pools testing positive, find the intersections and test those samples. To illustrate, let N = 29, r = 5, c = 6, and k = 1. Given that the pool of 29 is positive make the following array of the 29 samples:

						result of row pool
1	2	3	4	5	6	+
7	8	9	10	11	12	-
13	14	15	16	17	18	+
19	20	21	22	23	24	-
25	26	27	28	29	*	-
+	-	+	-	-	-	result of column pool

Positive results for samples (1 and 15), (3 and 13), (1,3, and 15), (1,13, and 15), or (1,3,13, and 15), could be obtained from these row and column tests. Therefore, it would be necessary to test all four samples to determine which ones are positive (+). This would have resulted in 1 + 5 + 6 + 4 = 16 analyses, or a savings of 13 analyses in detecting which samples tested positive.

6.9 THE ULTIMATE POOLED SAMPLING PROCEDURE

In certain situations, only the number N_p of individuals in a sample of N, or the fraction N_p / N, possessing a characteristic is desired. The identification of individuals with the characteristic, denoted as positives, is not required and anonymity is assured. If also the mean levels of both positives, M_p, and negatives, M_n, is known, then *one* sample analysis suffices for determining N_p or N_p / N. For many characteristics, values for M_p and M_n are known. The mean level of individuals who have had no alcohol, drugs, or other substances, M_n will be zero. The mean titer level of disease-free cows will not be zero but will be small and near zero for *Bovine leucosis*. The mean level M_p of all drivers picked up and suspected of driving under the influence of alcohol or drugs can be obtained from police or medical records. M_p for drug levels may also be obtained for people entering rehabilitation centers.

Given that the test is accurate in measuring the amount of the substance or titer at the dilution rates used, i.e., $1 / N$, the estimated number N_p or the fraction N_p / N may be obtained with one laboratory analysis as follows.

    a. Obtain a pool of all N samples with equal amounts from each sample.

    b. Test the pool and obtain the amount in the pooled sample, say M_s.

N_p equals the number of positives and $N - N_p = N_n$ equals number of negatives in the sample of N. Then,

$$N_p M_p + (N - N_p)M_n = NM_s \qquad (6.4)$$

or, since the only unknown is N_p,

$$N_p = N(M_s - M_n)/(M_p - M_n) \qquad (6.5)$$

If M_n equals zero, then

$$N_p/N = M_s / M_p \text{ or } N_p = NM_r / M_p. \qquad (6.6)$$

Because an equal amount of each sample may not be obtained or M_p and M_n may not be known exactly. N_p or N_p/N should be considered only as an estimate of the true number. However, for many purposes, an estimate of N_p is all that is required to achieve the goal of an investigation.

The following is a procedure suggested by Federer (1989) for an organization to determine if drug use is of frequent enough occurrence to warrant remedial action. It would be especially useful for an Army or Navy unit as well as for some businesses and schools. Drug tests that can be performed on urine samples would be appropriate for this sampling procedure. Let us consider an Army unit to illustrate the procedure, since one of the first things a member does upon arising is to urinate. If the urinals are all connected by a trough as in Figure 6.3, collection of all urine can be accomplished as shown.

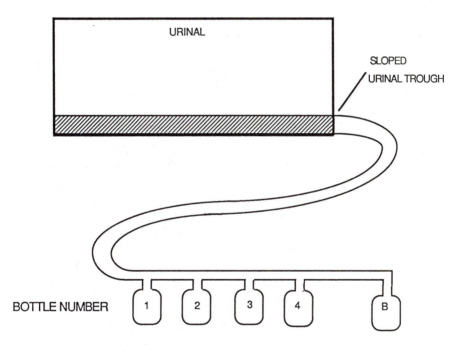

Figure 6.3. Urinal with sloped trough and B collection bottles.

Each bottle could be closed manually or automatically when it is full. Suppose that bottles (containers) are collected from N individuals as determined by the number entering (or leaving) a latrine. On each bottle there would be, on the average, $N/B = N_b$ individuals per bottle. One analysis on each bottle would yield estimates (from equation 6.6) of

$$N_{sb} = N_b\, M_{sb}\, /M_p. \qquad (6.7)$$

Then, the sum of the N_{sb} values would be an estimate of the number of individuals using drugs. Alternatively, an equal amount could be taken from each of the B bottles and pooled. Then, one analysis on the pool would suffice to estimate the number N_p.

The sampling time for schools, businesses, manufacturers, medical institutions, and others would need to be longer, say four to six hours, than for a regimented organization like the Army where the time of sampling may be only thirty minutes. The assumption is that an equal amount of urine is obtained from each individual and not that the frequency of urination is the same. If the amount of urine and drug use is unrelated, the above assumption could still be tenable. Otherwise, the estimate would be an approximation.

Detection of cocaine, arsenic, or other materials is possible using hair samples. In fact, use of cocaine in the past, say two months, can be detected by this method. In order to obtain an idea about drug use, barbers could be asked to deposit a small amount of hair from each customer in a pool. The pools for each day could be tested and an estimate of the number of users may be determined from equation (6.7). Then, if use is considered high enough, samples from each individual would be collected and analyzed to determine which people use drugs. This procedure would miss individuals who did not have their hair cut during the sampling period.

6.10 DOUBLE SAMPLING

Suppose that one characteristic X_1 can be easily and cheaply obtained on samples but that another characteristic X_2 is difficult and/or costly to obtain. The investigator would really like to obtain X_2 for every one of his N samples but his time and budget do not allow this. He could take a smaller sample size N but he could also measure X_1 on every one of the N samples and could measure X_2 on a sub-sample of size N_1. Thus, X_1 and

X_2 are both measured on N_1 samples. This procedure is called **double sampling**. If X_1 and X_2 are related, then a more accurate estimate of population parameters can be achieved, using appropriate statistical methodology, than would be obtained from N_1 samples and using only the X_2 measurements.

Many situations arise where this condition occurs. There might be a cheap, easy, and not too accurate procedure, for obtaining X_1 measurements available for detecting a disease and an accurate and costly method for obtaining measurement X_2. An experienced and highly accurate technician might be associated with a laboratory; it would be desirable to have this person do all the sample analyses to obtain the X_2 measurements. However, the volume of samples will not permit this, so several inexperienced and not so accurate technicians are employed to obtain measurements X_1. In the tobacco spray-arsenic in hair samples survey, the measurements X_1 equals stated exposure to tobacco sprays and could be taken on a large sample N. A sub-sample of size N_1 could be taken where X_1 measurements and X_2 measurements, which are laboratory analyses on hair samples for arsenic content, are both obtained. In ascertaining how much money is spent for specific items or activities of a business, the owner could be asked for this information, X_1 measurements, or his records could be studied to determine the X_2 measurements. Obtaining the latter would be a time-consuming and tedious task.

6.11 CHECKS AND QUALITY CONTROL OF LABORATORY ANALYSES

The accuracy of any laboratory procedure should be assessed from time to time to ascertain that the method is still under control. One method of doing this is to insert samples of known constitution at a low, a medium, and a high level of the substance being analyzed. For example, for the 300 biochemical analyses performed on the 260 pooled samples from the dietary-cancer mortality study, it was recommended that for each analysis, thirty checks (ten samples at a low level, ten at a medium level, and ten at a high level) be randomly and blindly (unknown to the analyst) inserted in with the pooled samples to check on variation among analyses at each of three levels and to check for biases in the laboratory method and/or laboratory analyst. This was done for some of the analyses but not all. In some cases, the analyst analyzed the 30 check samples separately from the others and this person knew that they were check samples. Results may not be the same when the analyst knows and when he does not know the identity of the samples being analyzed.

A quality control program is one that continually assesses the quality of the analyses being performed by an analytical laboratory. It should be an on-going program for *every* analytical laboratory as investigators need to know the quality of the analyses made by a laboratory.

6.12 SAMPLING PLANS FOR DISASTERS

Every nation experiences a major disaster at one time or another. In 1989, the United States was hit by the tropical hurricane Hugo, which caused devastating damage in some of the Carribean Islands and the Carolinas; and a major earthquake hit the San Francisco area, toppling the upper span of the Bay Bridge as well as damaging numerous buildings. Another major disaster was in South Africa where a large paper company spilled a large amount of a dangerous chemical in a river from which several towns obtained their water supply. This list could go on and on but that is not our point. In nations where a disaster occurs, it is essential that the leaders obtain quick and factual information in order to set emergency relief operations in motion. In many cases, it is essential to start relief operations within hours after a disaster happens.

In the case of Hugo, officials knew that the storm would hit various islands and when as well as where it would hit the Carolinas. Plans should have been made to assess the damage wrought within hours after Hugo stopped, to have the results summarized, and to provide for making decisions about relative courses of action. These should have been implemented within hours, not weeks, after the disaster. In order to do this, it is recommended that:

   a. A *small* Disaster Planning Unit with direct access to the leaders
   of a country be established for each nation. (The Unit should be
   small enough to act quickly and promptly without the bureaucratic
   wranglings of larger organizations and should consist of highly
   trained professionals in every area of possible disasters.)
   b. Sampling plans be devised for all possible disasters so that it will
   not be necessary to devise sampling plans after a disaster has
   happened (see Federer, 1984).
   c. A group of trained data collectors be accessible at the time any
   disaster occurs so that they can immediately obtain data on the
   damage from a disaster.

d. Methods and computing facilities for immediately summarizing the data be ready for use as soon as the data in (c) are collected.

e. The Disaster Planning Unit have immediate access to the leaders of a country who will act on their recommendations for relief.

In cases where the Disaster Planning Unit has notice of an impending disaster such as a hurricane or flood, they should begin planning immediately for collection of data relating to damage wrought by the disaster. This would allow quicker collection of pertinent data and of course quicker recommendations for emergency relief measures. They should set up plans of action and possible relief measures for every possible disaster. Such planning could save many lives and perhaps property from damage by disasters. It definitely would alleviate some of the suffering by humans in a disaster. A possible sampling scheme for a nuclear disaster or for a chemical dump is presented in Figure 6.4.

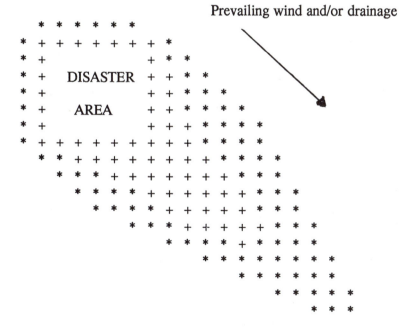

Figure 6.4. Sampling plan for a disaster area like a nuclear explosion or chemical dump with a prevailing wind or drainage pattern, with the plus (+) samples being analyzed first.

6.13 A SURVEILLANCE AND CONTROL PROGRAM

Many of the ideas presented thus far are appropriate for a surveillance and control program for a disease like *Bovine leucosis* (see Federer *et al.*, 1985). This disease can be controlled by management procedures in individual herds, as it is mainly contracted by transfer of blood from animal to animal via a hypodermic needle, dehorning apparatus, or other instrument which may pass blood. The program described by Federer *et al.* (1985) is presented below.

In stage one, it is essential to determine which cows have the disease *Bovine leucosis* (BL). Since the prevalence of BL is high, say 40 to 50%, in New York State, it was recommended that dairy cows be tested individually using a quick and an inexpensive testing procedure. It may be that the cows testing positive should be isolated (say 60 to 100 feet away from BLV negatives) but this may not even be necessary. The cows testing positive for the BL virus (BLV) would be the first ones culled from the herd. It is standard procedure in dairy herds to cull and replace cows on a continuing basis. No BLV positive new cows should be introduced into the dairy herd. Also, calves from BLV positive cows should not be kept as replacements for culled cows unless they test free of BLV.

The testing interval should be three to six months. As soon as the prevalence of BL diseased cows becomes relatively low, say less than 20 to 25%, pooling of blood samples for laboratory analysis should be instituted. Continuation of the process of testing and culling cows will reduce the prevalence of BLV positive cows to a very small percentage, say less than one percent. At this stage, pools of up to 75 blood samples could be pooled to screen for BLV positive samples. The ELISA test available is accurate enough to detect the presence of BLV even if only one cow in 75 has BLV. At this point, the testing interval should probably be lengthened to one year. When no more BLV positive cows are found in a herd, the testing inteval may be extended to 18 or 24 months.

In order to make the program effective, it is essential that the full cooperation of the dairy herd owner be obtained. Timely information of test results along with a full education about management procedures useful in preventing the spread of BL, will go far in obtaining the desired level of cooperation. Another factor that must be considered is the privacy of test results for each dairy herd owner. As of now the Freedom of Information Act would allow any person access to the test results for any dairy herd when the analyses are made by a New York State laboratory such as the

New York State Diagnostic Laboratory (NYSDL) at Cornell University. The way to assure privacy of test results for a dairy herd owner is to set up a private corporation or agency, say the No-Tell Corporation. Samples would be sent to No-Tell who would simply number the samples and identify the herd with a code number. No-Tell would send the samples to the NYSDL for analysis. When the test results are returned, No-Tell would decode them and send the test results to the dairy herd owner. Since No-Tell is a private organization, no unauthorized person would have access to their records.

The exact size of pools and length of interval between testing in each herd would need to be determined for each herd. The services of a statistician would need to be employed for this.

6.14 SAMPLING TO ASSURE UNIFORM DISTRIBUTION

In many situations a small amount of material (vitamin, drug, chemical) is added to a large quantity of other material (salt, animal feed, breakfast cereals). If the material in a small amount is evenly distributed throughout the large quantity of material, the resulting mixture is considered to be beneficial to the consumer. If, on the other hand, the vitamin, drug, or chemical is not evenly distributed, the consumer may receive an overdose or an underdose, depending upon which sample is consumed. For example, salt is treated with selenium in areas of China as a measure to lower incidence of cancer. Salt crystals are not uniform and during transportation the small crystals rise to the top of salt containers and the larger crystals to the bottom. Consumers receiving their supply of salt from the top of the container receive an overdose of selenium. Those obtaining their salt from the bottom of the container receive less than the desired amount of selenium. Since selenium in too high a concentration is toxic, undesirable consequences would result. Methods have been devised to assure uniform mixing of salt crystals of all sizes.

In adding growth hormones or antibiotics to animal feed, the ratio of the added material may be of the order of one to 10,000. How does one assure uniformity of mixing the two materials? It will be necessary to do continuous sampling and to check each sample for deviation from the desired ratio. As soon as a run of samples are obtained which are all at or near the desired ratio, the sampling could stop and the mixing would be considered uniform. This is an application of the ideas of sequential sampling design.

6.15 REFERENCES

Bush, K. A., W. T. Federer, H. Pesotan, and D. Raghavarao (1984). New combinatorial designs and their applications to group testing. *J. Statistical Planning and Inference* 10, 335-343.

Dorfman, B. (1943). The detection of defective members in large populations. *Annals of Mathematical Statistics* 14, 436-440.

Federer, W.T. (1984). Cutting edges in biometry. *Biometrics* 40, 827-839.

Federer, W.T. (1989). Anonymity, disease, drugs, and design. Technical Report no. BU-1036-M, Biometrics Unit Series, Cornell University, June.

Federer, W. T., L. C. Clark, E. J. Dubovi, and A. Torres (1985). A surveillance and control procedure for *Bovine leucosis* and other diseases. Technical Report no. BU-877-M, Biometrics Unit, Cornell University, September.

Feller, W. (1950). *Probability Theory and Its Applications*, John Wiley & Sons, Inc., New York, p. 189.

Raghavarao, D. and W. T. Federer (1973). Group testing - A combinatorial approach. Technical Report no. BU-473-M, Biometrics Unit, Cornell Univ., July.

Sobel, M. (1967). Optimum group testing. *Proc., Colloquium on Information Theory, Bolyai Mathematical Soc.*, Debrecen, Hungary, pp. 411-488.

Sobel, M. and P. A. Groll (1959). Group testing to eliminate efficiently all defectives in a binomial sample. *Bell System Technical J.* 38, 1179-1252.

Sobel, M. and P. A. Groll (1966). Binomial group testing with an unknown proportion of defectives. *Technometrics* 8, 631-656.

6.16 PROBLEMS

Problem 6.1. For the fractions of samples required in Table 6.1, prepare a graph with the group size as the abscissa and the fraction of total samples needed for p = 10%, 5%, 1%, and 0.1%. For p = 0.1% the fractions for group sizes 26 to 35 are: (26, 0.06414), (27, 0.06369), (28, 0.06334), (29, 0.06308), (30, 0.06290), (31, 0.06280), (32, 0.06276), (33, 0.06278), (34, 0.06286), and (35, 0.06298).

Problem 6.2. Suppose p were 4%. Using equation (6.1), compute the fraction of work required for groups of size g = 5, 6, 7, 8, and 9. Which group size is optimal, i.e., has the smallest fraction? Use equation (6.3) to verify this.

Problem 6.3. Suppose that the mean level of alcohol recorded for drivers picked up by a police department for 1989 was 0.14. Suppose further that a large number, N = 100, of blood samples was available but there was insufficient money to run analyses on all samples. Show how to estimate the number of individuals who used alcohol with one sample analysis {Hint: Use equations (6.4) to (6.6)}.

Problem 6.4. Suppose the average level of cocaine found in urine samples of drug users was 10 mg. Suppose Fraternity A wishes to monitor the number of individuals using the drug and obtains a pooled sample of urine from all members, N = 50, once a week. Suppose the sample results are 2 mg for week 1, 3 mg for week 2, and 1 mg for week 3. Estimate the fraction of the members using cocaine each week.

Problem 6.5. Make a list of laboratory analyses where you believe group testing could be used effectively to reduce the number of laboratory analyses. Give your reasons for each analyses.

Problem 6.6. Make a list of situations which you believe would lend themselves to sequential sampling procedures and explain why you think so.

Problem 6.7. Make a list of situations where you believe that double sampling procedures may prove useful. Justify your selections.

Problem 6.8. Suppose equal amounts of hair from the same position on an individual's head were obtained by barbers in a barbershop and deposited in a pool of hair each week. The sampling was to continue for a two month period. For the pool of hair from each barbershop, a single laboratory analysis for the presence of arsenic was conducted. Suppose that for one shop 200 people had their hair sampled and that the laboratory analysis indicated 4 mg in the sample. Suppose it is known that the mean level of arsenic in hair of people exposed to arsenic sprays is 14 mg. Estimate the number of individuals in this shop that were exposed to arsenic sprays.

Problem 6.9. Prepare a list of key words for this chapter and describe relationships among them.

CHAPTER 7. EXPERIMENT DESIGN

7.1 INTRODUCTION

The experiment designs considered in this chapter are for comparative experiments involving two or more treatments, where the object of the investigation is to obtain information on the treatments relative to each other. In other words, the interest is on differences between treatment averages (or effects) rather than on the averages (or effects) *per se.* One set of characteristics of designing the experimental arrangements or procedures we shall consider is:

1. Arrangement of the procedure to increase the **efficiency** of the experimental investigation. One design is said to be more **efficient** than a second if the variation in average response for any treatment or of the difference between two treatment effects is smaller than in the second design.

2. Grouping of the experimental material in such a manner that the units within a group are more alike than are units in different groups. This kind of grouping is called **blocking** or **stratification**.

3. Fairness to each treatment by subjecting all treatments to as nearly equal conditions as possible and by utilizing chance allotments or **randomization** thereafter.

First, these characteristics will be exemplified with three illustrative examples. Then we shall present a number of types of experiment designs which control various types or sources of heterogeneity among the individual items or units in the investigation.

It is necessary to understand the nature, size, and shape of the smallest unit of observation, **the observational unit**, and also about the smallest unit to which one treatment is applied, **the experimental unit**. In some cases the observational and experimental units are the same; in others the experimental and/or the observational unit size is fixed and cannot be varied. When the size and shape can be varied according to certain criteria, optimum size and shape needs to be considered but this is a topic beyond the scope of this text. We shall assume that the size and shape of the observational and experimental units are given. Examples of investigations wherein the unit is fixed are the animal in physiological and nutritional studies, the plant in physiological studies, the individual person in learning and educational studies, a cake or pie in baking studies (a whole cake or whole pie must be baked, even if size and shape can be varied), the classroom for teaching methods (the number of students can be varied, but the classroom is fixed), the automobile or the tire for road endurance tests, a piece of equipment used to produce or evaluate a product, fixed farms or pastures in certain management investigations, and so forth.

The first experiment design we shall consider deals with the weighing of very light objects on a scale, for instance a spring scale. Suppose that seven objects (a,b,c,d,e,f,g) are to be weighed. One experiment design for weighing them would be (The order of weighing should be randomized, but they are ordered here for easy reading.):

Weighing	Object weighed
1	determination for zero-correction
2	a
3	b
4	c
5	d
6	e
7	f
8	g

The weight of any object is the scale reading for its weight minus the scale reading for the first weighing. Such a design would require eight weighings, and each object would be weighed only once.

Alternatively, let us consider a different "weighing design" which was first suggested by Yates (1935) of the Rothamsted Experimental Station in England. Suppose that the following experiment design is utilized to weigh the seven objects:

Weighing	Object weighed
1	a, b, c, d, e, f, g

2	a, b, d
3	a, c, e
4	a, f, g
5	b, c, f
6	b, e, g
7	c, d, g
8	d, e, f

The same number of weighings is used here as for the previous weighing design, but each object has been weighed four times rather than only once as in the previous design. It turns out that the variation in weights from the above design is only one-third that of the first design and three times as efficient. The weight of each of the objects is obtained as follows:

Weight of object	Coefficients for weights from weighing							
	1	2	3	4	5	6	7	8
a	+	+	+	+	-	-	-	-
b	+	+	-	-	+	+	-	-
c	+	-	+	-	+	-	+	-
d	+	+	-	-	-	-	+	+
e	+	-	+	-	-	+	-	+
f	+	-	-	+	+	-	-	+
g	+	-	-	+	-	+	+	-

If the object is present in the weighing, that particular weight receives a plus sign and if not, a minus sign. The sum of the first four weighings minus the sum of the last four weighings gives the weight of object a; the sum of weighings 1, 2, 5, and 6 minus the sum of weighings 3, 4, 7, and 8 gives the weight of object b; and so on.

A number of scientific papers on weighing designs for a chemical balance, a spring balance, and other weighing devices have been published in statistical journals. Reference to these papers may be found in Section XV-4 of Federer (1955) and in Section VIII of Federer and Balaam (1972). Our purpose here is not to discuss weighing designs but to illustrate the pay-off in efficiency that is sometimes possible when an appropriate experiment design is used. Characteristic (1) above is exemplified by this example.

As a second example used to illustrate characteristics (2) and (3) above, let us suppose that the investigator is comparing four nutritional treatments (for example, standard ration = S, S + vitamin A, S + vitamin B, S + vitamin D), and is using a rat as the experimental animal or unit. Suppose that for design 1, he randomly selects ten rats for each treatment without any regard to the rat's parentage. This allows all four treatments fair and equal chance to be allotted any ten of the 40 rats. Suppose that another investigator takes account of the rat's parentage and uses ten litters of four male rats each. (The word litter is used to designate the members born to a mother within a short period of time, say one day. Thus, twins in humans would be a litter of size two, triplets would be a litter of size three, etc.. In certain types of animals such as rabbits, dogs, cats, swine, and so forth, the members of a litter are brothers and sisters, or half sibs, and therefore not identical in genetic composition but they are related.) The four treatments are then allotted by chance to the four male rats of each of the ten litters to form design II. This is "fair" to all four treatments as each has an equal chance of being allocated to any rat in the litter. In this design, the comparison among treatments is within a litter (on members of the same litter) and on rats of the same sex. The variation for many characteristics including nutritional response among members of the same litter or family is less than it is among members of different litters. Hence, design II would be expected to yield treatment means which are less variable than the corresponding means from design I. In fact, it was found from nutritional experiments on swine that the variation of treatment means compared on individuals of the same litter was about one-half of that obtained when the animals were not grouped or stratified into litters. Practically, this means that investigators using design II would require only one-half as many animals to obtain the same degree of variation among treatment means as those using design I. A simple change of design from I to II would cut the cost of experimentation by one-half; alternatively, for a fixed amount of experimentation, it would decrease the variation among treatment means by one-half.

The use of blocking or stratifying experimental material into relatively homogeneous groups can greatly increase the efficiency of experimentation, since total variation is equal to that due to assignable causes plus bias plus random error. By blocking, a portion of the random error is placed into the assignable or controllable category, thereby reducing the amount of variation in the chance or random category.

As a third example, we shall utilize an illustration demonstrating characteristics (2) and (3); it is adapted from one described by W. J.

Youden (formerly of the National Bureau of Standards and now deceased) in a lecture at Cornell University a number of years ago. The owner of a large fleet of cars wished to compare the effect of four brands of motor oil on the performance of cars in the fleet. His first experiment was conducted as follows. A new car was purchased; motor oil K was used in the car from zero to 20,000 miles; motor oil C was used from 20,000 to 40,000 miles; motor oil P was used from 40,000 to 60,000 miles and finally motor oil M was used from 60,000 to 80,000 miles. From the measures of performance used, it appeared that motor oil M was definitely inferior to the other three motor oils. The sales representative was called in, and the fleet owner told him that he would not be purchasing their product any longer. Upon inquiring about the reason for this decision, the salesman was informed about the experiment performed to compare the four brands of motor oils and about the poor performance of his company's product. After thinking about the experiment a moment, the salesman retorted, "Your experiment was unfair to our product. The car was all worn out and the poor performance you observed was not due to the oil used but to the dilapidated condition of the car. You were fair in driving the car 20,000 miles for each of the four oils but you were unfair for the order in which the oils were tried."

The fleet owner promised to take this into consideration before he made a final decision. It happened that four new cars for the fleet were purchased at this time. The owner decided to use these cars to perform an experiment which would be fair to all four motor oils. He decided to assign the numbers 1, 2, 3, and 4 in a random manner to the four oils by rolling a six-sided die and ignoring 5's and 6's. The number on the die obtained from the first roll would be assigned to oil K, the number obtained on the second roll, excluding the number used for oil K, would be assigned to oil P, and so on. He obtained the following code K=3, P=1, C=4, M=2. Since he knew nothing about the performance of each of the four new cars, he allotted the numbers 1, 2, 3, and 4 to the cars in an arbitrary manner. Also, the numbers 1, 2, 3, and 4 were allotted to the mileage groups in a random manner such that

    3 = 0 to 20,000 mile group,
    1 = 20,000 to 40,000 mile group,
    4 = 40,000 to 60,000 mile group, and
    2 = 60,000 to 80,000 mile group.

He then drew up the following plan:

Mileage group number			Car number		
		1	2	3	4
(20,000 to 40,000 miles)	1	1 (P)	2 (M)	3 (K)	4 (C)
(60,000 to 80,000 miles)	2	2 (M)	1 (P)	4 (C)	3 (K)
(zero to 20,000 miles)	3	3 (K)	4 (C)	1 (P)	2 (M)
(40,000 to 60,000 miles)	4	4 (C)	3(K)	2 (M)	1 (P)

When rearranged in the order of mileage, the grouping becomes:

Mileage group	Car number			
	1	2	3	4
0 to 20,000	K	C	P	M
20,000 to 40,000	P	M	K	C
40,000 to 60,000	C	K	M	P
60,000 to 80,000	M	P	C	K

The experiment was run according to the above experiment design, and oil M still ranked considerably lower in performance tests than the other three motor oils. The salesman for brand M motor oil was told that his oil would not be used any longer. Again he wanted to know how the experiment had been conducted. He was informed of the experimental procedures. He was quite disappointed, but after some time his face brightened, and he said, "Ah, yes, but how do I know that you didn't assign me a horrible driver, and this is what is causing the low performance for my company's product?" Well, the owner of the fleet had considered this variable as well, and he replied, "I took care of that, too. The drivers were assigned to the cars according to the following plan:

Mileage group	Car number			
	1	2	3	4
(zero to 20,000)	K S	C R	P D	M G
(20,000 to 40,000)	P R	M S	K G	C D
(40,000 to 60,000)	C G	K D	M R	P S
(60,000 to 80,000)	M D	P G	C S	K R

where drivers were assigned to the letters as follows: S = Smiley, R = Red,

D = Demon, and G = Grumpy. The plan was fair in all respects, and extraneous variation was controlled. Your oil just doesn't stand up in comparison with the other oils."

The salesman went away very depressed about losing a customer; however, in a few days he was back again, because his company had developed a new oil called Super Special M. The new oil had been tested in the manner described above; it was found to be equal or superior in all respects to oils K, C, and P. Everyone lived happily ever after, until a better oil was developed and tested!

From Problem 3.2, the "Georgia Peach Squeezing Experiment", we noted that the experimenter had not been fair in all respects to the four treatments (the four quarters of the peach). He could have been fair to all treatments by picking four random samples of 250 peaches, assigning a treatment (quarter) to each sample of 250 peaches, and measuring the pressure necessary to puncture the skin on a given quarter of the peach. Thus, only one treatment would be performed on each peach. More efficient procedures of comparing all four treatments on each peach will be discussed later in this chapter.

7.2   COMPLETELY   RANDOMIZED   DESIGN---ZERO-WAY CONTROL OR ELIMINATION OF HETEROGENEITY

If the experimental material available to an investigator contains only non-assignable variation, then it is impossible to block or to group the material into sub-groups such that the variation among sub-groups is larger than among individuals within sub-groups with regard to the response being considered in the investigation. Any grouping would be no more effective than a random assignment of individuals to the sub-groups·and, hence, would be useless. This means that there is a *single* population of experimental units and that a simple random sample of experimental units is obtained for each treatment. The resulting design is denoted as a **completely randomized design**. The term completely randomized means that there is no blocking or grouping of the experimental units into sub-groups. In most experiments the treatments are usually allocated the same number of experimental units. If more information is desired on some of the treatments, they are allotted additional experimental units. The number of experimental units that a treatment receives is denoted as the number of **replicates** for the treatment.

To illustrate the above, suppose that our treatments consist of eight different kinds of cooking fats, one of them a standard. The characteristic to be observed is the quantity in grams of fat absorbed by six doughnuts during cooking. The experimental material consists of one large batch of doughnut mix which is enough to make more than $8 \times 6 \times 5 = 240$ doughnuts. The experimental unit is six doughnuts, as these will all be baked at one time. The observational unit is also the set of six doughnuts, since no data are available on the grams of fat absorbed by an individual doughnut. In order to be fair to all treatments, let us assign consecutive batches of six doughnuts to treatments in a random fashion, until each treatment is observed on five sets, or replicates, of six doughnuts each. We can do this by putting numbers 1 to 8 on round tags which are as nearly alike as possible. These numbers are put in a hat or covered jar and thoroughly mixed. A number is drawn blindly, and this is the treatment number to receive the first set of six doughnuts. The number is returned to the hat, the tags are thoroughly mixed, and a tag is again drawn blindly. This second number represents the treatment to receive the second set of six doughnuts. This process is continued until the $8 \times 5 = 40$ sets of six doughnuts have been allocated to the $v = 8$ treatments and each treatment has received five experimental units of six doughnuts each. As soon as $r = 5$ sets of doughnuts have been allotted to a given treatment, the tag with that number may be removed from the hat. Note that this is sampling with replacement until there are r experimental units allotted to a treatment. After that, it is sampling without replacement for that treatment.

Relative to the three characteristics to be considered in designing experiments, fairness is exhibited by allowing any treatment to receive any set of six doughnuts. No selection or "intelligent selection" by the experimenter is practiced. The material is relatively homogeneous in that a single batch is used and the order of cooking the doughnuts has no effect. Therefore, no blocking or stratification is required. Relative to the first characteristic, it does not appear that the procedure could be made more efficient by utilizing another procedure.

As a second example of the completely randomized design, suppose that 100 chicks from a single hatch of eggs of a single strain of dams and one sire are randomly divided into four groups of 25 chicks each. Suppose that four types of single-dose vitamins in capsule form represent the $v = 4$ treatments. The four treatments are randomly allocated to the $r = 4$ groups of 25 chicks, and a capsule is given to each of the 100 chicks. The response is weight at eight weeks of age. The 100 chicks are treated alike in all other respects except for type of capsule, that is, they are all in the same pen and have the same food and water sources. The chicks intermingle and so all are subjected to the same elements of the environment in the enclosure. The treatments are compared in as nearly equitable manner as possible.

As a third example, suppose that a large oven is available for baking purposes. Suppose further that there are no gradients or heat pockets in the oven and the heat remains at the designated temperature once the oven has been heated to this point, and therefore temperature fluctuations are minor. Suppose that five different amounts of thickening in pies are to be used, and that these represent the $v = 5$ treatments, say A, B, C, D, E. Suppose further that four pies are to be baked for a given amount of thickening and that the 20 pies can be baked at one baking. The treatments are randomly allocated to the 20 places in the oven. One possible arrangement is shown in Figure 7.1.

If the 20 pies cannot be baked at one time, we might use a single oven to bake the 20 pies in the following sequence:

Order	Treatment	Order	Treatment	Order	Treatment	Order	Treatment
1	E	6	A	11	B	16	E
2	E	7	D	12	C	17	D
3	C	8	D	13	A	18	A
4	B	9	B	14	C	19	D
5	E	10	A	15	B	20	C

For this situation there should be no gradients in the successive bakings in order for the completely randomized design to be the appropriate one. The above layout for an experiment in a completely randomized design might be appropriate for 20 pots on a greenhouse bench or for a series of soil analyses involving five treatments.

The completely randomized design is the simplest of all experiment designs to design and to analyze the results. It involves zero-way or no elimination of heterogeneity in the experimental material. The total variation in the experiment may be written as:

Total variation = variation among treatment means + error variation

= assignable cause + non-assignable cause.

The yield of any experimental unit may be written as treatment mean plus

E	E	C	B	E
A	D	D	B	A
B	C	A	C	B
E	D	A	D	C

Figure 7.1. Arrangement of 20 pies in an oven for v = 5 treatments and r = 4 replicates of each treatment.

an error or discrepance term. It is permissible to use the above form when the different components of variation are additive in their effects. We may need to partition the assignable cause in the variation among the treatment means and a bias factor. An ordinary arithmetic mean is biased by the amount of the bias factor. Now let us consider an individual observation from an experiment and consider the various sources or contributing factors to the variation among observations. From the above, we may write the ijth observation Y_{ij} as:

An observation = estimated treatment mean + random error, or

$$Y_{ij} = \bar{y}_i + (Y_{ij} - \bar{y}_{i.} = e_{ij}),$$

where $\bar{y}_{i..}$ is the arithmetic mean of the r observations $Y_{i1}, Y_{i2}, \dots , Y_{ir}$. Also, an observation equals the overall average of all treatment means plus a deviation of an estimated treatment mean from the estimated over-all average plus random error, or $Y_{ij} = \bar{y}_{..} + (\bar{y}_{i.} - \bar{y}_{..}) + e_{ij}$. The symbols are given above to indicate that we can and will use a shorthand notation in writing equations about variation. A fairly universal convention in statistical writings is to use capitol letters to denote observations and totals, to use small letters with a bar over the letter to denote means, to use Greek letters to denote values of parameters, and to use small Roman letters (or Greek letters with carat or hat over the top of the letter) to denote estimates of the parameters in the response equation. The dot in the total $Y_{i.}$ or the dots in the total $Y_{..}$ indicate a summation over the subscript(s) where the dot appears. Thus,

$$Y_{1.} = Y_{11} + Y_{12} + Y_{13} + \ldots + Y_{1r}.$$

To illustrate these ideas, consider the following experiment: Let treatment A = no crust on the pie top = " open-faced", let treatment B = crossed strips of pie dough on the top = "cross-hatched", and let treatment C = pie dough on the top = "kivered". Let $\bar{y}_{A.}$, $\bar{y}_{B.}$, and $\bar{y}_{C.}$ = arithmetic mean scores on consistency for treatments A, B, and C, respectively. Let Y_{ij} = jth score on the ith treatment for j = 1, 2, 3, 4 pies. Suppose that there is a bias factor of 10 because the scoring started at 10 instead of zero, that is all scores are read 10 too large. Let the 12 observations, or scores, be those given in Table 7.1. The various estimated treatment effects $\bar{y}_{i.} - \bar{y}_{..}$ are computed in the top part of the table. The computed random error deviations are given in the middle part of the table while the various components of each observation are given in the bottom part of the table.

In order to obtain some intuition about an equation with additive effects for the score or yield of an observation, consider that we would have a situation wherein the average of the three treatment means is 22, and to this we add a bias factor of 10. All 12 of our observations would have a score of 32. Then, suppose that we add -3 to the first four, zero to the next four, and +3 to the last four observations. Lastly, suppose that we assign random error components of -5, +1, 0, and +4 to the first 4 observations, +4, 0, -4, and 0 to the second four, and +3, -3, -1, and +1 to the last four to obtain the 12 observations. The addition of effects, or sources of variation, rather than their multiplication for example, produces an additive effects equation for an observation.

The above additive effects equation for the yield of an observation has been presented in terms of averages of observations, that is in terms of statistics from the sample. In terms of the parameters of an entire population we may refer to Table 7.1.

Table 7.1. Observations, means, and error deviations for scores of consistency for 12 pies.

	Treatment A	Treatment B	Treatment C	All scores
	$Y_{A1} = 24$	$Y_{B1} = 36$	$Y_{C1} = 38$	
	$Y_{A2} = 30$	$Y_{B2} = 32$	$Y_{C2} = 32$	
	$Y_{A3} = 29$	$Y_{B3} = 28$	$Y_{C3} = 34$	
	$Y_{A4} = 33$	$Y_{B4} = 32$	$Y_{C4} = 36$	
Total	116	128	140	384
Treatment mean	$\bar{y}_{A.} = 29$	$\bar{y}_{B.} = 32$	$\bar{y}_{C.} = 35$	$\bar{y}_{..} = 32$
	$\bar{y}_{A.} - \bar{y}_{..} = -3$	$\bar{y}_{B.} - \bar{y}_{..} = 0$	$\bar{y}_{C.} - \bar{y}_{..} = 3$	0

$\bar{y}_{..}$ = 384 / 12 = 32 = average of all observations = bias + estimate of true population mean.

Estimated error deviations = e_{ij} = $Y_{ij} - \bar{y}_{i.}$.

	Treatment A	Treatment B	Treatment C
	24 - 29 = -5	36 - 32 = 4	38 - 35 = 3
	30 - 29 = 1	32 - 32 = 0	32 - 35 = -3
	29 - 29 = 0	28 - 32 = -4	34 - 35 = -1
	33 - 29 = 4	32 - 32 = 0	36 - 35 = 1
Total	0	0	0

Observation = Y_{ij} = bias + estimate of true mean + ($\bar{y}_{i.} - \bar{y}_{..}$) + e_{ij}

Treatment A	Treatment B	Treatment C
24 = 10 + 22 - 3 - 5	36 = 10 + 22 + 0 + 4	38 = 10 + 22 + 3 + 3
30 = 10 + 22 - 3 + 1	32 = 10 + 22 + 0 + 0	32 = 10 + 22 + 3 - 3
29 = 10 + 22 - 3 + 0	28 = 10 + 22 + 0 - 4	34 = 10 + 22 + 3 - 1
33 = 10 + 22 - 3 + 4	32 = 10 + 22 + 0 + 0	36 = 10 + 22 + 3 + 1

$$Y_{ij} = \mu_{i.} + \varepsilon_{ij}, \text{ for } i = 1, 2, ..., v; j = 1, 2, ..., \infty,$$
$$= \mu_{..} + (\mu_{i.} - \mu_{..}) + \varepsilon_{ij}$$
$$= \mu_{..} + \tau_i + \varepsilon_{ij}$$

In the above, $\varepsilon_{ij} = Y_{ij} - \mu_{i.} = $ a random error, $\tau_i = \mu_{i.} - \mu_{..} = $ a treatment effect, and $\mu_{..} = (\mu_{1.} + \mu_{2.} + ... + \mu_{v.}) / v$. $\mu_{i.}$ is the true population mean associated with the ith treatment. Now, $Y_{ij} = \mu_{..} + \tau_i + \varepsilon_{ij} = \bar{y}_{..} + (\bar{y}_{i.} - \bar{y}_{..}) + e_{ij}$ does *not* mean that $\bar{y}_{..} = \mu_{..}, \tau_i = (\bar{y}_{i.} - \bar{y}_{..})$, and $\varepsilon_{ij} = e_{ij}$ but that the sum of three components equals the sum of three other components, for example $4 + 5 + 6 = 15 = 13 + 2 + 0$.

7.3   RANDOMIZED COMPLETE BLOCK DESIGN--ONE-WAY ELIMINATION OF HETEROGENEITY

If it is possible to group the experimental material or conditions in a manner such that the variation among experimental units within a group is less than the variation would have been without grouping, this should be done in order to compare treatments on the less variable material or under less variable conditions. The second illustrative example in the introduction demonstrates this point. Suppose that the four nutritional treatments were labeled A, B, C, and D. The four rats in each litter would be randomly allocated to a treatment. One possible arrangement for design II would be:

Litter Number	Rat Number and treatment number			
1	1- B	2- A	3- D	4- C
2	5- B	6- C	7- A	8- D
3	9- C	10- A	11- B	12- D
4	13- A	14- B	15- D	16- C
5	17- D	18- C	19- A	20- B
6	21- D	22- C	23- A	24- B
7	25- B	26- A	27- D	28- C
8	29- C	30- B	31- A	32- D
9	33- D	34- C	35- A	36- B
10	37- D	38- A	39- C	40- B

The rats could be tagged in some manner in order to retain their identity, or they might be housed in individual cages with the cages exposed to as nearly equal environments as possible. If this is not possible, then the four caged rats of a litter should be put in one environment, those from a second litter in a second environment, and so forth; then the observed variation among the ten groups is composed of variation among litters plus variation among environments. However, as far as the treatments are concerned, they are compared within a group, and the variation among treatment means is less than it would have been without grouping by litter plus environment.

A design such as the above is denoted as a **randomized complete block design**. The block is the litter plus environment, all treatments occur within *each* block or litter and hence is **complete**, and a *randomization* must be used in laying out the design of the experiment. The number of times each treatment occurs in a block may vary for each treatment but all treatments must occur in each of the blocks in order to be complete.

In order to make any useful inferences from an experiment designed as a randomized complete block design, it is necessary to precisely and completely describe the population for which the design in question is representative of the population. The purpose of an experiment is to relate the experiment facts to parameters of the population. A description of the results from an experiment that considers the experiment as its universe is of little or no use in scientific experimentation. The population structure for a randomized complete block design has a block as a sub-population with the universe consisting of all the possible sub-populations. A simple random sample of blocks (sub-populations) is made to obtain the blocks for an experiment. Then within each of the sub-populations (blocks), a simple random sample of experimental units is made. The treatments are then randomly allotted to the selected experimental units. As explained in Chapter 5, samples drawn in this way are representative of the population.

To illustrate the above, consider that we are interested in a particular breed of animal. Our population would consist of all possible litters (sub-populations) for this breed. A simple random sample of litters would be made. For each of the selected litters (blocks), there would be, say, v individuals. Since the particular conception for an individual could be considered a random event, the members of a litter could be considered to be a simple random sample from that particular sub-population. For a nutritional experiment with the individual as the experimental unit, a randomly selected individual would be allocated to each of the v nutrition treatments.

As a second illustration, consider that an experimenter wishes to compare v different varieties of green beans and to make recommendations for all farmers in a specified section of New York State. The population would consist of all farms (sub-populations) in the area. Each farm would be made up of experimental units on which the varieties of green beans are to be grown. Thus each farm is a sub-population. A simple random sample of farms (blocks) is made to obtain the blocks for the experiment. Then, within each selected block a simple random sample of experimental units is made and the v varieties of green beans are randomly allotted to these experimental units. Experiments conducted in this manner are representative of the population under consideration.

If ten of the above 40 rats are randomly allocated to the four treatments without any regard to parentage, it is possible that an arrangement of the following form could be obtained:

Litter and environment	Rat number and treatment letter			
1	1- A	2- A	3- A	4- A
2	5- A	6- A	7- A	8- A
3	9- B	10- B	11- B	12- B
4	13- B	14- B	15- B	16- B
5	17- C	18- C	19- C	20- C
6	21- C	22- C	23- C	24- C
7	25- D	26- D	27- D	28- D
8	29- D	30- D	31- D	32- D
9	33- A	34- A	35- B	36- B
10	37- C	38- C	39- D	40- D

The difference between the arithmetic means of treatments A and C would be: $\bar{y}_{A.} - \bar{y}_{C.}$ = [effect of treatment A + (effects of litters and environments 1, 2, and one-half of 9)] - [effect of treatment C + (effects of litters and environments 5, 6, and one-half of 10)]. In this case, it cannot be determined whether the difference is due to differences in treatments, in litters, or in environments. Such a complete mixing of effects is known as **complete confounding**. The difference between the means of treatments A and B is **unconfounded** or unmixed in only one, number 9, of the 10 litter and environment groups. Likewise the difference between the means of treatments C and D is obtainable free of other effects in only litter 10. The differences for A versus C or D and B versus C or D cannot be separated from the litter and environment effects. If these four pairs of treatments had occurred together in some of the other litters, the effects and treatments would have been only partially mixed up with the effects of the litters. Such an arrangement is denoted as **partial mixing or partial confounding** of effects. When solutions for treatment and litter effects are possible, the design is denoted as **connected** and when they are not the design is **not connected**. The design above is not connected whereas the previous one is. The lack of confounding such as in the designs before the last one leads to informative and less variable treatment responses.

As a second example, suppose that the relative effectiveness of nine different herbicides in eliminating dandelions from home lawns is to be tested, and that 12 different lawns have been selected for the investigation; the 12 lawns are relatively uniform in topography, grass cover, and dandelion infestation. Each lawn forms a relatively uniform *block* of land. (It was for situations like this that the randomized complete block design was first described, used, and named by Sir Ronald A. Fisher.) Each of the 12 blocks, or lawns, is divided into nine experimental units which are as alike as possible. Then, the nine treatments are randomly allocated to the nine experimental units in each of the 12 blocks or lawns, with the numbers 1, 2,..., 9 representing the treatments. One possible arrangement is shown in Figure 7.2. The characteristic to be measured is number of dandelions in a **plot** or experimental unit, at two months after an application of the herbicide. In such an experiment there could be considerable variation in dandelion count among the 12 lawns, but this would not affect the differences between treatments since all treatments are compared with each other on each of the 12 lawns.

If the treatments had been randomly allocated to the $12 \times 9 = 108$ experimental units or plots, and if the lawns differed in dandelion count, all 12 plots of some treatments could by chance have been allocated to lawns with a low dandelion count and other treatments to lawns with a high dandelion count. The comparison between treatments would then be mixed up with differences between lawns. In a randomized block design, each of the nine treatments appears on each of the 12 lawns. Given that lawns differ in dandelion count, this arrangement makes the differences between means less variable than if there had been no stratification or blocking.

The count of dandelions in a plot may be expressed as the sum of the block and treatment estimated means minus the overall mean plus an error term, thus: Count = block mean + treatment mean - overall mean + error, or symbolically as

$$Y_{ij} = (\bar{y}_{.j} = \text{block mean}) + (\bar{y}_{i.} = \text{treatment mean}) - (\bar{y}_{..} = \text{overall}$$

$$\text{mean}) + e_{ij} = \bar{y}_{..} + (\bar{y}_{i.} - \bar{y}_{..}) + (\bar{y}_{.j} - \bar{y}_{..}) + e_{ij}.$$

In terms of the parameters of the population, we may write the above as:

$$Y_{ij} = \mu_{i.} + \mu_{.j} - \mu_{..} + \varepsilon_{ij}$$
$$= \mu_{..} + (\mu_{i.} - \mu_{..}) + (\mu_{.j} - \mu_{..}) + \varepsilon_{ij}$$
$$= \mu_{..} + \beta_j + \tau_{i.} + \varepsilon_{ij},$$

where $\mu_{..} = \Sigma_{j=1 \text{ to } 12} \mu_{i.} / 9$, $\mu_{i.}$ = population mean of treatment i over all blocks or environments in the population, Σ is a symbol denoting a summing or totaling process, $\mu_{.j}$ = the population mean of block j summed over all individuals in the sub-population for block j, $\varepsilon_{ij} = Y_{ij} - \mu_{i.} - \mu_{.j} + \mu_{..}$ = random error, τ_i = the true effect for treatment i, and β_j = true effect for the jth block. Of course, if these parameters were known, there would be no need to conduct an experiment.

The sum of all counts for two given treatments, say 1 and 2, in the experiment described above is:

treatment one sum + sum of 12 block means - 12 (overall mean) + 12 error terms associated with treatment one

treatment two sum + sum of 12 block means - 12 (overall mean) + 12 error terms associated with treatment two

The difference between two treatment means is:

{treatment one sum - treatment two sum + 12 error terms - 12 other error terms} / 12

$$= \bar{y}_{1.} - \bar{y}_{2.} + \Sigma_{j=1 \text{ to } 12}(e_{1j} - e_{2j}) / 12,$$

$$= (\bar{y}_{1.} - \bar{y}_{..} = t_1) - (\bar{y}_{2.} - \bar{y}_{..} = t_2) + \Sigma_{j=1 \text{ to } 12}(e_{1j} - e_{2j}) / 12$$

$$= \tau_1 - \tau_2 + \Sigma_{j=1 \text{ to } 12}(\varepsilon_{1j} - \varepsilon_{2j}) / 12,$$

where $\Sigma_{j=1 \text{ to } 12} \varepsilon_{ij}$ is equal to $\varepsilon_{i1} + \varepsilon_{i2} + \varepsilon_{i3} + \varepsilon_{i4} + \varepsilon_{i5} + \varepsilon_{i6} + \varepsilon_{i7} + \varepsilon_{i8} +$
$\varepsilon_{i9} + \varepsilon_{i10} + \varepsilon_{i11} + \varepsilon_{i12}$. Here we may note that the effects of the true
overall mean and of the true block means do not appear in the difference
between two treatment means.   Since we are comparing herbicide
treatments for effectiveness of dandelion control, the differences between
means are the statistics of interest. All designs having the property that
differences between all pairs of arithmetic means for any of the categories
do not contain any effects other than the category effects and random
errors, are known as **orthogonal designs**. Also, if the differences
between arithmetic means of treatments contain only differences due to
true treatment effects plus differences of random errors, then the treatment
effects are said to be **orthogonal** to the other sources of variation in the
experiment. This is an important property of experiment designs in that
the differences between two treatment means is unaffected by all other
sources of variation in the experiment *except* that due to random error
variation or non-assignable cause variation. Also, being able to use
arithmetic means greatly simplifies the computational procedures, and
orthogonal designs are an efficient class of experiment designs.

The above definition of orthogonality relates to the parameters in an
experiment design.   An alternate definition of orthogonality is a
combinatorial one which is: "If n_{ij} is the number of times that the ith
treatment occurs in the jth block and if the ratio $n_{1j}:n_{2j}:n_{3j}: \ldots :n_{vj}$ stays
constant for *every* value of j = 1, 2,..., r, then the treatment effects are
said to be orthogonal to the block effects." To illustrate this, consider the
first example in this section. Each treatment A, B, C, and D occurred once
in each of the 10 blocks or litters. Hence, $n_{Aj}:n_{Bj}:n_{Cj}:n_{Dj}$ = 1:1:1:1 for
every value of j = 1, 2,..., 10 = r. As a second illustration, consider the
following design consisting of r = 5 litters of six rats each with v = 3
treatments, A, B, and C, and with treatment A occurring on three rats in
each litter, B occurring on two rats in each litter, and C occurring on one
rat in each litter. With the rats randomly allotted to each letter in each
litter, the design would be:

Litter number	Rat number and treatment (letter)					
1	1- A	2- B	3- A	4- A	5- B	6- C
2	7- A	8- A	9- A	10- C	11- B	12- B
3	13- B	14- A	15- A	16- A	17- C	18- B
4	19- A	20- A	21- A	22- B	23- C	24- B
5	25- A	26- C	27- B	28- B	29- A	30- A

Thus, $n_{A1} = n_{A2} = n_{A3} = n_{A4} = n_{A5} = 3; n_{B1} = n_{B2} = n_{B3} = n_{B4} = n_{B5} = 2$; and $n_{C1} = n_{C2} = n_{C3} = n_{C4} = n_{C5} = 1$. The ratio $n_{Aj}:n_{Bj}:n_{Cj} = 3:2:1$ for every value of j = 1, 2, 3, 4, 5, and hence treatment effects are orthogonal to litter (block) effects.

Note that the individual response in the preceding experiments involving blocks and treatments is assumed to be the sum of four terms. This need not be the case, as some responses may be the product of these terms rather than their sum. The appropriateness of the assumption of additive effects must be questioned for every type of experiment. If the response is the product of terms instead of the sum, one could use another function of the responses to obtain additive effects. In this case, the response could be transformed to log of response. Thus, if Y = abcd, then log Y = log a + log b + log c + log d. One might wonder why additivity of effects is desirable. The answer is that computations and interpretations are simpler on the additive scale. Despite the desirability of simplicity, it may be necessary to work on a non-additive scale; this can greatly complicate the statistical and computational procedures, but it may be necessary to do this in order to attain reality.

In field and laboratory experimentation on biological material, the randomized complete block design is probably the most frequently used experiment design. Ease of construction, layout, and analysis of results contribute heavily to its frequent use. Also, it has been found to be considerably more efficient for the above types of experiments than the completely randomized design. Summarization of several hundred field experiments over a period of years indicates that six blocks or replicates of a randomized complete block design are approximately equivalent to ten replicates of a completely randomized design in attaining the same degree of variability associated with a treatment mean. For field experimentation, the blocking or stratification into blocks reduces the variability among treatment means to six-tenths of the variation without blocking. The value of blocking experimental material is dependent upon the type of material under consideration. Each type of experimentation requires individual evaluation. One can always block as a form of insurance against heterogeneity, but over-stratification results in some disadvantages which will be discussed later. As a rule, *a minimum amount of blocking* should be used to control the heterogeneity or the suspected heterogeneity present in the experimental material.

As an illustration of the above consideration, suppose that one were interested in only three herbicides instead of nine and that the size of the experimental unit was fixed, in that the lawns were divided into nine plots or experimental units instead of three. Then, blocks of size three could be used, and there would be three blocks per lawn. However, if the nine experimental units were relatively homogeneous, a completely randomized design of three treatments and three replicates on each treatment for *each* lawn should be used. This would result in minimum blocking which would control the lawn to lawn variability. It should be pointed out, however, that the lawn would probably be divided into thirds and larger experimental units would be used.

To illustrate another variation on the randomized complete block design, suppose that only five herbicides were of interest with four of these (1, 2, 3, and 4) being of more interest than the fifth one, number 5, and suppose that nine experimental units were available on each lawn. Treatments 1, 2, 3 and 4 could be included twice on each lawn and treatment 5 could be put in once. If we let numbers 1 and 6 be the plots for treatment 1, numbers 2 and 7 be the plots for treatment 2, numbers 3 and 8 be the plots for treatment 3, numbers 4 and 9 be the plots for treatment 4, and number 5 be the plot for treatment 5 in the original dandelion design, then the arrangement in the first three blocks or lawns would appear as shown in Figure 7.3.

Both of the above variations on the randomized complete block design are orthogonal designs. That is, differences between treatment means do not involve the block effects. As long as the orthogonality of block and treatment effects is a property of the design, the analysis remains simple.

As an example of checking orthogonality of effects suppose that we have a randomized complete blocks design of three treatments A, B, and C such that A is included three times, B twice, and C once in each of two blocks. We then express the yields symbolically and in a systematic manner in terms of the estimated effects of the parameters in the model as follows:

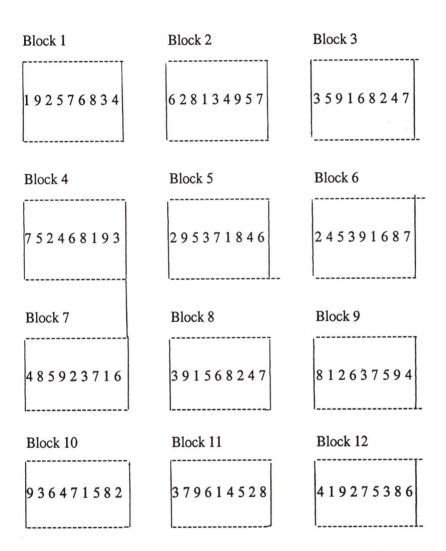

Block 1

`1 9 2 5 7 6 8 3 4`

Block 2

`6 2 8 1 3 4 9 5 7`

Block 3

`3 5 9 1 6 8 2 4 7`

Block 4

`7 5 2 4 6 8 1 9 3`

Block 5

`2 9 5 3 7 1 8 4 6`

Block 6

`2 4 5 3 9 1 6 8 7`

Block 7

`4 8 5 9 2 3 7 1 6`

Block 8

`3 9 1 5 6 8 2 4 7`

Block 9

`8 1 2 6 3 7 5 9 4`

Block 10

`9 3 6 4 7 1 5 8 2`

Block 11

`3 7 9 6 1 4 5 2 8`

Block 12

`4 1 9 2 7 5 3 8 6`

Figure 7.2.   Randomized complete block arrangement for v = 9 treatments and r = 12 replicates.

block 1	block 2
$Y_{A11} = \bar{y}_{...} + t_A + b_1 + e_{A11}$	$Y_{A21} = \bar{y}_{...} + t_A + b_2 + e_{A21}$
$Y_{A12} = \bar{y}_{...} + t_A + b_1 + e_{A12}$	$Y_{A22} = \bar{y}_{...} + t_A + b_2 + e_{A22}$
$Y_{A13} = \bar{y}_{·..} + t_A + b_1 + e_{A13}$	$Y_{A23} = \bar{y}_{...} + t_A + b_2 + e_{A23}$
$Y_{B11} = \bar{y}_{...} + t_B + b_1 + e_{A11}$	$Y_{B21} = \bar{y}_{...} + t_B + b_2 + e_{A21}$
$Y_{B12} = \bar{y}_{...} + t_B + b_1 + e_{A12}$	$Y_{B22} = \bar{y}_{...} + t_B + b_2 + e_{A22}$
$Y_{C11} = \bar{y}_{...} + t_C + b_1 + e_{C11}$	$Y_{C21} = \bar{y}_{...} + t_C + b_2 + e_{C21}$

where $t_i = \bar{y}_{i..} - \bar{y}_{...}$, $b_j = \bar{y}_{.j.} - \bar{y}_{...}$, $e_{ijh} = Y_{ijh} - \bar{y}_{i..} - \bar{y}_{.j.} + \bar{y}_{...}$, $\bar{y}_{...}$ = overall mean, $\bar{y}_{i..}$ = mean of ith treatment for i = A, B, C, and $\bar{y}_{.j.}$ = mean of jth block for j = 1, 2. The treatment means are:

$$\bar{y}_{A..} = \bar{y}_{...} + t_A + 3(b_1 + b_2)/6 + (e_{A11} + e_{A12} + e_{A13} + e_{A21} + e_{A22} + e_{A23})/6$$

$$\bar{y}_{B..} = \bar{y}_{...} + t_B + 2(b_1 + b_2)/4 + (e_{B11} + e_{B12} + e_{B21} + e_{B22})/4$$

$$\bar{y}_{C..} = \bar{y}_{...} + t_C + (b_1 + b_2)/2 + (e_{C11} + e_{C21})/2$$

where the symbols are as previously defined. Comparing all possible differences among arithmetic means of treatments, we note that treatment effects t_i are orthogonal to the $\bar{y}_{...}$ and the b_j effects, and also that the population parameter values may be used in place of the sample values to illustrate orthogonality.

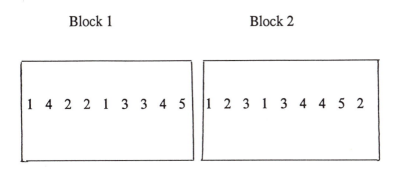

Block 1   Block 2

| 1 | 4 | 2 | 2 | 1 | 3 | 3 | 4 | 5 |

| 1 | 2 | 3 | 1 | 3 | 4 | 4 | 5 | 2 |

Block 3

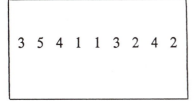

| 3 | 5 | 4 | 1 | 1 | 3 | 2 | 4 | 2 |

Figure 7.3. A randomized complete block design arrangement for $v = 5$ treatments with six replicates on treatments 1 to 4 and three replicates on treatment number 5.

7.4 BALANCED INCOMPLETE BLOCK DESIGN--ONE-WAY ELIMINATION OF HETEROGENEITY

In the previous section, the block size was large enough to accommodate all treatments at least once. Suppose that we have more treatments to test than the number of experimental units in a relatively homogeneous block of material. To be specific, suppose that we are interested in five mosquito repellents; the experimental unit is an arm of an enlisted army private who has "volunteered" to serve as a guinea pig in the experiment. Obviously enlisted men only have two arms! Thus, we have blocks of size two, but there are five treatments. We solve this dilemma by the following scheme, as "volunteers are plentiful":

Block number or "volunteer"	Left arm	Right arm
1	A	B
2	A	C
3	D	A
4	E	A
5	B	C
6	B	D
7	E	B
8	C	D
9	C	E
10	D	E

where the five treatments are numbered A, B, C, D, and E. Let us suppose that the left arm is no more or less susceptible to mosquitoes than the right arm. (If not, then we could include the mirror images of each of the above pairs on another ten "volunteers", or on the same "volunteers" at a later time.) Then the designation of right or left arm is omitted. The experiment is conducted as follows. There are ten pairs of cages each containing the same number of mosquitoes. Each of the ten "volunteers" is randomly allocated a block number or pair number from one to ten; this amounts to randomly assigning a pair of treatments to the "volunteer". Then, a coin is flipped to determine which member of the pair of treatments falls on the left arm and which on the right arm. The specified treatment is applied to the arm of the "volunteer" who puts the left arm into one cage and the right arm into a second cage of hungry mosquitoes! The arms are left in the cage for a specified length of time, say one hour, the number of landings and bites by hungry mosquitoes are recorded.

It should be noted that we were fair in allocating a pair of treatments to each volunteer and the arms to the members of the pairs. This experiment design is fair in another respect. It should be noted that every treatment occurs an equal number of times, once in this case, with every other treatment in one of the ten blocks. To check this, note that treatment D, for example, appears with A in block 3, with B in block 6, with C in block 8, and with E in block 10. This same type of relationship holds for any other treatment.

For designs such as the one above, statisticians have coined a special name. If there are v treatments in b blocks of size k, and if every pair of treatments occurs together the *same* number of times, say λ, in the b blocks, the design is called a **balanced incomplete block design**, which is often shortened to BIBD or bibd. When v = k, i.e., the number of treatments is equal to the number of experimental units in a block, and the treatment pairs occur together once in each block, we have a randomized complete block design, denoted as rcbd. Thus, we see that this rcbd is a special case of the bibd. The bib designs are balanced but are not in general orthogonal. The statistical computations are more complicated than those for the rcbd, but they still are not overly difficult. The population structure for balanced incomplete block designs is the same as for randomized complete block designs, i.e., the blocks form the sub-populations and the entire population is made up of the sub-populations.

For blocks of size two and for v > 2 treatments, this design is known in some literature citations as a **paired comparison design**. A listing of bib designs for v treatments in b blocks of size k may be found in several places (for example, Cochran and Cox, 1957, Cox, 1958, and Federer, 1955).

As a second example, suppose that we wish to compare v = four brands of shoes. In order to be fair to each brand of shoes, we select a two-footed subject, the block, and we have him wear a left shoe of Brand A, say, and a right shoe of Brand B for a specified period, say six weeks. Then, the shoes are reversed in that the second member of the pair of Brand A is worn on the right foot, and Brand B on the left; these are also worn for six weeks. The wear measurements are recorded for both pairs of shoes. Six subjects are required to obtain a bib design. The design is:

		Treatment		
Subject		A	B	C	D
1		x	x		
2		x		x	
3		x			x
4			x	x	
5			x		x
6				x	x

If 12 subjects were available, the above design could be repeated or could be conducted in one half the time by letting a second individual wear the remaining left and right shoes of two pairs. The randomization scheme described for the first example should be followed here also.

The design with blocks of two units may be used in many types of investigations, since there are many instances where blocks of size two occur naturally. Besides the two arms and the two legs of an individual the following come to mind:

1. identical twins,
2. opposite leaves on a plant,
3. roasts and other cuts of meat from opposite sides of an animal,
4. double-yolked eggs,
5. opposite halves of a leaf,
6. opposite halves of a fruit, and
7. two eyes, eyelids, ears, etc. of an individual.

The natural pairing of entities in the biological world is of importance in designing experiments so that treatments are compared on relatively homogeneous material.

An example of a bib design was obtained from J. C. Moyer, Geneva Experiment Station, Cornell University. He was interested in comparing the flavor of juice obtained from hand-harvested and from mechanically harvested grapes held for various periods of time after harvest. Five treatments were used. To evaluate the flavor of the juice, a panel of judges was available. Since grape juice is rather tart, each judge could taste no more than three, or at most four samples, at one sitting. If four samples could be tasted, a bibd for $v = 5$, $k = 4$, $b = 5$, $\lambda = 3$, and $r = 4$ replicates is:

Blocks = Judges

1	2	3	4	5
1	1	1	1	2
2	2	2	3	3
3	3	4	4	4
4	5	5	5	5

However, if only three samples could be tasted at one sitting, then blocks of size three must be utilized. A bibd for this case is:

Blocks = Judges

1	2	3	4	5	6	7	8	9	10
1	1	1	1	1	1	2	2	2	3
2	2	2	3	3	4	3	3	4	4
3	4	5	4	5	5	4	5	5	5

Here $v = 5$, $b = 10$, $k = 3$, $\lambda = 3$, and $r = 6$ replicates on each treatment.

It was decided to use the above design. The five treatments for one series of experiments were:

1 = mechanical-harvest held for 5 hours,
2 = hand-harvest held for 0 hours,
3 = hand-harvest held for 2 hours,
4 = hand-harvest held for 12 hours, and
5 = hand-harvest held for 21 hours.

Each judge scored the grape juice samples from 1 to 10, with 1 being at the top of the scale and 10 at the bottom of the scale relative to desirable taste. The results of one of these experiments are given below:

Score of block = panelist or judge

Treatment	1	2	3	4	5	6	7	8	9	10	Sum
1	5			1	8		8	7		4	33
2		2		4		3	3		5	7	24
3			3		6	1		3	4	8	25
4	9	6	9	9	9	8					50
5	8	9	8				10	10	10		55
Sum	22	17	20	14	23	12	21	20	19	19	187

To determine the effect on flavor of the resulting grape juice, 34 such sets of data as the above were obtained for different times and methods of harvesting grapes. Many other examples may be found in published literature.

The estimated treatment means in a balanced incomplete block design are not simply the arithmetic averages. This is because there is non-orthogonality between the incomplete block and the treatment effects. The formula for computing the treatment mean adjusted for block effects is

treatment mean adjusted = k(sum of all observations for the treatment - sum of the means of the blocks in which the treatment occurred) / (kr - r + λ) + $\bar{y}_{..}$. For treatment 1 above, the adjusted mean is computed as:

{3(5 + 1 + 8 + 8 + 7 + 4) - (22 + 14 + 23 + 21 + 20 + 19)} / (3[6] - 6 + 3) + 187 / 30

= (99 - 119) / 15 + 187 / 30 = - 20 / 15 + 187 / 30 = 147/30 = 4.9,

whereas the unadjusted arithmetic mean is 33 / 6 = 5.5. The other adjusted treatment means are computed in a similar manner. The more non-orthogonal a design becomes, the more complex the computation of effects.

As might be suspected, an investigator may find that balanced incomplete block designs are not available for his particular situation. In such cases, it may be necessary to use a **partially balanced incomplete block design** which is an incomplete block design wherein the number of times the individual pairs of treatments occur together is not a constant but may take on two, three, or more different values. There are many classes of such designs and much literature has been published on the subject. There are several methods for constructing such incomplete block designs (See, e.g., Patterson and Williams, 1976, and Khare and Federer, 1981).

7.5 SIMPLE CHANGE-OVER DESIGN--TWO-WAY ELIMINATION OF HETEROGENEITY

Suppose that one were to compare two merchandising treatments, say two different displays, simultaneously, in stores where the two treatments are on the same counter and the line of traffic moves from left to right past the counter. If the two treatments were identical, then the one in the first position would often be purchased more frequently than the one in the second position. In order to be fair to both treatments, say A and B, one could set up the following design for ten stores (Conditions of this nature were found in an actual marketing experiment.):

	Store number									
Position	1	2	3	4	5	6	7	8	9	10
1	B	B	A	B	B	A	A	A	B	A
2	A	A	B	A	A	B	B	B	A	B

In the above design, both treatments appear in each store once, and each treatment appears five times or one-half of the time in each position. Thus, we have been fair to both treatments, and variation between positions and among stores has been balanced out or controlled. This removal of variation from two sources from the differences between pairs of treatment means decreases the variation between treatment means. To illustrate this, let a single observation or measurement be expressed in terms of estimated means or effects, as follows:

Observation = position mean + store mean + treatment mean - 2(overall mean) + error

= overall mean + position effect + store effect + treatment effect + error

$$= \bar{y}_{...} + (\bar{y}_{h..} - \bar{y}_{...}) + (\bar{y}_{.i.} - \bar{y}_{...}) + (\bar{y}_{..j} - \bar{y}_{...}) + e_{hij}.$$

Using this form, the first response or observation Y_{11B} may be written as

$$Y_{11B} = \bar{y}_{...} + (\bar{y}_{1..} - \bar{y}_{...}) + (\bar{y}_{.1.} - \bar{y}_{...}) + (\bar{y}_{..B} - \bar{y}_{...}) + e_{11B}$$

The sum of the ten responses or observations for treatment A is:

$$Y_{21A} + Y_{22A} + Y_{13A} + Y_{24A} + Y_{25A} + Y_{16A} + Y_{17A} + Y_{18A} + Y_{29A} + Y_{10A}$$

$$= 10\bar{y}_{...} + 5(\bar{y}_{1..} - \bar{y}_{...}) + 5(\bar{y}_{2..} - \bar{y}_{...}) + 10(\bar{y}_{..A} - \bar{y}_{...}) + \text{the}$$
sum of the ten treatment effects + the sum of the ten error terms associated with these ten observations.

and the sum of the ten observations for treatment B is:

$$Y_{11B} + Y_{12B} + Y_{23B} + Y_{14B} + Y_{15B} + Y_{26B} + Y_{27B} + Y_{28B} + Y_{19B} + Y_{20B}$$

$= 10\bar{y}_{...} + 5(\bar{y}_{1..} - \bar{y}_{...}) + 5(\bar{y}_{2..} - \bar{y}_{...}) + 10(\bar{y}_{..B} - \bar{y}_{...}) +$ the sum of the ten store (column) effects + the sum of the ten error terms associated with these ten observations.

The means are obtained by dividing by 10. Then, the difference between the mean of treatment A and the mean of treatment B is:

treatment A effect - treatment B effect + (the sum of the ten e_{hiA} - the sum of the ten e_{hiB}) / 10.

From the definition of orthogonality given previously, we see that the above design is orthogonal. This design is known as a **simple change-over design** for two treatments. The schematic plan of the simple change-over design for three treatments (A,B,C) in three rows and in 3s = 12 columns is:

```
                          Columns
Row    1   2   3   4   5   6   7   8   9   10  11  12
 1     A   A   A   A   B   B   B   B   C   C   C   C
 2     C   C   C   C   A   A   A   A   B   B   B   B
 3     B   B   B   B   C   C   C   C   A   A   A   A
---------------------------------------------------------------
```

The design for v treatments in v rows and vs columns may be constructed in a manner similar to that described for two and for three treatments. Simple change-over designs may be used in many situations. For example, suppose that we wish to compare v foods in a cafeteria line in v different positions on vs different days. As another example, consider the comparison of v programs or subjects at v different hours of the day in vs schools (teachers, years, etc.). It is emphasized again that the rows and columns refer to two sources of non-treatment variation and not necessarily to a row and column spatial arrangement, and the two sources of variation are to be eliminated in the experiment.

In the simple change-over design, all treatments appear once in each column and s times in each row. To randomize a given plan, allot the letters (treatments) to the first row in the same manner as for a completely randomized design. Since each letter must appear once in each column, this completes the randomization for two treatments, and we simply write in the remaining letter in the second row. For three treatments, randomly allot the letters to the 3s different positions such that each letter appears s times in each of the first two rows and no letter appears more than once in a column; treatments in the third row are inserted so that all treatments appear once in each column. For more than $v = 3$ treatments, simply extend the above process.

There are several population structures possible for row by column designs such as the simple change-over design and the latin square design discussed in the next section. The first one considered is of the following nature. There is a single population of sampling units. A simple random sample of the sampling units is selected. These units are the columns of the row by column design. Then, on each of these sampling units (columns), the treatments will appear in one of the orders of applying the treatment to the sampling unit. A randomization procedure is applied to the simple change-over and latin square designs to determine which treatment appears in which order. As an example, consider that the population consists of all the grocery stores in a given section of the United States. A simple random sample of grocery stores is made to form the columns. Suppose the v treatments are to appear in a store for one month and that the experiment is to run for v successive months. The months form the rows of the row by column design.

As a second illustration of population structure for row by column designs, consider that the universe consists of sampling units which are farms. The individual farm consists of fields which vary in two directions. One randomly selected farm is obtained. Then, on this single sampling unit, a row by column design is constructed to control and eliminate heterogeneity encountered in the field. This population structure holds for a number of field experiments conducted by Agriculturists.

Several other types of population structure exist for row-column designs. The particular population structure needs to be determined and defined for each experiment laid out or conducted using a row-column design such as the simple change-over or latin square design.

7.6 LATIN SQUARE DESIGN--TWO-WAY ELIMINATION OF HETEROGENEITY

The **latin square** is a plan of k rows and k columns of a square with k symbols arranged such that each symbol appears once in each row and once in each column. If the symbols are Latin letters, we could, as Sir Ronald A. Fisher did, call this a Latin square plan. If the symbols used were Greek letters, we could call the plan a Greek square. If the symbols used were Arabic symbols, we could call the plan an Arabic square, and so on. By common usage, this plan is used with Latin letters and when properly randomized is called a latin square design. Furthermore, this design for the removal of row and column variation from treatment differences controls variation from two sources and not necessarily from rows and columns.   The row and column designation merely refers to the two sources. To illustrate, suppose that three different pie recipes represent the treatments, that a large oven is available for baking the nine pies simultaneously, and that the treatments are arranged in the oven as follows:

$$
\begin{array}{ccc}
C & B & A \\
B & A & C \\
A & C & B \\
\end{array}
$$

The latin square design controls variation in two directions. Now suppose that only one pie can be baked at one time, that three pies can be baked on a given day, and that the order of baking has an effect. Then the following plan of baking would be useful:

	Day								
	1			2			3		
	order			order			order		
	1	2	3	1	2	3	1	2	3
Treatment	C	B	A	B	A	C	A	C	B

which when rearranged looks "like a square" or row-column design:

	Day		
Order of baking	1	2	3
1	C	B	A
2	B	A	C
3	A	C	B

Each treatment appears once on each day and once in each order of baking.
     Considerable use has been made of the latin square design for studying merchandising innovations as they affect the sale of grocery store products.

We shall consider an experiment (Dominick, 1952) involving the following four treatments on McIntosh apples:

A = regular apples,
B = apples 2.25 inches in diameter,
C = apples 2.50 inches in diameter, carefully selected, and
D = apples 2.50 inches in diameter, highly colored and uniform.

Four stores from the same chain of grocery stores were to be used. This is often a necessity for experiments of this type in order to disentangle management factors and building arrangement factors from their differential effects on the treatments. Stores from different chains may introduce difficulties in assessing treatment responses; it is therefore preferable to run the experiment on stores of a single chain. The stores are the columns and the first four days of the week are the rows in the following latin square design:

		Store		
Day of week	1	2	3	4
Monday	A	B	C	D
Tuesday	B	A	D	C
Wednesday	D	C	B	A
Thursday	C	D	A	B

It was felt that this experiment was too small to estimate properly the differences in sales of apples per 100 customers for the four treatments. Therefore, four latin square designs for these four treatments were used as follows:

	Week 1				Week 2			
Day or part of day	Store				Store			
First part of week	1	2	3	4	1	2	3	4
Monday	A	B	C	D	B	D	C	A
Tuesday	B	A	D	C	D	A	B	C
Wednesday	D	C	B	A	A	C	D	B
Thursday	C	D	A	B	C	B	A	D

Second part of week

Friday a.m.	B	A	D	C	D	C	B	A
Friday p.m.	C	D	B	A	B	A	C	D
Saturday a.m.	D	C	A	B	C	D	A	B
Saturday p.m.	A	B	C	D	A	B	D	C

The week was divided into two parts as described above; the sales of apples in the first four days of the week were approximately equal to the sales in the last two days of the week. Friday and Saturday were split to equalize sales in the two parts of a day as nearly as possible. Also, since the purchase of apples is generally not a daily but rather a weekly event, the purchase of apples from a given treatment on Monday, for instance, would not affect sales of apples during the rest of the week; that is, the purchase of apples from any given treatment would eliminate the purchase of apples by that customer for another week. If the purchase of apples one week were to affect the purchase of apples the following week, all treatments would be affected equally. Also, the shopping habits of the customers from the four experimental stores were similar with respect to such variables as frequency of shopping, volume of purchases, and proportion and number of customers per day.

A possible randomization procedure for the above design is to

  a. construct a latin square,
  b. randomly allot the letters to the treatments,
  c. randomly allot the column numbers to stores, and
  d. randomly allot the row number to the days.

To illustrate, suppose that we have three treatments. We number three circular tags of the same size and shape as 1, 2, and 3 and place these into a hat (or jar). Let our square be

$$1 \quad 2 \quad 3$$
$$3 \quad 1 \quad 2$$
$$2 \quad 3 \quad 1$$

Shake the hat with the tags, draw out one tag blindly, and assign that number to the first treatment; draw a second number from the hat and let that be the second treatment number; the remaining number in the hat is assigned to the third treatment. Put the tags back into the hat, shake, and again draw out the three tags, for example, 2,1,3 which is the allotment of the stores to the columns, do likewise for the rows, such as 3,1,2. Following the last two steps, we have:

Days	Store 2	1	3
3	1	2	3
1	3	1	2
2	2	3	1

to produce the plan

Days	Store 1	2	3
1	1	3	2
2	3	2	1
3	2	1	3

The observation or measurement in the orthogonal latin square design is of the same form as the simple change-over design, that is

Observation = overall mean + row effect + column effect + treatment effect + error

$$= \bar{y}_{...} + (\bar{y}_{h..} - \bar{y}_{...}) + (\bar{y}_{\cdot i \cdot} - \bar{y}_{...}) + (\bar{y}_{\cdot \cdot j} - \bar{y}_{...}) + e_{hij}$$

$$= \bar{y}_{...} + r_h + c_i + t_j + e_{hij},$$

where $\bar{y}_{...}$ is the mean of all observations, $\bar{y}_{h..}$ is the mean of the hth row, $\bar{y}_{\cdot i \cdot}$ is the mean of the ith column, $\bar{y}_{\cdot \cdot j}$ is the mean of the jth treatment, and where each term in one equation is replaced by its alternative form in the other equation. Special analyses are necessary when the above form of additivity of effects does not hold.

As another illustrative example, the Georgia peach squeezing experiment in Problem 3.2 could have been set up as follows:

Group of peaches	Order of measurement 1	2	3	4
1 (250 peaches)	left front	left back	right back	right front
2 (250 peaches)	right front	left front	left back	right back
3 (250 peaches)	right back	right front	left front	left back
4 (250 peaches)	left back	right back	right front	left front

The order of performing the measurements would be orthogonal to the treatments (quarters of a peach) in this design, whereas in the design used the treatment effects and the order of performing the measurement effects were completely confounded or mixed-up.

W. J. Youden, formerly of the National Bureau of Standards, presented a lecture entitled "How statistics improves physical, chemical, and engineering measurement" to United States Department of Agriculture personnel on 12/14/49. In his lecture, he gave many examples in various fields on the use of the latin square design. The following is an excerpt from a mimeographed copy of his lecture (with the permission of W. J. Youden):

I am going to stop here in my discussion of how to estimate errors. Everyone is much more interested in how you reduce them. This is much more challenging; and really, we are more useful as statisticians, I think, at this phase of the work. I would not dare to claim that statisticians will help reduce the error of measurements if I was not fortified by my own personal experience and by the experience of scientists I have worked with on their projects. Let me enter this phase of it with a momentary digression and tell you about a farmer who had four sons. He offered a prize to that son who got the best yield with some crop. The boys entered this contest with enthusiasm, but when the farmer had set aside a field for this contest, a question came up immediately. How should they divide the field to make sure the various portions allotted were as closely equal as possible in their fertility?

Suppose we divide the field into a checkerboard by marking off 4 horizontal strips with 4 vertical strips. That gives us 16 plots or 4 plots for each boy. Something like this would do:

A	B	C	D
C	D	A	B
D	C	B	A
B	A	D	C

This is an attempt to make sure that each of the 4 boys has a fair sample of the field. Son A gets 4 plots and samples every vertical and every horizontal strip. This is also true for every other boy. After the harvest is in, each boy takes the average yield of the 4 plots assigned to him. A scientist would immediately ask if the differences between the averages for the boys are great enough to indicate a real difference in farming ability for the 4 boys. Suppose I defer the answer to that for a moment. I will only tell you that this particular arrangement is very widely used in experimental agriculture. It is very successful in reducing the error of the comparisons in spite of the fact that there is something artificial and arbitrary, something almost hopeful, in the idea that the fertility can be considered to go by strips.

For the rest of the talk I am going to show you that this same arrangement is even more successful in the physical, chemical, and engineering laboratories. Indeed, I think I was one of the first to take it out of the field and bring it in as far as the greenhouse. I found pathologists were studying tobacco mosaic virus, and, in order to compare the toxicity of different solutions, they would smear the virus solutions over the leaves of tobacco plants. In 3 or 4 days little spots came out on the leaves; the stronger the virus, the more spots. To compare the solutions then, they would smear them on the leaves and count the spots.

For some reason or other most of these tobacco plants were grown to the point where they had about five leaves. By smearing the same solution on all leaves, for several plants, it was revealed immediately that there were certain natural groupings. The leaves from the same plant, as might well be expected, had a common quality of susceptibility to the production of spots. Another plant would be resistant; all the leaves on that would give smaller counts. The total count on the five leaves of one plant might be one-fifth or one-third what it was on another plant. But even more striking was the fact that there was a positional effect. The top leaves tended to be alike (as did the second, the third, and the fourth, and the bottom leaves) in the sense that all the top leaves might give about half the count of their corresponding bottom leaves from the same plant.

Here, then, you see the familiar rows and columns made to order. Nothing hopeful about this regularity, it is there. And to compare five virus solutions, we will simply make sure, if we label them A, B, C, D, E, that they are allotted to the leaves in the same kind of pattern I had a moment ago for the farmer's sons.

Leaf	Plant Number				
Position	1	2	3	4	5
Top	A	B	C	D	E
2nd	B	E	D	C	A
3rd	C	A	E	B	D
4th	D	C	A	E	B
5th	E	D	B	A	C

Note the arrangement of the 5 letters: all 5 on every plant, all 5 in each leaf position. The net result of this was to so improve the accuracy of the comparisons, that it is quite conservative to say it was like presenting the pathologist with an extra greenhouse. He did not need to test as many plants with each solution. This was a case where these strips really paid off.

Now I am going to just briefly run over some other cases where this same design -- which is called the latin square because Latin letters are used in it -- has been used.

It is being used in the measurement of standards of radioactivity and in rating the samples that are sent in to be compared with those standards. These measurements are made by first putting the known standard in front of a Geiger Counter and getting a count, and then in turn placing the unknowns and getting counts, and comparing these counts. It takes a certain time to make these measurements. During this time the voltage changes, and conditions change. That's one of the troubles in doing experiments. If a standard and three unknowns are each measured four times the familiar arrangement in a latin square makes it possible to consider each column as a period of time. The rows, which correspond to the 1st, 2nd, 3rd, and 4th measurement in each time period, correspond to positional effect on the enamel panels. This is a precision type of measurement.

Another case that is rather interesting comes from physical chemistry where they were comparing sources of temperature and have some standard cells which will set up a temperature with tremendous faithfulness, probably even to four decimal places. It stretches the best thermometers in the land to the uttermost to try to detect differences among these cells. They want to know whether they can make a series of cells that are really all alike. One trouble is that the resistance thermometer has to be married to a cell for a whole day to come to equilibrium. So to compare two cells using the same thermometer you must make measurements on successive days. Or if you want to compare them on the same day, you must use two different thermometers. Another study revealed that there were day to day effects, for example, comparing two cells using the same thermometer on different days brought in an error from the different days. Then they wanted to compare two cells and avoid this error by doing both measurements on the same day, they had to use two different thermometers. Then they had to take somebody's word for it that the thermometers were the same. They were right at that borderline where they were making such precise measurements that this assumption seemed to be questionable. Let cell I, cell II, cell III, and cell IV correspond to rows in the latin square and thermometer A, thermometer B, thermometer C, thermometer D refer to the columns. On the first day every cell gets a thermometer. The assignment of thermometers to cells for each day is shown in the latin square.

Cell Number	Thermometer			
	A	B	C	D
I	1	2	3	4
II	3	4	1	2
III	2	3	4	1
IV	4	1	2	3

The numbers refer to days. We are not interested so much in the difference between days. They appear where treatments usually appear in an agricultural design. In agricultural work we are not interested in the differences in fertility between the horizontal strips and the vertical strips; but in this experiment it is the differences among the row averages and the differences among the column averages in which we are interested. Now we want to remove from the problem the difference between days. This design does it.

The latin square has also been used in studies of electroplating where the rows and columns correspond to positions on the metal plates and the letters to different laboratories (in an interlaboratory test). It has been used in comparing makes of tires, and this is another case where it is a natural. Let us take four makes of tires from each manufacturer, and take four automobiles. How are you going to compare them fairly? You should not assign all 4 tires of one make to one automobile because there are differences in drivers. It is so easy to set up this same latin square arrangement. Let each column equal an automobile, each row a wheel position, each letter a tire manufacturer.

The latin square was used in the worst possible way once in studying 8th Air Force Bombing. Rows became targets, and columns represented the order in which the bomb groups went over the target, and the entries in the square were the bomb groups themselves. We learned about bombing that way.

7.7 LATIN RECTANGLE DESIGNS--TWO-WAY ELIMINATION OF HETEROGENEITY

In a marketing investigation by Dr. Max E. Brunk, Cornell University, seven different merchandising practices were studied to determine their effects on sales of sweet corn. When the experiment was started, it was not known how long the sweet corn season would last, although it was fairly certain to last at least four weeks. It was therefore necessary to set up a design such that as much balance as possible would be retained if some of the rows in a 7 × 7 latin square design were to be deleted or added. The following design was used:

Week	Store 1	2	3	4	5	6	7
1	E	F	G	A	B	C	D
2	B	C	D	E	F	G	A
3	D	E	F	G	A	B	C
4	F	G	A	B	C	D	E
5	G	A	B	C	D	E	F
6	A	B	C	D	E	F	G
7	C	D	E	F	G	A	B
8	A	B	C	D	E	F	G

Every treatment (A,B,C,D,E,F,G) appears once in each row. Hence treatments and rows are orthogonal. If seven weeks are used, the design is an ordinary latin square design, and rows, columns, and treatments are all orthogonal to each other. If the design ends at four weeks, columns and treatments are not orthogonal, but it should be noted that the design of $v = 7$ treatments in $b = 7$ blocks of size $k = 4$ forms a bib design with each treatment pair occurring together twice in the seven blocks. Thus, a latin rectangle design of four rows by seven columns can be obtained from the first four rows of the above plan, the treatments and row effects are orthogonal, the row and column effects are orthogonal, and the treatment and column effects are associated in the same manner as in a bib design when the treatment pairs occur together in the columns the same number of times. Such a design is dubbed a **Youden design** after W. J. Youden who created a number of these designs while working at the Boyce Thompson Research Institute. This is one of two experiment designs bearing the name of a man. The other is a special type of latin square called the Knut Vik square. It appears that the influence of Sir Ronald A. Fisher is responsible for the names in both cases.

Likewise, the first three rows of the above rectangle design is also a Youden design with $v = b = 7$, $r = k = 3$, and $\lambda = 1$. The first four rows given above also produce a Youden design with $v = b\ 7$, $r = k = 4$, and $\lambda = 2$. Any six rows of a 7×7 latin square also form a Youden design such that treatment pairs occur together in the columns $\lambda = 5$ times. The deletion of any row from a $k \times k$ latin square design produces a Youden design. Also, the addition of any row of the square to the latin square produces a design with the same balanced properties as the Youden design.

Several additional types of latin squares and latin rectangles have been constructed and are available in published literature. The grouping of treatments in rows and/or columns may not be balanced, resulting in non-orthogonality of effects. Consequently, the statistical analyses will be more complicated than those for orthogonal designs such as the rcbd, the latin square and the simple change-over. The requirements of the experiment and *not the ease of analysis* determine the appropriate design.

The equation for yield of an observation and the randomization procedure follows that for the latin square design. It should be noted that the simple change-over design is also a latin rectangle design and that the latin square design is a special case of the latin rectangle design.

7.8 *LATIN CUBE DESIGNS--THREE-WAY ELIMINATION OF HETEROGENEITY*

Although latin squares were first studied in the latter part of the eighteenth century by Professor L. Euler, the famous Swiss mathematician, the concept of latin cube arrangements is rather recent. K. Kishen, Indian statistician, and Sir Ronald A. Fisher independently presented latin cube designs in the early 1940s. A practical use for these designs has been found by M. E. Brunk and his co-workers at Cornell University in setting up a series of latin square designs. The following design for three merchandising treatments (A,B,C) was set up in three stores in Ithaca, New York:

		Week							
	1			2			3		
Day of	store			store			store		
week	1	2	3	1	2	3	1	2	3
Monday	A	B	C	B	C	A	C	A	B
Tuesday	B	C	A	C	A	B	A	B	C
Wednesday	C	A	B	A	B	C	B	C	A

If we set up a design with the three axes representing a cube, we obtain the diagram shown in Figure 7.4. It should be noted that a plane perpendicular to any axis results in a latin square design, and so in the above, the design in each store, the design on any day of the week, and the design in any week each forms a latin square. The three sources of variation removed from the differences between treatment means are days of the week, weeks, and stores. The statistical analyses for latin cube designs have been partially worked out; the analysis can take many forms depending upon the purposes of the experiment. An exhaustive evaluation of all analyses has not yet been made in published literature.

The latin cube design has many practical applications. A study on rotating practices in field experiments has been designed in a 5 × 5 × 5 latin cube design for a soil conservation study. The 4 × 4 × 4 latin cube design has been used in a merchandising study on paper products for which several statistical analyses were developed.

The full randomization procedure has not been studied, but an approximate randomization procedure after the construction of a latin cube is the random allotment of the letters to the treatments, of the weeks to the planes on the x_1 axis, of the stores to the planes on the x_2 axis, and of the days to the planes on the x_3 axis.

7.9 MAGIC LATIN SQUARE DESIGN--THREE-WAY ELIMINATION OF HETEROGENEITY

In addition to latin square plans, Professor L. Euler discussed the magic latin square plan. The 4 × 4 and 6 × 6 magic latin squares are:

A	B	C	D
C	D	A	B
B	A	D	C
D	C	B	A

A	B	C	D	E	F
D	E	F	A	B	C
C	A	B	E	F	D
F	D	E	C	A	B
B	C	A	F	D	E
E	F	D	B	C	A

In the above 4 × 4 square all four treatments (A,B,C,D) appear once in each row, once in each column, and once in each 2 × 2 square. In the above 6 × 6 square all six treatments (A,B,C,D,E,F) appear once in each row, column, and 2 × 3 rectangle. The variation controlled is row, column, and the 2 × 2 and 2 × 3 rectangles in the above designs. The randomization procedure and analysis is described by Federer (1955). Although the magic latin square design has been utilized in field experiments, one wonders whether it is not a case of over-stratification.

7.10 DESIGNS FOR TREATMENTS APPLIED IN SEQUENCE WHEN THE TREATMENT EFFECT CONTINUES INTO THE NEXT PERIOD

Suppose that three merchandising treatments,
        A = a display of 4-pound Polythene bags,
        B = a display of 6-pound Polythene bags, and
        C = a display of 8-pound Polythene bags,

were used to determine the effect of size of bag on the sale of McIntosh apples in grocery stores. Six stores from a given chain in central New

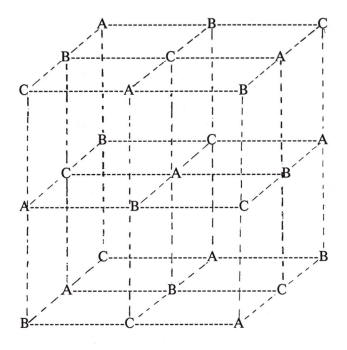

Diagrammatic representation

	I			II			III	
A	B	C	B	C	A	C	A	B
B	C	A	C	A	B	A	B	C
C	A	B	A	B	C	B	C	A

Schematic representation

Figure 7.4. 3 × 3 × 3 Latin cube of first order.

York are selected for the study. The period of observation on sales of apples is one week. If a person purchases two four-pound or one eight-pound bag of apples, it is possible that this would affect their purchase of apples during the following week, that is, the effect of a treatment might last for more than the treatment period; this is called a **residual effect** of the treatment. The sale of apples for a given treatment above or below the mean during the treatment period (the week the treatment is in the store) is the **direct effect** of the treatment. The following design, known as a **double change-over design**, would be used for this situation (see Federer, 1955, example XIV-2):

Week	Store 1	2	3	4	5	6
1	A	B	C	A	B	C
2	B	C	A	C	A	B
3	C	A	B	B	C	A

In this design, each treatment follows and is followed every other treatment. The number of times this occurs is often equal for every pair of treatments, in which case the design is said to be balanced for residual effect. In the above, it will be noted that treatment B follows A twice and treatment A follows B twice. The same balance is attained for pairs A and C and B and C.

Here the population is a single population of sampling units, the columns. A simple random sample of columns is made. Then, the randomization procedure is to randomly allot the stores to the columns and the letters to the treatments. The rows are not randomized, because the sequence of treatments in each column must be maintained. The statistical analysis is somewhat complicated, since not all effects are orthogonal. The design is useful in many types of experiments; the dairy cow, the patient, the worker, the rat, the hospital, and so on, replace the store category, and the period of treatment replaces the week. This design is also known as a **repeated measures design** as the sampling unit is measured repeatedly through time.

There are several variants of designs for measuring direct and residual effects of treatments. Designs can be constructed to obtain measurements for residual effects that last for more period and even for measuring the continuing effect of a treatment throughout the entire length of the experiment and beyond. One design variation is:

| | | Store | | |
Period		1	2	3
1		A	B	C
2		B	C	A
3		C	A	B
4		A	B	C
5		C	A	B
6		B	C	A
7		A	B	C

The randomization is the same as that for the previous design. The yield
equation is of the same form as the latin square design except that is
contains an additional term for a residual effect of a treatment.

7.11   AUGMENTED   DESIGNS--ONE-   AND   TWO-WAY ELIMINATION OF HETEROGENEITY

In certain areas of investigation, an experimenter may desire to use an
unequal number of experimental units for each of the treatments. This
may be desired because of lack of material or because of the desire to
screen out undesirable treatments with minimum effort. For example, the
Hawaiian Pineapple Research Institute was interested in soil fumigants as
they affect the growth of the pineapple plant. Chemists can produce many,
say 300 to 500, new possible soil fumigants each year. Since a satisfactory
soil fumigant was already available and since any fumigant superior to the
present one would have to be quite exceptional, the researcher decided that
he would allocate only one experimental unit for the trial of any new
possible soil fumigant; this was necessary since an ineffective soil fumigant
would produce no pineapples and a fumigant that was too strong would kill
the plants. The economic loss resulting from plots with no pineapple fruit
could not be tolerated on a large scale. Also, the effective dose of a
chemical could be determined fairly closely by reference to known levels
of a similar chemical compound. Any new soil fumigant that was not as
good as the standard or better would be rejected on the basis of its
performance from one experimental unit. If it was not rejected, the new
soil fumigant was a candidate for further testing and would go into the
group that would be entered in the second stage of screening or testing.

Each year, in addition to the 300 to 500 new possible soil fumigants available, the investigator has 2, 3, 4, or more promising new soil fumigants, the standard commercially used soil fumigant (say A), and an extremely effective soil fumigant that was not practically useful (say B). Suppose that the three promising new soil fumigants are labelled C, D, and E and that the 300 new soil fumigants to be tested are numbered 1, 2, ..., 300. The investigator decides that he needs 20 replicates of A,B,C,D and E, and only one of treatments 1 to 300. He sets up the following schematic arrangement:

				Block					
1	2	3	4	5	6	7		19	20
A	A	A	A	A	A	A		A	A
B	B	B	B	B	B	B		B	B
C	C	C	C	C	C	C		C	C
D	D	D	D	D	D	D		D	D
E	E	E	E	E	E	E		E	E
1	16	31	46	61	76	91		271	286
.	.	.	.	.		.	.		
.	.	.	.	.		.	.	.	.
15	30	45	60	75	90	105		285	300

In the blocks of 20 experimental units (plots) and 20 treatments, he randomly allots the letters to the 20 experimental units in each of the 20 blocks. Then he randomly assigns the numbers 1, 2,..., 300 to the 300 possible new soil fumigants and fills in the remaining 15 plots in blocks of 20 as indicated above. The treatments A, B, C, D, and E, which may be called **standard treatments**, are arranged as in a randomized complete block design. The name **new treatments** or the name **augmented treatments** could be applied to the 300 possible new soil fumigants. Such a design has been dubbed an **augmented randomized complete block design**. Each block of a standard rcb design has been augmented with the new treatments. (In Hawaiian the word for augmented is "**hoonuiaku**". Hence, we could, as the author has done (Federer, 1956), call this design a **hoonuiaku rcb design**.)

If the investigator had wished to control variation in two directions, he could have set up four 5 × 5 augmented latin square designs as follows:

A	B	C	D	E		A	B	C	D	E
1-	28-	31-	58-	61-		226-	253-	256-	283-	286-
3	30	33	60	63		228	255	258	285	288
B	C	D	E	A		D	E	A	B	C
4-	25-	34-	55-	64-		229-	250-	259-	280-	289-
6	27	36	57	66		231	252	261	282	291
C	D	E	A	B		B	C	D	E	A
7-	22-	37-	52-	67-	...	232-	247-	262-	277-	292-
9	24	39	54	69		234	249	264	279	294
D	E	A	B	C		E	A	B	C	D
10-	19-	40-	49-	70-		235-	244-	265-	274-	295-
12	21	42	51	72		237	246	267	276	297
E	A	B	C	D		C	D	E	A	B
13-	16-	43-	46-	73-		238-	241-	268-	271-	298-
15	18	45	48	75		240	243	270	273	300

Instead of there being only one experimental unit in each row and column intersection, there are four experimental units or plots. The standard treatment is randomly allocated to one of the four plots. The latin square design and randomization follows that given in Section 7.6 and the new treatments are numbered in the same manner as for an augmented randomized complete block design and allocated to the remaining plots.

An alternative method of a schematic layout of an augmented latin square design with 20 replicates of the standard treatments and 300, say, of the new treatments each appearing once, would be as indicated on the following page. This design is not as flexible as the previous one, but it may be more efficient in controlling extraneous variation among treatment effects in the experiment. The randomization procedure for a latin square design may be used for this design. All five standard treatments appear once in each row and once in each column and hence treatment effects are orthogonal to both row and column effects. Each of the new treatments

| | Column | | | | | | | | | | | | | | | | | | | |
Row	1	2	3	4	5	6	7	8	9	10	11	12	13	14	15	16	17	18	19	20
1	A	B	C	D	E	1	2	3	4	5	6	7	8	9	10	11	12	13	14	15
2	16	A	B	C	D	E	17	18	19	20	21	22	23	24	25	26	27	28	29	30
3	31	32	A	B	C	D	E	33	34	35	36	37	38	39	40	41	42	43	44	45
4	46	47	48	A	B	C	D	E	49	50	51	52	53	54	55	56	57	58	59	60
5	61	62	63	64	A	B	C	D	E	65	66	67	68	69	70	71	72	73	74	75
6	76	77	78	79	80	A	B	C	D	E	81	82	83	84	85	86	87	88	89	90
7	91	92	93	94	95	96	A	B	C	D	E	97	98	99	100	101	102	103	104	105
8	106	107	108	109	110	111	112	A	B	C	D	E	113	114	115	116	117	118	119	120
9	121	122	123	124	125	126	127	128	A	B	C	D	E	129	130	131	132	133	134	135
10	136	137	138	139	140	141	142	143	144	A	B	C	D	E	145	146	147	148	149	150
11	151	152	153	154	155	156	157	158	159	160	A	B	C	D	E	161	162	163	164	165
12	166	167	168	169	170	171	172	173	174	175	176	A	B	C	D	E	177	178	179	180
13	181	182	183	184	185	186	187	188	189	190	191	192	A	B	C	D	E	193	194	195
14	196	197	198	199	200	201	202	203	204	205	206	207	208	A	B	C	D	E	209	210
15	211	212	213	214	215	216	217	218	219	220	221	222	223	224	A	B	C	D	E	225
16	226	227	228	229	230	231	232	233	234	235	236	237	238	239	240	A	B	C	D	E
17	E	241	242	243	244	245	246	247	248	249	250	251	252	253	254	255	A	B	C	D
18	D	E	256	257	258	259	260	261	262	263	264	265	266	267	268	269	270	A	B	C
19	C	D	E	271	272	273	274	275	276	277	278	279	280	281	282	283	284	285	A	B
20	B	C	D	E	286	287	288	289	290	291	292	293	294	295	296	297	298	299	300	A

appears once in the experiment. The statistical analyses for these designs is not too difficult.

Any experiment design with a standard set of treatments may be augmented with a set of new treatments, for instance, the augmented balanced incomplete block design, the augmented latin rectangle design, etc.. Furthermore, some treatments may be included once, some treatments may be included twice, some three times, etc.. A statistical analysis can be devised for any augmented design consisting v_1 treatments included once, v_2 treatments included twice, v_3 treatments included three times, ..., v_r treatments included r times. Such designs allow considerable freedom in the use of all experimental material in the desired proportions.

Although there are numerous other types of experiment designs created by statisticians, the above should be sufficient to convey to the reader some of the principles for construction and some properties of experiment designs.

7.12 SUMMARY OF PRINCIPLES OF EXPERIMENT DESIGN

At the beginning of this chapter we listed three desirable characteristics
for designing experiments; we discussed the concepts of randomization,
blocking, efficiency, orthogonality, confounding, and balancing. Sir
Ronald A. Fisher defined and developed these concepts of experiment
design. Figure 7.5 is a small adapted replica of one that is said to have
hung on the wall of his office at the Rothamsted Experiment Station in
England; it illustrates three of the basic principles of design, viz.
replication, randomization, and local control.   Randomization and

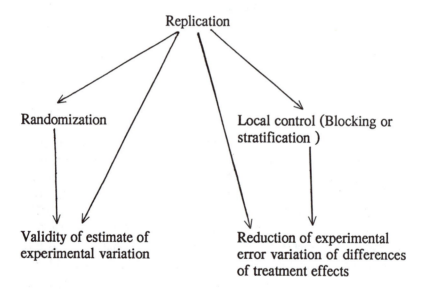

Figure 7.5. Relationships of three basic principles of experiment design.

replication are necessary to obtain a valid estimate or measure of the experimental variation. Replication and "local control" (blocking or grouping) are necessary to achieve a reduction in the random variation among treatment effects in the experiment. Use of "local control" has been made throughout this chapter in blocking or grouping to eliminate or to control the various sources of variation.

We stated that randomization was used in order to be fair to the treatments. If the variation among the experimental units within a group is due entirely to random error with no known method of grouping, then the random allotment of treatments to the experimental units is entirely fair. Any treatment then has an equal chance of receiving the highest or the lowest observation; any other method of assignment would lead to unfairness to some treatments in that there would be discrimination. Also, any device of allotting treatments that tends to make them more alike or more unlike than they would have been by a random allocation leads respectively to larger or smaller estimates of error variation than for the random allotment of treatments. We may summarize the purpose of randomization as follows:

1. to obtain unbiased estimates of differences among treatment responses (means or effects) and
2. to obtain an unbiased estimate of the random error variation in the experiment.

With respect to the latter, a valid estimate of the experimental variation may be obtained if there is sufficient replication and additivity of effects (See example 5.6, pages 77 to 78, in Cox, 1958.). There are several procedures for achieving a random allocation of treatments to the experimental units within the block. We have discussed the use of identical and numbered balls or tags in a jar or hat, the use of coins, and the use of a die or a pair of dice. There are also random number tables available from many sources, and some of these were discussed in Section 5.4.1.

Replication is the repetition of treatments in different blocks or sources of variation in the experiment. For example, in the randomized complete block design, we need at least two blocks in order to distinguish between the estimated overall mean $\bar{y}_{..}$ and the estimated block effect, say

$\bar{y}_{.j} - \bar{y}_{..}$. If there is only one block, then $\bar{y}_{.1} = \bar{y}_{..}$. This is true also for treatments in a comparative experiment, since two or more treatments are required in order to compare items. If there is only one treatment, then the estimated treatment mean, say $\bar{y}_{1.}$, and the overall mean $\bar{y}_{..}$ are indistinguishable and the estimated treatment effect $\bar{y}_{1.} - \bar{y}_{..}$ cannot be obtained. The **number of replicates** for a treatment is the number of experimental units assigned to that treatment.

Also, an increase in the number of replicates of a treatment tends to decrease the variation in the estimate of a difference between two treatment means from orthogonal designs. This is the manner in which replication leads to a reduction in the experimental error of differences of treatment effects.

In addition to the three basic principles of experiment design (randomization, replication, and blocking), the fourth principle of orthogonality is important in order to ensure that the estimate of the random variation between treatment means is the same for all pairs of treatments having equal replication and having the same degree or magnitude of random error variation. Statistical analyses are simpler than those for non-orthogonal designs, and orthogonal designs are the most efficient of all designs.

If orthogonal designs are not possible, then we strive for balanced designs which still ensures that differences between pairs of treatment effects all have the same variance. In the balanced designs discussed, all treatment pairs occur with each other equally frequent in the b blocks of size k for the v treatments. Since bk = total number of experimental units and since there are v treatments each repeated r times, bk = vr in balanced designs. Orthogonal designs may be balanced designs, but the reverse may not be true. The randomized complete block design is usually a balanced design, but the balanced incomplete block design is generally not an orthogonal design. Figure 7.6 illustrates the relationships of six (the above five plus confounding) Fisherian principles of experiment design.

Partial confounding in incomplete block designs may lead to more efficient designs than complete block designs, depending upon the experimental variation and the blocking. The double-headed arrow indicates that blocking and orthogonality are related but that one does not necessarily lead to the other. If we have blocks with an appropriate design, for instance, the rcb design, we can have orthogonality between treatment and block effects. The interrelationships between the first three criteria listed at the beginning of this chapter and the six in Figure 7.6 may be represented as shown in Figure 7.7.

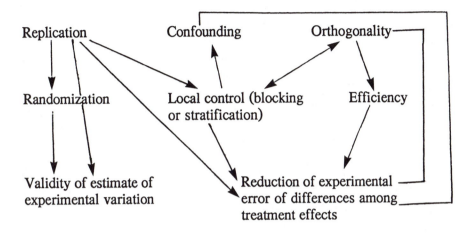

Figure 7.6. Interrelations among six principles of design.

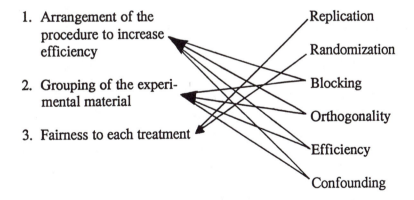

Figure 7.7. Interrelationships of principles of design and three criteria.

The above discussion is an introduction to properties, construction, and layout of experiment designs. Various designs may be compared in order to determine which are the better designs for experimental work. One of the items to be considered in the fourth principle of scientific inquiry, the design of the investigation, is the selection of the appropriate experiment design. A number of experiment designs have been presented in this chapter to illustrate the kinds of designs available for different types of variation among the experimental units.

In Problem 7.2 of this chapter you are asked to design an experiment for comparing the speed of two calculating machines in performing a series of calculations, specifically the computation and accumulation of the squares of a set of numbers. The calculation is known as the computation of the "sum of squares" of a set of numbers. In a smaller class or with laboratory sections in the class, this experiment could be performed with pairs of students. One student could do the computing, and the other student could perform the timing operation and record the time for calculation. The conduct of this experiment can be demonstrated with only one pair of students to determine which of two calculating machines is faster in performing the stated statistical computations (also, see Problem 12.5). This experiment has been performed routinely in the introductory statistics course at Iowa State University for many years. Such an experiment provides each pair of students with a set of data on which to perform a variety of statistical computations. This experiment with a particular set of data is also discussed in Chapter 1 of Cochran and Cox (1957) in connection with principles of scientific inquiry.

7.13 BOOKS ON THE DESIGN AND ANALYSIS OF EXPERIMENTS

There are numerous books on statistical methods or analyses of the Snedecor and Cochran (1980) type, for instance, Steel and Torrie (1980), etc.. These books, in general, stress analyses after the data are available rather than methods of procuring the data; examples of the rcbd and latin square designs are given, but the reasons for using these designs receive little or no discussion. Some of the books which stress the purposes, construction, and layout of experimental arrangements or designs and treatment designs and the corresponding analyses are listed below. Unfortunately, all are necessarily written assuming some knowledge of statistical analyses on the part of the reader; elementary facts of statistical analyses of data may be obtained from later chapters of this book.

Cochran, W. G. and Cox, G. M. (1957). *Experimental Designs, 2nd edition* (lst edition, 1950), John Wiley and Sons, Inc., New York, xiv + 617 pp.
(This book requires a fair knowledge of the statistical methods in Snedecor and Cochran (1980). It is a comprehensive book on plans for designs; analyses and numerical examples are presented for a number of designs.)

Cox, D. R. (1958). *Planning of Experiments*, John Wiley and Sons, Inc., New York, London, and Sydney, vii + 308 pp..
(Professor Cox has attempted to avoid statistical and mathematical technicalities. The book is an elementary discussion of experiment and treatment design, with some discussion of the reasons and objectives behind each.)

Federer, W. T. (1955). *Experimental Design -- Theory and Application*, Macmillan Company, New York, xix + 593 pp.. (Republished in 1974 by The Oxford and IBH Publishing Co., New Delhi, India.)
(A somewhat advanced text on the construction, layout, and analyses of many types of designs. Emphasis is placed on the reasons for using a design as well as on the statistical analysis. A considerable amount of generality is achieved. Many numerical examples covering a variety of fields are presented to illustrate the various designs and analyses.)

Finney, D. J. (1955). *Experimental Design and Its Statistical Basis*, The University of Chicago Press, Chicago and London, xi + 169 pp..
(Professor Finney has written a book at about the same level as Professor Cox, but it is slanted more toward biological topics. The design topics are treated with different emphasis and in a different manner than are those in Cox's book. Both books could be read to advantage.)

Fisher, R. A. (1926). The arrangement of field experiments, *Journal of the Ministry of Agriculture*, 33, 503.
(This is a paper which is listed because it is the first publication on designing experiments using randomization in designs; it forms an excellent introduction to the subject of experimental designs.)

Fisher, R. A. (1935) . *The Design of Experiments*, lst edition in 1935 with several subsequent editions. Oliver and Boyd, Edinburgh, xi + 236 pp..
(Nearly all the general ideas in experiment design stem from the pioneer

work of Sir Ronald A. Fisher as expressed in his writings, lectures, consultations, informal conversations, and reflections. Fisher's works could profitably be read prior to other accounts of experiment and treatment designs, and again after each book read; this applies even to the most advanced and the most mathematical books. Do not expect to understand all of Fisher at first reading, at second, or even at the third reading. Some of the ideas are complex enough that people are still struggling to understand them.)

Ghosh, S. (editor) (1990). *Statistical Design and Analysis of Industrial Experiments.* Marcel Dekker, Inc., 440 pp..

Kempthorne, 0. (1952). *The Design and Analysis of Experiments,* John Wiley and Sons, Inc., New York, ix + 631 pp..
(An advanced and general text on experiment and treatment designs, stressing the theoretical approach to the construction, randomization procedure, and analysis, few numerical illustrations are used.)

Yates, F. (1937). The design and analysis of factorial experiments, *Imperial Bureau of Soil Science, Technical Communication No. 35,* Harpenden, England, 95 pp..
(This pamphlet is a classical writing on the subject of the treatment design known as a factorial. Many writers have taken their material from this excellent publication, which discusses many aspects of experiment design, including as yet unsolved problems.)

There are many other books which contain the words "design of experiments" or "experimental design" in their titles but except for the more mathematical ones, they contain little that is not in the above books or in statistical methods texts. If an individual were to comprehend fully the principles and theory contained in the books listed above, he/she would be considered a competent statistician in the field of experiment and treatment design. They would also be able to do research in this field and to work on some of the unsolved problems.

7.14 RESEARCH WORK IN EXPERIMENT DESIGN

The fields of treatment design and experiment design are being actively investigated by numerous statisticians and mathematicians. A bibliography of papers published in these fields has been prepared for the period 1950 to 1967; nearly 7000 references are listed therein (See Federer and Balaam, 1972). Before 1950, nearly 2000 citations relate to work in these fields.

7.15 REFERENCES AND SUGGESTED READING IN ADDITION TO THOSE LISTED IN SECTION 7.13

Dominick, B. A., Jr., (1952). Merchandising McIntosh apples under controlled conditions -- customer reaction and effect on sales. Ph.D. Thesis, Cornell University, Ithaca, New York.

Federer, W. T. (1956). Augmented (or hoonuiaku) designs. *The Hawaiian Planters' Record* 50, 191-208.

Federer, W. T. and L. N. Balaam (1972). *Bibliography on Experiment and Treatment Design Pre 1968.* Published for the International Statistical Institute by Oliver & Boyd, Edinburgh, 769 pp..

Jowett, G. H. and Davies, H. M. (1960). Practical experimentation as a teaching method in statistics. *Journal of the Royal Statistical Society, Series A,* 123, 10-35.

Kahan, B. C. (1961). A practical demonstration of a needle experiment designed to give a number of concurrent estimates of π. *Journal of the Royal Statistical Society, Series A,* 124, 227-239.

Khare, M. and W. T. Federer (1981). A simple construction method for resolvable incomplete block designs for any number of treatments. *Biometrical Journal* 23(2), 121-132.

Patterson, H. D. and E. R. Williams (1976). A new class of resolvable incomplete block designs. *Biometrika* 63(1), 83-92.

Snedecor, G. W. and Cochran, W. G. (1980). *Statistical Methods, 7th edition.* The Iowa State University Press, Ames, Iowa, xiv + 507 pp..

Steel, R. G. D. and Torrie, J. H. (1980). *Principles and Procedures of Statistics, second edition.* McGraw-Hill Book Company, Inc., New York, Toronto, and London, xvi + 481 pp..

Yates, F. (1935). Complex experiments. *Journal of the Royal Statistical Society, Series B*, 2,181-247.

Yates, F. (1937). The design and analysis of factorial experiments. *Imperial Bureau of Soil Science, Technical Communication* 35, 1-95.

Youden, W. J. (1962). *Experimentation and Measurement.* National Science Teachers Association and National Bureau of Standards, Scholastic Book Services, New York, 127 pp..

7.16 PROBLEMS

Problem 7.1.   Set up a randomized complete blocks design for an experiment involving five dental cleaning preparations developed by Dr. Iva Payne. Use the five preparations on one patient as the block and use ten different patients. The characteristic measured is "cleanliness" of teeth after cleaning. Describe the randomization procedure and a method of conducting the experiment.

Problem 7.2.   Design an experiment for evaluating fairly the speed of computing sums of squares on two calculating machines, brand A and brand B. One operator and ten different sets of numbers will be used. Each of the ten sets of sums of squares will be computed on each of the machines. We know from past experience that an operator computes a sum of squares for a given set of numbers faster the second time than he does the first time.

Problem 7.3.   In Section 36 of Fisher's (1935) book, entitled *The Design of Experiments*, the following puzzle is given:

> Sixteen passengers on a liner discover that they are an exceptionally representative body. Four are Englishmen, four are Scots, four are Irish, and four are Welsh. There are also four each of four different ages, 35, 45, 55, and 65, and no two of the same age are of the same nationality. By profession also four are lawyers, four soldiers, four doctors, and four clergymen, and no two of the same profession are of the same age or of the same nationality. It appears, also, that four are bachelors, four married, four widowed, and four divorced, and that no two of the same marital status are of the same profession, or

the same age, or the same nationality. Finally, four are conservatives, four liberals, four socialists, and four fascists, and no two of the same political sympathies are of the same marital status, or the same profession, or the same age, or the same nationality. Three of the fascists are known to be an unmarried English lawyer of 65, a married Scots soldier of 55, and a widowed Irish doctor of 45. It is then easy to specify the remaining fascist. It is further given that the Irish socialist is 35, the conservative of 45 is a Scotsman, and the Englishman of 55 is a clergyman. What do you know of the Welsh lawyer?

For the problem in this class you may answer the above questions or you may answer the simpler problem given below; it is an adaptation of the above puzzle.

Nine passengers on an airliner discover that they are an exceptionally representative group. three are Englishmen, three are Scots, and three are Irish. By profession, three are lawyers, three are doctors, and three are clergymen. No two of the same profession have the same nationality. Furthermore, three are married, three are widowed, and three are divorced. No two of the same marital status are of the same profession or the same nationality. Finally, three are laborites, three are conservatives, and three are socialists. No two of the same political sympathies are of the same marital status, of the same profession, or of the same nationality. If the married English clergyman and the widowed Irishman are conservatives, specify the remaining seven individuals if you can. If not, what additional information do you need?

Problem 7.4. An investigator knows that there are two sources of variation inherent in his experimental material. He wishes to compare two treatments and to use eight experimental units for each treatment. Give three experiment designs for controlling two sources of variation among the 16 experimental units.

Problem 7.5. Construct a schematic arrangement of a balanced incomplete block design for $v = 6$ treatments in blocks of size $k = 4$. How many times, λ, do the individual pairs of treatments occur together in the b blocks?

Problem 7.6. Construct a schematic arrangement of a balanced incomplete block design for v = 7 treatments in blocks of size k = 4. How many times, λ, do the individual pairs of treatments occur together in the b blocks? Repeat but use incomplete blocks of size k = 3. How many times do the individual pairs of treatments occur together in the b blocks?

Problem 7.7. Write down the linear and additive yield equations using estimated effects for the completely randomized design, the randomized complete block design, and the latin square design. Which elements in each of the yield equations represent variation due to assignable causes, bias, and random error?

Problem 7.8. Devise a 7 × 7 latin square such that the first three rows form a Youden design and the last four rows also form a Youden design. (See your solution in Problem 7.6.)

Problem 7.9. Use the telephone directory to obtain a randomization plan for the dandelion experiment described in Section 7.3. Describe your procedure, including the place started in the directory.

Problem 7.10. Use the same telephone directory used for Problem 7.9 and obtain a randomization procedure for a 5 × 5 latin square design. Describe the procedure you used, including the place in the directory where you started.

Problem 7.11. Conduct a comparative experiment wherein you follow the principles of scientific investigation. Select an experiment and treatment design and take a set of measurements or observations. Some topics for investigation might be to compare v individuals as the treatments and to measure the distance from a target or line that a penny falls when tossed, compare v pairs of individuals in a series of bridge games, compare v different brands of ping-pong balls for height of bounce when dropped from a height of, say, five feet, compare v different kinds of seed for time from planting until germination occurs, compare weighings by v individuals as described on page 96 of Youden's (1962) book, or compare any other v treatments of interest.

Problem 7.12. Twenty-eight Falcon dealers in Chicago were interviewed by two of eight interviewers (*Applied Statistics*, pages 93-97, 1961, A. F.

Dealer	A	B	C	D	E	F	G	H	Mean
			Interviewer (Price minus $2,000)						
1	100	125							113
2	235		95						165
3	50			30					40
4	133					80			107
5	50				30				40
6	25						88		57
7	140							150	145
8		41	50						46
9		180		195					188
10		65			75				70
11		50				100			75
12		100					96		98
13		170						150	160
14			75	95					85
15			25			55			40
16			132	50					91
17			145					96	121
18			100				152		126
19				99	235				167
20				100		100			100
21				50			50		50
22				35				50	43
23					150	163			157
24					135		150		143
25					70			138	104
26						50	100		75
27						75		65	70
28							100	89	95
Range	25-235	41-180	25-145	30-195	30-235	50-163	50-152	50-150	

Interviewer

Mean	105	104	89	86	106	89	105	105	99
Adjusted									
Mean	111	99	87	84	99	94	114	102	

Jung). The design used was a balanced incomplete block design with $v = 8$ interviewers and $b = 28$ dealers. Falcon prices (minus $2,000) quoted to the interviewers by the dealers are given above.

Compute the overall range (largest value minus the smallest value) for this set of data. How many replicates are there? What is the block size? Ascertain whether or not the design is a bibd. If the design is a balanced incomplete block, what is the value of λ? (Note that the adjusted means have been adjusted as described in the text.)

Problem 7.13. A number of technicians are asked to rate a particular kind of shoe according to the following scale:

1--extremely unsatisfactory   6--slightly satisfactory
2--very unsatisfactory   7--moderately satisfactory
3--moderately unsatisfactory   8--very satisfactory
4--slightly unsatisfactory   9--extremely satisfactory
5--not good, not bad

In the experiment, there were six different courses on which the technician will judge the shoe. Twenty technicians were to be used. The design used for the 20 technicians and six courses, was:

	Technicians (numbered 1 to 20)			
	1, 5, 9,	2, 6, 10	3, 7, 11	4, 8, 12
Course	13 & 17	14 & 18	15 & 19	16 & 20
1	x	x		
2	x		x	
3	x			x
4			x	x
5		x		x
6		x	x	

The following represent the data collected from this experiment:

Technician	1	2	3	4	5	6	Total	Mean
1	3	8	9				20	6.67
2	1				1	5	7	2.33
3		9		4		4	17	5.67
4			9	5	3		17	5.67
5	3	6	6				15	5.00
6	4				7	1	12	4.00
7		2		3		9	14	4.67
8			5	6	9		20	6.67
9	6	2	9				17	5.67
10	4				5	9	18	6.00
11		6		8		1	15	5.00
12			9	5	8		22	7.33
13	6	5	1				12	4.00
14	2				6	6	14	4.67
15		4		8		7	19	6.33
16			9	5	2		16	5.33
17	5	9	7				21	7.00
18	5				1	3	9	3.00
19		9		6		7	22	7.33
20				6	7	9	22	7.33
Total	39	60	70	57	51	52	329	

(← Course and score)

What is the design? What are the values for the number of treatments v, of blocks b, of block size k, of replicates r, and of times treatment pairs occur together in the blocks?

Problem 7.14. The following results are for three judges judging three different items (A, B, C) in three orders where the experiment design was a latin square:

Order	Judge and score 1	2	3	Sum	Mean	Treatment Mean
1	B 1	A 0	C 5	6	2	A 2
2	C 0	B 5	A 4	9	3	B 4
3	A 2	C 4	B 6	12	4	C 3
Sum	3	9	15			
Mean	1	3	5	27	3	

Show how to compute the effects for judges, for orders, and for treatments. Also, show how to compute the nine residuals for these data.

Problem 7.15 Given that the overall mean is 10, that the treatment means for A, B, and C are 15, 12, and 3 respectively, and that the four block means are 12, 10, 11, and 7, show how to construct the 12 responses for this situation. What is the value of the 12 residuals and why?

Problem 7.16. Construct a set of responses for a randomized complete block design and a latin square design when the residuals are not all zero.

Problem 7.17. Construct a balanced incomplete block design for $v = 10$ and $k = 3$. Obtain a randomized plan for this design and describe in detail all the steps involved in the randomization.

Problem 7.18. Given the following three designs, i. e., the schematic arrangement before randomization:

Block number	1	2	3	4	5	6
Design 1	A	A	A	A	A	A
	B	B	B	B	B	B
	C	C	C	C	C	C
	D	D	D	D	D	D

Design 2

A	A	A	B	B	C
A	A	A	B	B	C
B	C	D	C	D	D
B	C	D	C	D	D

Design 3

A	A	A	A	A	C
B	B	B	B	A	C
C	C	C	C	B	D
D	D	D	D	B	D

Discuss and compare the blocking, replication number, block size, balancedness and orthogonality for the three designs.

Problem 7.19. Investigate the following two plans for balance and orthogonality:

Block number	1	2	3	4	5	6	7
Plan 1	A	A	A	B	C	D	
	B	B	E	E	E	E	
	C	C					
	D	D					
Plan 2	A	B	C	D	E	F	
	B	C	D	E	F		A
	D	E	F		A	B	C

What are the values for the block sizes and replication numbers?

CHAPTER 8. TREATMENT DESIGN AND SELECTION OF CONDITIONS FOR THE EXPERIMENT

8.1. INTRODUCTION

As described under the fourth principle of scientific investigation in Chapter 4, the treatment design or the selection of the treatments in an experiment may be one of the more important aspects of an experimental investigation; in many experiments the success of the experiment is vitally connected to the selection of the treatments. In this chapter we shall discuss

1. the presence of conditions in an experiment as related to a treatment design,
2. the relation of treatment design, the conditions of the experiment, and the population in which inferences are to be made, and
3. a number of treatment designs.

First, we shall consider an article written by Jellinek (1946). Excerpts from the article are reproduced below to illustrate several concepts (reprinted with the permission of H. A. David, Editor of *Biometrics*).

A headache remedy, designated here as drug A, is composed of ingredients a, b and c. Ingredient b was running short, and the manufacturers wished to know whether or not the efficacy of this drug would be lowered through the omission of this ingredient. In order to answer this question, 200 (only 199 subjects completed the tests) subjects suffering from frequent headaches were to be treated for two weeks on each occurrence of headaches with drug A, two weeks with drug B which contained ingredients a and c, two weeks with drug C which contained ingredients a and b, and two weeks with drug D which was a placebo consisting of ordinary lactate which is pharmacologically inactive.

The four drugs were made to appear identical in color, shape, size, and taste. Neither the subjects nor the physicians administering the drugs were aware of the differences in the composition of the four drugs, i.e., the study was doubly blind. Because of possible progressive sensitization or desensitization to the drugs, they were administered in different sequences as follows:

Group I	50 subjects
Group II	49 subjects
Group III	50 subjects
Group IV	50 subjects

A full account of the selection of subjects, type of records kept, instructions and mode of administration as well as psychological implications of the experience with the placebo will be given elsewhere.

The subjects took the tablets whenever a headache occurred. At the end of each two week period they reported to the physician the number of headaches they had in the course of that period and how many of these were relieved satisfactorily by the drug. They also reported the dosage taken on each occasion and the time elapsing between administration of the drug and the onset of relief from pain. Observations on psychological, gastric and heart reactions were noted. For each subject his "success rate" for each of the four drugs was computed as follows:

$$\frac{\text{Number of headaches relieved}}{\text{Number of headaches treated in the two week period}}$$

The potency of the drugs is expressed in terms of the arithmetic means of these individual "success rates". The analyses were carried out on the individual rates.

Some subjects had only three headaches in the course of a two week period while others had up to ten attacks in the same period. Thus, the individual rates are based on a varying number of headaches. This introduces an undesirable element into the analysis, but the great consistency of the data shows that the results may have been affected only to a small degree by this aspect of the tests. In other surveys, however, it may be desirable to stipulate the testing of each drug on four or five occasions rather than during a fixed period of time. (The design of the experiment was a latin square design with four columns (the two week periods), four rows (the four groups of 50 patients), and four treatments A, B, C, and D as given below.).

Group of patients	First 2 Weeks	Second 2 Weeks	Third 2 Weeks	Fourth 2 Weeks
I	A	B	C	D
II	B	A	D	C
III	C	D	A	B
IV	D	C	B	A

The mean success rates of the three analgesics, A, B, and C, and the placebo, D, compared as follows:

	A	B	C	D
Mean Success Rate	0.84	0.80	0.80	0.52

Inasmuch as placebos have been used at all in clinical tests of drugs, the procedure was to express the efficacy or non-efficacy of a drug in terms of "how much better" the drug was than the placebo. Thus in the present instance, it would have been said that drugs A, B, and C were only 53 to 62 per cent better than placebo. That such statements are meaningless and misleading will be seen from the further analysis of the data.

The success rate of .52 on placebo was due to 120 out of the 199 subjects. No relief whatever was reported by 79 subjects although they had three to ten headaches treated with placebo. On the other hand, these same 79 subjects when treated with one of the three analgesics reported from one third to all of their headaches relieved. The 120 subjects who reported relief at all through placebo did not do so only on one or two occasions, but rather consistently. The nature of response to placebo is seen best from the distribution of the number of headaches reported as relieved by placebo in subjects who had a constant number of attacks. In Table 8.1, the distribution of the number of relieved headaches is given for 59 subjects of this study who were exposed on five headache attacks to placebo. The distribution is given also for 121 subjects including these 59 subjects and another 62 subjects from later studies who also had five exposures to placebo on the occasion of headache attacks.

Examples of the rare U shaped distribution are seen here. Thus there are individuals who definitely tend to respond to placebo. This difference in response to placebo must reflect a difference in the nature of headaches. The sample is drawn from at least two broad populations of sufferers from headaches. If subjects never report relief through a pharmacologically inactive substance but always report at least some attacks relieved through bona fide analgesics, it must be assumed that they represent a "pure culture" of physiological headaches not accessible to suggestion, while the 120 subjects who either always or most of the time responded to placebo represent, perhaps predominantly, psychogenic headaches and to some extent also milder physiological headaches coupled with a tendency toward suggestibility.

Table 8.1. Distribution of the number of headaches reported as relieved by subjects who had been treated with placebo on 5 attacks of headaches

Number of Relieved Headaches	Present Study Number of Subjects	Present and Later Studies Number of Subjects
0	22	49
1	1	1
2	5	6
3	7	12
4	8	18
5	16	35
Total	59	121

Evidently persons suffering from psychological headaches lack the prerequisite condition for discrimination of potency among drugs, as any substance of the appearance of a drug and prescribed or administered by a physician will serve the purpose. This finding suggested a separate analysis of the "success rates" of the three analgesics on subjects who did not and on subjects who did react to placebo. The mean "success rates" of the analgesics are shown in Table 8.2 for these two classes of subjects in each of the four groups of different sequences of drug administration.

In spite of the small number of individuals in any of the four groups, the order of mean "success rates" in the class of subjects not reacting to placebo shows great consistency. In each of the four groups drug A occupies the first, drug C the second, and drug B the third place. The mean "success rates" of the entire class of subjects not reacting to placebo suggest definitely the importance of ingredient b which was lacking in drug B and a minor importance of ingredient c which was lacking in drug C as the mean "success rate" of the full formula, A, was much superior to that of B and somewhat superior to that of C.

Table 8.2. Mean "success rates" on 3 analgesic drugs. 79 subjects not reacting to placebo and 120 subjects reacting to placebo.

Group No.	Subjects not reacting to placebo				Subjects reacting to placebo			
	Number of Subjects	A	B	C	Number of Subjects	A	B	C
		"Success rates"				"Success rates"		
1	14	.90	.65	.86	36	.76	.87	.83
2	26	.88	.66	.70	23	.76	.84	.87
3	20	.85	.60	.71	30.	.89	.85	.76
4	19	.91	.82	.86	31.	.86	.90	.83
All groups	79	.88	.67	.77	120	.82	.87	.82

No consistency of mean "success rates" is seen in the class of subjects reacting to placebo; each of the three analgesics occupies first, second, and third places in one or the other of the four groups. As a matter of fact, the placebo, which is not shown in Table 8.2, occupied the first place in Group 1 with a mean "success rate" of .89, and for the entire class of 120 subjects the mean "success rate" of placebo was .86.

The sequences in which the drugs were administered had apparently no effect on their "success rates". The highest rates of drug A occurred when it was the first and the fourth in order of administration. The highest rate of B was seen when it was third in sequence and of C when it was the third and second of the drugs administered in the respective periods.

In the class of subjects reacting to placebo, the "drugs" variance was not significant, and it was only a fraction of the corresponding variance in the other class of subjects. The variation of the overall response to drugs was, however, significant even among those subjects who did react to placebo. This probably does not reflect a true response to the drugs but rather a difference between two types of sufferers from psychological headaches, namely, an erratic type who wish to impress the physician through the great variations in their condition, and those whose psychological headaches are not complicated by hypochondrasis. These types can be distinguished from the presence or absence of reports by the subjects on extremely minute detail. The subjects reporting such minute detail reported only one third to one half of their headache attacks relieved by placebo as well as by the bona fide analgesics, while the subjects not reporting minute observation as a rule reported complete success with placebo and the analgesics. In addition, there may have been some subjects with true physiological headaches but accessible to suggestion. The net result of these factors is a significant variation of overall susceptibility to drugs among the reactors to placebo.

Banal as it may sound, discrimination among remedies for pain can be made only by subjects who have a pain on which the analgesic action can be tested. The imagined pain, the psychological headache, may be a source of great discomfort to the subject, but it does not form the prerequisite condition for drug discrimination. Through the use of placebo subjects who lack the basis of drug discrimination can be screened out, and the relative potency of drugs can be determined on the subjects in whom the essential condition for discrimination of analgesic action is given.

8.2. CONDITIONS UNDER WHICH THE EXPERIMENT IS CONDUCTED

In order to assure success in an investigation, the conditions must be such as to allow treatment differences to be expressed if, indeed, they are present. The most elaborate and elegant treatment design can be rendered useless if the conditions are such that it is impossible to assess treatment differences. This fact was very much in evidence in the Jellinek (1946) experiment described in the preceding section. In order to compare the effectiveness of headache remedies, it is necessary for subjects to have headaches. Discrimination among pain remedies can be made only by subjects having a pain on which analgesic action can be tested. The screening of subjects by use of a placebo leads to the desired group. Also, if an investigator had obtained only the number of headaches not relieved over a two-week period and if the subjects had had no headaches, all headache remedies would have been equally effective.

In the dandelion experiment in the randomized complete block design discussed in Chapter 7, all herbicides would receive a "perfect score" if there were *no* dandelions present in the lawns at the beginning of the experiment. Likewise, in fumigation and insect spraying experiments, organisms of the specified type *must* be present before it can be determined whether or not the fumigant and the sprays are effective.

As an example of the above, it was desired to learn whether soil fumigation with soil fumigants used on pineapples would be effective in increasing the yields of sugarcane. The use of soil fumigants for pineapples resulted in greatly increased yields of pineapple fruit. The pineapple fields were adjacent to fields on which sugarcane was grown. If the soil fumigant was killing a detrimental organism or was creating soil conditions for better plant growth in the pineapple fields, it was possible that the same result could be obtained on sugarcane. Consequently, it was decided to set up experiments to investigate this possibility; the sugarcane plantation managers offered fields which were high producing ones for use in the experiment. Their contention was that they wanted higher yields from these fields. There is no doubt that this was a desirable motive, but there is considerable doubt that the effectiveness of soil fumigation could be assessed on such fields. Presumably, high producing fields have nothing wrong with them and hence would be useless in assessing the effectiveness of a soil fumigant. Instead, the investigator asked for land which, because of unknown causes, produced the lowest yields. He wanted to conduct his experiment on so-called "sick" fields. Since it was not known what soil fumigation does for the pineapple plant, it was postulated that the highest chance of success for soil fumigation on sugarcane would be on "sick" fields.

In an experiment involving the determination of the nature of the ability of the tomato plant to set fruit shortly after fertilization, two tomato varieties, Porter and Ponderosa, were selected to be grown in Oklahoma. When grown side by side in fields there, Porter set fruit immediately, but Ponderosa did not do so until six weeks later even though both varieties began flowering at the same time. Grown in another environment, Porter and Ponderosa would have set fruit in the same length of time after flowering; it would have been impossible to study the reason for the difference under these conditions. It had been stated that the reason for the differential reaction of the two varieties was physiological, but the investigator found the difference to be genetic. He determined the nature of inheritance of this characteristic and showed that a Ponderosa type variety could be produced which would set fruit immediately under the environmental conditions of Oklahoma.

8.3 EXPERIMENTAL CONDITIONS AS RELATED TO APPLICATION

The conditions of the experiment must be the same, or at least similar, to those to which the treatments are to be applied in practice. To illustrate this point we shall use an example in marketing research. Prior to 1949, when Professor M. E. Brunk and his colleagues started their marketing research investigations at Cornell University, it was common practice for market researchers to set up experiments in which all treatments appeared in the store at the same time. In practice, however, store managers would use only one of the treatments. Under the conditions of the experiment the customer was given a choice of all treatments; the commodity most often preferred by customers was reported to the store managers with the suggestion that they adopt this treatment. The fact that one treatment is preferred over another when the customer is given a choice is no assurance that this treatment will increase sales over the other treatment when the customer has no choice. This illustrates a cardinal point in experimentation involved in selecting a treatment for general use by farmers, teachers, businessmen, etc., and that is to *conduct the experiment under the same conditions that are to be used in practice and for which recommendations are to be made.*

As a second illustration, in biological investigations an experiment conducted in a greenhouse or in a laboratory may not be comparable to an experiment conducted in a farmer's field and under the usual cultural conditions used by a farmer. The ease of obtaining results in a greenhouse, laboratory, clinical, small animal, or other experiment should not be construed to mean that these results are applicable under practical conditions. The results obtained by a trained set of panelists or judges may not be useful for the general public. Professor G. W. Snedecor, formerly of Iowa State University and now deceased, was working with a trained set of butter tasters, some of whom had won international fame. He wondered whether the general public agreed with their tastes. To obtain some information on the question, he took a number of butter samples that had been rated by the "experts" to a meeting of a ladies club. The ladies did not agree with the experts and rated their top choice at the bottom. The experts may have picked the best butter using specified criteria, but this is no guarantee that the general public will agree with them.

Many other examples can be given. Some of our Cornell food experts, for example, had certain criteria for what constituted good applesauce. Their panelists rated the samples according to these criteria. The samples were tried on a sample of the general public that, it appeared, had become used to the strained applesauce fed to babies and ulcer sufferers; of course, they disagreed with the panelists. In addition, when a panel is set up from the general population, the panel itself may become unrepresentative of the population. This is especially true for such items as farm business records, nutritional studies (e.g. on milk), social habits, reading habits, and so forth. When people know that they are under observation, they frequently act quite differently than they do when they do not know it. For example, a person who works with the College Extension Service, keeping farm records, may make different decisions than when he keeps no records. When a person keeps a record of material read, he may automatically and consciously be more selective in his reading material than when no record is kept. As a member of the sample, he may read the *New York Times*. whereas he would read the local newspaper if he were not a member of the panel.

As another example of how people may change or may answer questions, a survey was conducted in a town in Western New York where the researchers had access to records of the total amount of milk sold, where it was sold, and the person to whom it was sold. There were a number of dairies in the town, and they had lists of all their customers as well as of the amounts sold to each customer. The surveyors also knew the amount of milk bought and sold by all the stores in the town. A promotional campaign was launched to determine whether milk usage could be increased. A set of large tumblers with the word milk on the glass was given to each cooperating member. A questionnaire was taken before, during, and after the campaign, to ascertain the amount of milk reportedly consumed by families. It became evident that the mother, then the father, then the grandparents, and then an unrelated person living in the house were increasingly more reliable in reporting milk consumption by the children of the family. Evidently, the mother reported the amount she thought her children should use rather than what they actually used. The father was almost as biased as the mother. The grandparents and an unrelated occupant of the house were much more reliable than the parents in reporting the actual milk consumption of the child. Some mothers reported that they bought from dairy A, which reportedly sold high quality milk, when dairy B has just delivered milk to their door! Also, even though the people interviewed indicated that the milk promotional campaign had been a great success in increasing the consumption of milk, the records indicated that the campaign did not change milk usage in this town. If people are at an optimum usage level as far as they are concerned, a campaign will be useless in changing their habits unless it changes their opinions about optimum quantity. If milk usage is at full capacity, there can be no increase.

8.4. CONTROLS

The necessity of using a control was amply illustrated by the Jellinek experiment. There the control or placebo was necessary to divide the participants of the study into two groups, the group that responds to a placebo and the group that does not. The type of control is also illustrated very well in this experiment, in that the placebo was like the drugs in all external appearances, but was lacking the active ingredient for relieving headaches.

In preceding sections, there were illustrations of the necessity of having a control in determining whether a specified condition existed. This is true in the "squirt and count" type of experiments involving sprays and counting the number of insects remaining alive. If there are no organisms present or if they do no damage, it is necessary that this should be known in order to assess the effectiveness of the sprays being used. The control could be no spray or it could be the commercial or standard spray that is used for this type of material. Unless it is known that organisms are always present in sufficient numbers to damage the crop under question, it is essential that one control involving no spray and one control involving the standard spray should be included. Two such controls allow information on the number of organisms, or the amount of damage present, and on the effectiveness of new sprays as compared to the standard spray.

As an example of the above, sprays were being compared for their effectiveness in controlling damage caused by a specified insect to the potato plant. Damage is measured by the reduction in the weight of potato tubers obtained from a plant. The weights of tubers from 100 sprayed plants and from 100 unsprayed plants were compared. If the weights were the same within sampling variation, then we could say that

1. no damage resulted from not spraying,
2. the act of spraying was detrimental enough to offset the beneficial effect of the spray, or
3. no insects were present.

It was found that the act of running spraying equipment over the ground between potato plants caused a reduction in the weight of tubers because the sharp lugs in the sprayer wheels destroyed roots near the surface. Therefore, it was necessary to include two control treatments, one representing no spray and no machinery run over the ground, and the second representing no spray but in this case the sprayer would be run over the ground in the same manner as for the sprayed plots. With two such controls the investigator can assess the damage caused by running the sprayer over the ground and the effectiveness of the spraying treatment. Note that a spraying treatment could be both detrimental by damaging the plant due to the chemical compounds in the spray and beneficial by killing insects; these two effects could offset each other.

For organoleptic tests (e.g., taste tests, odor tests, etc.), it is necessary that a standard or control treatment should be included in order to compare new treatments with a standard. Likewise, for fertilizer trials a treatment with no fertilizer and perhaps a treatment involving the standard fertilizer treatment are usually necessary. In variety trials one or more standard varieties are included as controls for the new varieties in the experiment. A standard laboratory technique needs to be included in an experiment involving a comparison of new laboratory techniques. The inclusion of appropriate controls in medical and educational experiments is vital for their successful conclusion. The type of controls to be used often requires considerable thought in order to make the correct comparisons. In most cases, the best medical treatment and educational practice should be one of the controls.

The many diverse examples listed above indicate the need for and beneficial effects from the inclusion of a control, a *point of reference,* in the experiment. Without this point of reference, the experiment may turn out to be only an experience rather than a procedure for obtaining factual and useful information. In some experiments, the problem of establishing standards can become a statistical one, as was the one described by Youden, Connor, and Severo (1959); their study resulted from an inquiry from the United States Geological Survey concerning the choice of standards for estimating the amount of uranium in soil samples.

8.5. SINGLE FACTOR WITH SEVERAL LEVELS

The first item to be considered in single factor investigations is the factor or variable to be investigated. For example, let us suppose that we are interested in investigating the effect of inorganic chemicals on plant growth over the life period of the plant. Since many inorganic chemicals are involved, we will select one of them, say nitrogen, in the nitrate form, NO_3. Thus, we need select not only the factor, say nitrogen, and the amounts or levels to be used, but also the form NO_3 of the factor to be used; quite different results may be obtained with a different form, say from NH_3 (ammonia). The same amount of nitrogen could be used for the two forms, but the effect on plant response or growth could be quite different for the two forms of nitrogen. Another example would be to have three, say, forms of administering medicine. The three forms could be vaccination, oral, or intravenous.

After selecting the inorganic chemical and its form, we need to determine the range over which nitrogen will be studied. Suppose that we decide to use a sand culture experiment. Should we start at zero nitrate nitrogen in the sand culture? What should be the upper limit of amount of nitrate nitrogen used? The answers to these questions depend upon the objectives of the experiment, the type of plant used, and the stage of growth of the plant. Let us suppose that we are using the celery plant, that cuttings are to be used, and that growth measured as increase in weight will be studied over a 12-week period. We may decide to use a zero addition of nitrate nitrogen as the lowest level and to use 500 gms of nitrate nitrogen per ten kilograms of other material as the highest level. The other material will contain identical amounts of sand, water, and all other elements considered necessary for plant growth. Sufficient amounts of all other elements will be included. The only item that varies will be amount of nitrate nitrogen applied. Do we want the complete response picture or curve over the entire range? If we do, then we utilize this range; if not then we either shorten the range or lengthen it to obtain the desired coverage.

After the range has been established, we have to determine the number of levels to be used. If we know the form of the response curve and the range, then we can say something definite about the choice of levels. For example, let us suppose that the response is linear, that is,

$$\text{response} = Y_i = a + bN_i + \text{error}$$

where N_i is the amount of nitrate nitrogen applied in grams per 10 kilograms of other material. This may be pictured as shown in Figure 8.1. If the response is a straight line, we need only the zero application and the upper level, say 500 grams. These two different levels of the amount applied would be sufficient to estimate the linear plant response to amount of nitrate nitrogen. Likewise, if we know that $\log Y_i = a + bN_i + \text{error}$, or $Y_i = a + b \log (N_i+1) + \text{error}$, is the form of the response, we need only two levels, the lowest and the highest, in order to estimate the two constants a, the intercept, and b, the slope, of the response function.

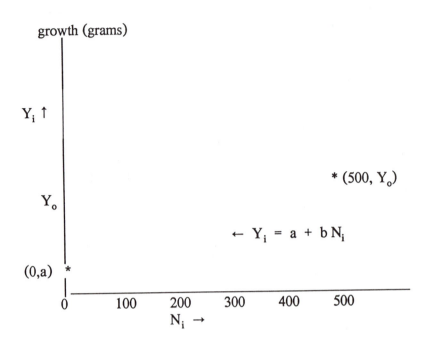

growth (grams)

Figure 8.1. Grams of nitrate nitrogen (N_i) applied versus growth.

If the response is curved, then more than two levels of nitrate nitrogen may be required. Suppose that the response is of the form $Y_i = a + bN_i + cN^2_i +$ error. In this case the response may be similar to one of the forms shown in Figure 8.3. In order to estimate the constants a, b, and c, we must use three or more levels of nitrate nitrogen. One procedure would be to put 4 / 9 of our observations at each end of the range and 1 / 9 in the middle. Given that the curve is second degree, or quadratic, the only need for observations in the middle is to determine whether c is negative or positive. The reason for a higher proportion of observations at the ends of the range is to tie down the ends of the curve.

The design of the number or proportion of observations at each point is the subject of past and current statistical research. The problem of efficiently allocating proportions of observations to the various levels of the variable becomes increasingly difficult as the degree, k, of the polynomial,

$$Y = a + \sum_{j=1 \text{ to } k} b_j N^j + \text{error}$$

$$= a + b_1 N + b_2 N^2 + b_3 N^3 + ... + b_k N^k + \text{error}$$

increases. However, an equal allocation of proportions of observations at equally spaced levels should be followed if the degree, k, of the polynomial is unknown. Also, an equal allocation of proportions and spacings may not be as efficient as some other procedure, but information is still obtainable just as it was in the first weighing experiment discussed in Section 7.1 of the previous chapter.

So far we have been concerned with selecting levels at any point that we wish. In certain cases, however, the levels will not be determined by the investigator but may appear at random or be obtained as a result of other experimental conditions. As long as the level of the factor can be measured or determined and the error of determining the level is small compared to the range of the levels, the procedures discussed above all apply. For example, it would be simple to pre-select levels of nitrate nitrogen applied to a greenhouse potted plant or to a group of plants in a four-meters-square plot of ground, *but* it would be difficult if not impossible to do this with groups of *plants* having different pre-specified levels of nitrate nitrogen. The mere fact that nitrate nitrogen is added to a plot of ground is no assurance that the plants will absorb this chemical in the proportion in which it is applied. *Perhaps* fertilizer studies should be conducted in terms of the amount of the chemical found in the plant rather than in terms of the amount that is applied to the ground.

An example of the functional form W_i = dry weight of a chick embryo at ages 6 to 16 days = $A e^{bX_i}$, where X_i represents age in days, A and b are constants computed from data, and e is the base of the Naperian system of logarithms, is given by Snedecor (1946), page 376, and it is reproduced in Table 8.3 and Figure 8.2 (Reprinted by permission from *Statistical Methods*, 4th edition, by G. W. Snedecor, 1950 by The Iowa State University Press).

The plot of the data and of the response function W_i = $0.002046(1.57)W_i$ indicates quite close agreement of the data with the functional form. As another way of looking at the data the form $\log_{10} W_i$

Table 8.3. Dry weights of chick embryos from ages 6 to 16 days, together with common logarithms

Age in Days X	Dry Weight, W (grams)	Common Logarithm of Weight Y
6	0.029	-1.538*
7	0.052	-1.284
8	0.079	-1.102
9	0.125	-0.903
10	0.181	-0.742
11	0.261	-0.583
12	0.425	-0.372
13	0.738	-0.132
14	1.130	0.053
15	1.882	0.275
16	2.812	0.449

* From the table of logarithms, one reads $\log 0.029 = \log 2.9(10^{-2}) = \log 2.9 + \log 10^{-2} = 0.462 - 2 = -1.538$

$= \log_{10} 0.002046 + (\log_{10} 1.57)X_i = -2.689 + 0.1959X_i$.

There are many mathematical forms possible for an unknown response function. The investigator should try to obtain or to postulate the appropriate form prior to experimentation and then select an appropriate treatment design for the experiment. The emphasis in statistics textbooks is on the polynomial form, but this is not necessarily the most universal response function in real life. The polynomial form represents a rather easy and mathematically tractable form. Since this problem was solved before others, the results were the first to be incorporated into statistical writings, resulting, perhaps, in the emphasis placed on the polynomial response function.

In the main, we have briefly illustrated that

    1. the variable to be investigated must be selected,
    2. the range of levels of the variable must be determined,
    3. the levels must be selected, and

    4. the proportion of observations at each level of the variable of
    interest must be determined.

In general one could use equal spacings of levels of the variable and equal
proportions of observations at each level, but a more efficient treatment
design usually results if a higher proportion of observations are placed at
the end-points of the range and the remaining observations at points of

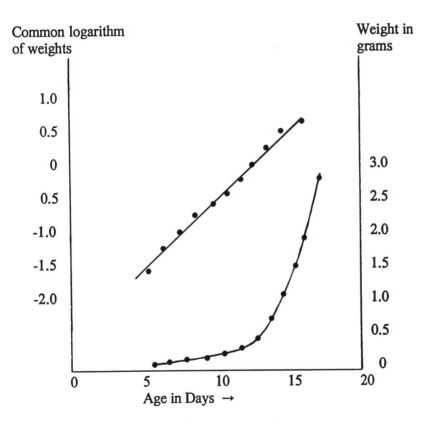

Uniform scale: W = 0.002046 (1.57)^X
Logarithmic scale: Y = 0.1959X - 2.689

Figure 8.2. Dry weights of chick embryos at ages 6 to 16 days with fitted
curves.

maxima and minima which are peculiar to the particular form of the response function being studied.

8.6 TREATMENT DESIGNS INVOLVING TWO OR MORE FACTORS AND THE FACTORIAL TREATMENT DESIGN

The experimenter may wish to study two or more variables jointly to observe the manner in which the response varies with the changing levels of the variables under study. Suppose that it is desired to study the effect on plant response of varying amounts of nitrogen and phosphorous. If the levels are selected, the same comments as discussed in the preceding section for a single variable apply here. Whether the levels are pre-selected or not, a level of each of the variables under study will be associated with each response observation. For example, N_1 and P_1 will be associated with Y_1, N_2 and P_2 with Y_2, N_3 and P_3 with Y_3, and so forth. The form of the response function will involve both variables and might be one of the following forms where N_i = i^{th} level of nitrogen, P_i = i^{th} level of phosphorous, and e_i is the error term associated with the ith observation:

$$Y_i = a + bN_i + cP_i + e_i$$

$$\log Y_i = a + bN_i + cP_i + e_i$$

$$\log Y_i = a + b \log N_i + c \log P_i + e_i$$

$$Y_i = a + bN_i + cP_i + dN_i^2 + fP_i^2 + gN_iP_i + e_i$$

$$Y_i = a\,N_i^b\,P_i^c\,e^{dN_i + fP_i + gN_iP_i} + e_i$$

$$Y_i = a\,N_i^b\,e^{cN_i} + d\,P_i^f\,e^{gP_i} + e_i\;.$$

In the last two equations, e is the base for natural logarithms, whereas e_i is an error deviation. The response in the last equation would be the sum of two exponential forms.

From the above, we note that the response in the experiment could be of various forms. From the laws and theories evolved for the subject matter in the area where the investigation is being conducted, it is often possible

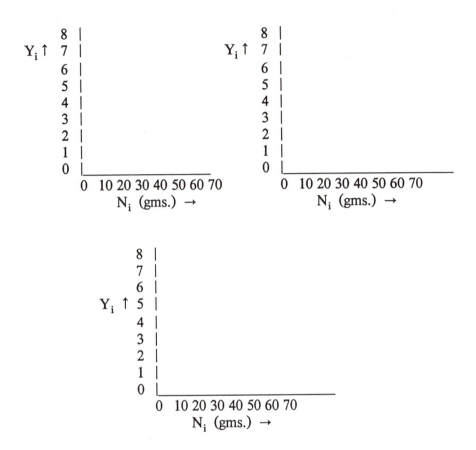

Figure 8.3. Some forms of response.

to eliminate one or more of the above forms and to choose a small subset of plausible mathematical forms. Sometimes it is possible to pick the exact form for the response function. Many examples of this occur in the physical and biological sciences. For example, the exponential growth law is written as $Y = a e^{bt}$ = weight at time t_i; the rate of increase of growth at a specified time, say t_0, is $dY / dt = ab e^{bt}$ evaluated at $t = t_0$; Mendel's law of segregation has the form of the binomial distribution, and so on.

When the form, or even an approximate form, is known, it should be utilized. When the mathematical function describing the phenomenon under observation becomes established beyond reasonable doubt then a law has been established; the goal of the empirical investigation has been achieved.

An example is given in Table 8.4; the nature of the measurements from various parts of the plant obtained from an Hawaiian Sugar Planters' Association sugarcane experiment in Hawaii is illustrated.

Table 8.4 Measurements obtained from a sugarcane experiment.

Observation number	Sugar (tons/acre)[1]	Total nitrogen in base	Total potash in 8-10th node	Total phosphorous in 8-10th node
1	$Y_{11} = 5.0$	.0568	.554	.0723
2	$Y_{12} = 4.3$	.0621	.608	.0802
3	$Y_{13} = 4.6$	.0528	.653	.650
4	$Y_{21} = 10.0$	.1143	.368	.0317
5	$Y_{22} = 9.1$	.1557	.463	.0322
6	$Y_{23} = 9.8$	.1359	.576	.0299
7	$Y_{31} = 12.3$	.2443	.190	.0184
8	$Y_{32} = 11.6$	.2738	.324	.0187
9	$Y_{33} = 14.3$	.2790	.473	.0258

[1] The first subscript of Y_{ij} refers to level of nitrogen application and the second refers to level of potash application.

As a second example, consider a set of hypothetical data where Y equals grade-point average for four years at S-W University for the 20 most recent graduates from fraternity øπð, X_1 equals high school grade average, and X_2 equals aptitude test scores as given in Table 8.3.

Suppose that it has been decided to utilize the data for the 20 most recent graduates of the fraternity to determine how well one could predict a four-year grade-point average from high school standing and aptitude test scores. Suppose that the fraternity "research" team consulted with a statistician who obtained for them a least squares regression equation of the form:

$$Y^*_i = a + b_{Y1.2} X_{1i} + b_{Y2.1} X_{2i} = -3.52 + 0.085 X_{1i} - 0.0014 X_{2i}$$

(He actually used more significant figures to compute Y^*_i above.) For example, the first estimated grade-point average was computed as $-3.52 + 0.085(75) - 0.0014(480) = 2.2 = Y^*_i$. From the equation, we note

Table 8.5. Hypothetical data for 20 students of fraternity øπ∂.

Student number	High School grade = X_1 in %	Number of points for aptitude test = X_2	4-year grade-point average = Y	Estimated grade-point aver. = Y*	Error of estimate = Y - Y*
1	75	480	2.0	2.2	-0.2
2	99	700	4.1	4.0	0.1
3	86	680	3.5	2.9	0.6
4	97	750	4.1	3.7	0.4
5	86	550	3.5	3.0	0.5
6	97	750	4.1	3.7	0.4
7	80	450	2.8	2.7	0.1
8	89	650	3.5	3.2	0.3
9	91	700	3.5	3.3	0.2
10	97	760	3.0	3.7	-0.7
11	94	700	2.8	3.5	-0.7
12	96	750	4.0	3.6	0.4
13	96	720	3.9	3.7	0.2
14	89	700	3.3	3.1	0.2
15	94	700	3.3	3.5	-0.2
16	93	720	3.1	3.4	-0.3
17	78	400	2.4	2.6	-0.2
18	60	350	1.0	1.1	-0.1
19	91	700	2.8	3.3	-0.5
20	97	780	3.5	3.7	-0.2

that an increase of one percent point in high school grade increases the four-year grade-point average by 0.085 grade-point for constant aptitude test scores, whereas an increase of one point in the aptitude test score decreases the four-year grade-point average by 0.0014 point when X_1 is held constant. Both X_1 and X_2 exhibit a positive relationship with Y when considered individually, but X_2 exhibits a negative effect when the effect of X_1 is removed or held constant. For example, consider the two students with a high-school grade of 96; here, the one with the higher aptitude test score total had a lower estimated average than the one who had the lower score (3.9 *vs.* 4.0). If these data were to be used to predict the performance of "pledges", the pledgee would be well advised to obtain as high a high school grade and as low a test score as possible! The truth of the matter would be the actual grade-point average obtained by a pledge. Many factors, not just the above two, enter into accounting for a student's performance in college.

Although it would have been possible to study pre-selected levels for the above set of data, it would have been difficult to obtain the prescribed levels of the variables of interest. For example, suppose it is desired to study levels of X_1 equal to 60, 70, 80, 90, and 100, and levels of X_2 equal to 500, 550, 600, 650, 700, and 750. We may note that there is no combination of X_1 and X_2 for any of these values in the preceding set of data. For our study and for most similar situations, it is not necessary to have pre-selected levels of X_1 and X_2. However, it is necessary that the range of levels of X_1 and X_2 cover the range desired by the experimenter. For the above combinations, a very large sample would be required in order to find such combinations and then the desired combinations must be selected from this sample, a time-consuming and data-wasting procedure.

Although the difficulties of obtaining selected levels for the above examples are apparent, we should emphasize that there are many types of investigation that allow selection of levels of the variables of interest to the investigator. The variables in a study are often denoted as **factors**. For example, suppose that the experimenter is investigating growth of celery plants in sand culture experiments where all the items except the two factors nitrate nitrogen, NO_3, and potash in the form of K_2O are held constant. Various amounts by weights of NO_3 and K_2O will be added to the experimental units. Suppose that it is decided to use the following levels of NO_3 and K_2O:

factor one: levels of NO_3: 10, 20, 40, 80 grams
factor two: levels of K_2O: 13, 39, 65, 91 grams

Now the question arises as to what combinations of the various levels of the two factors should be used in the investigation. The following combinations could be used:

Levels of K$_2$0 (grams)	Levels of NO$_3$ (grams)			
	10 = 0	20 = 1	40 = 2	80 = 3
13 = 0	$n_0k_0 = 00$	$n_1k_0 = 10$	$n_2k_0 = 20$	$n_3k_0 = 30$
39 = 1	$n_0k_1 = 01$	$n_1k_1 = 11$	$n_2k_1 = 21$	$n_3k_1 = 31$
65 = 2	$n_0k_2 = 02$	$n_1k_2 = 12$	$n_2k_2 = 22$	$n_3k_2 = 32$
91 = 3	$n_0k_3 = 03$	$n_1k_3 = 13$	$n_2k_3 = 23$	$n_3k_3 = 33$

(Note: Instead of using 0,1,2,3 to designate levels, 1,2,3,4 could have been used. This is an arbitrary convention.)

The combination of the ith level of NO_3 and of the jth level of K_2O is denoted as $n_ik_j = ij$. Once the order of the subscript has been defined, the subscript ij is sufficient to define the treatment corresponding to the combination n_ik_j of the two factors NO_3 and K_2O. The observation (yields or weights) for the ij^{th} combination of the two factors would be Y_{ij} as denoted in the following table:

Level of K$_2$0 (grams)	Levels of NO$_3$ (grams)				Sum	Mean
	0	1	2	3		
	(weight or yield in grams)					
0	Y_{00}	Y_{10}	Y_{20}	Y_{30}	$Y_{.0}$	$\bar{y}_{.0}$
1	Y_{01}	Y_{11}	Y_{21}	Y_{31}	$Y_{.1}$	$\bar{y}_{.1}$
2	Y_{02}	Y_{12}	Y_{22}	Y_{32}	$Y_{.2}$	$\bar{y}_{.2}$
3	Y_{03}	Y_{13}	Y_{23}	Y_{33}	$Y_{.3}$	$\bar{y}_{.3}$
Sum	$Y_{0.}$	$Y_{1.}$	$Y_{2.}$	$Y_{3.}$	$Y_{..}$	
Mean	$\bar{y}_{0.}$	$\bar{y}_{1.}$	$\bar{y}_{2.}$	$\bar{y}_{3.}$		$\bar{y}_{..}$

The treatment design for the two factors described above is known as a **factorial treatment design** which is a design that contains all combinations of two or more levels of two or more factors. From such a treatment design, with each combination repeated an equal number of times, we find that the arithmetic means may be used to estimate the various effects. For example, the estimated effects for each of the levels of NO_3 and of K_2O are:

Effect of 10 gms of NO_3 = 0 level of NO_3 : $\bar{y}_{0.} - \bar{y}_{..}$

      ”   ”   20   ”   ”   ”   = 1   ”   ”   ”   : $\bar{y}_{1.} - \bar{y}_{..}$

      ”   ”   40   ”   ”   ”   = 2   ”   ”   ”   : $\bar{y}_{2.} - \bar{y}_{..}$

      ”   ”   80   ”   ”   ”   = 3   ”   ”   ”   : $\bar{y}_{3.} - \bar{y}_{..}$

Effect of 13 gms of K_2O = 0 level of K_2O : $\bar{y}_{.0} - \bar{y}_{..}$

      ”   ”   39   ”   ”   ”   = 1   ”   ”   ”   : $\bar{y}_{.1} - \bar{y}_{..}$

      ”   ”   65   ”   ”   ”   = 2   ”   ”   ”   : $\bar{y}_{.2} - \bar{y}_{..}$

      ”   ”   91   ”   ”   ”   = 3   ”   ”   ”   : $\bar{y}_{.3} - \bar{y}_{..}$

This type of effect, or of linear combinations of these, has been denoted as a **main effect** or **one-factor effect** in a factorial experiment. In biological literature, these are often called **interactions** but should more properly be called **one-factor interactions**.

In addition to main effects, this two-factor factorial treatment design contains another type of effect, *viz.* **two-factor interaction effects** which are effects peculiar to a mutually beneficial or mutually detrimental effect of a particular combination. The interaction effect for the ith level of NO_3 with the jth level of K_2O is computed as

$$Y_{ij} - \bar{y}_{i.} - \bar{y}_{.j} + \bar{y}_{..} = Y_{ij} - \bar{y}_{..} - (\bar{y}_{i.} - \bar{y}_{..}) - (\bar{y}_{.j} - \bar{y}_{..})$$

If the above quantity is zero, the two-factor interaction effects would be zero, and $Y_{ij} = \bar{y}_{i.} + \bar{y}_{.j} - \bar{y}_{..}$. Graphically we may represent the zero two-factor interaction case as shown in Figure 8.4.

Alternatively, levels of K_2O may be plotted on the abscissa as shown in Figure 8.5. (Note that 20 grams and 40 grams of NO_3 gave identical values for all levels of K_2O in the former graph. Hence the points would be identical on the latter graph. This is an artificial example.) In the case of zero two-factor interaction, *the distance between any two curves remains constant throughout all levels of the factor represented on the abscissa.*

In an actual experiment, there will be sampling variation, and a zero two-factor interaction will not usually be observed even if the population interaction effects are truly zero. There are statistical procedures for setting bounds on sampling variations and to determine whether or not the

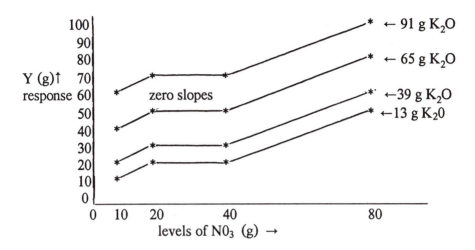

Figure 8.4 Representation of a zero two-factor interaction.

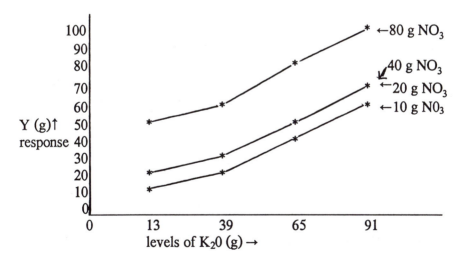

Figure 8.5. Alternative representation of zero two-factor interaction.

effects may reasonably be expected to fall between a designated set of values. Some of these procedures will be considered later under methods of summarizing data from experiments and surveys.

If the purpose of an experiment utilizing a factorial treatment design is to estimate main effects and interactions, then it has been shown (Kempthorne, 1952, page 425) that the factorial treatment design is optimum, that is, no other selection of treatments does this more effectively than a factorial with equal numbers of observations on each treatment. If the experimenter is given the particular form of the response function, then it is often possible to select a treatment design which is more efficient than the factorial.

The factorial treatment design listed above is called 4 levels of NO_3 × 4 levels of K_2O factorial, or alternatively, a $4 \times 4 = 4^2$ factorial of amounts of NO_3 and K_2O. In the term 4^2, the four refers to the number of levels, and the superscript 2 refers to the number of factors. For example, a 4^3 factorial would represent 4 levels of three different factors in all combinations. In general, we use symbols such as s^n to denote that there are s levels each of the n different factors. In the form 4×4, the first four represents the number of levels of the first factor and the second four those in the second factor. This form is convenient when the number of levels of the factors is not the same. For example, suppose that only the first three levels of NO_3 had been used above. Then the factorial would have been a 3 × 4 = 3 levels of NO_3 × 4 levels of K_2O factorial treatment design. Likewise, a p × q × r factorial would represent a factorial with p levels of factor one × q levels of factor two × r levels of factor three.

The 4×4 factorial design was selected to introduce the idea of factorials. There are smaller factorials, such as 2 × 2, 2 × 3, 2 × 4, 3 × 3, and 3 × 4, but the 4×4 is large enough to illustrate several ideas which could not be illustrated as well with some of the smaller factorials. In practice, however, these smaller factorials are extensively used. In fact, the 2^n (n factors each at 2 levels) is a very popular factorial series. Statisticians also like the 2^n series because of certain mathematical group theoretic properties.

As an example of a 2 × 2 factorial treatment design, let us suppose that we are interested in comparing two types of rubber, that obtained from the hevea rubber tree $= r_0$ and that produced synthetically $= r_1$, manufactured by two different processes p_0 and p_1. The yield characteristic is number of miles a truck may be driven until only 1/4 inch of the tread remains on the tire. Suppose that a latin square design is used on each of several trucks possessing eight sets of dual wheels with the rows of four tires corresponding to the rows of the latin square and the position of the tires from left to right corresponding to the columns. Suppose that the data in thousands of miles for the four treatments are of the form:

	Manufacturing process (miles / 1,000)			
Type of rubber	p_0	p_1	Sum	Mean = $\bar{y}_{...}$
hevea = r_0	$\bar{y}_{00..} = 15.1$	$\bar{y}_{01..} = 19.5$	34.6	17.3
synthetic = r_1	$\bar{y}_{10..} = 16.7$	$\bar{y}_{11..} = 14.3$	31.0	15.5
sum	31.8	33.8	65.6	-
mean = $\bar{y}_{.j..}$	15.9	16.9	-	$16.4 = \bar{y}_{....}$

rubber main effects: $r_0 = \bar{y}_{0...} - \bar{y}_{....} = 17.3 - 16.4 = 0.9$

$r_1 = \bar{y}_{1...} - \bar{y}_{....} = 15.5 - 16.4 = -0.9$

process main effects: $p_0 = \bar{y}_{.0..} - \bar{y}_{....} = 15.9 - 16.4 = -0.5$

$p_1 = \bar{y}_{.1..} - \bar{y}_{....} = 16.9 - 16.4 = 0.5$

interaction effects: $rp_{00} = \bar{y}_{00..} - \bar{y}_{0...} - \bar{y}_{.0..} + \bar{y}_{....} = 15.1 - 17.3$
$- 15.9 + 16.4 = -1.7$

$rp_{01} = \bar{y}_{01..} - \bar{y}_{0...} - \bar{y}_{.1..} + \bar{y}_{....} = 19.5 - 17.3$
$- 16.9 + 16.4 = 1.7$

$rp_{10} = \bar{y}_{10..} - \bar{y}_{1...} - \bar{y}_{.0..} + \bar{y}_{....} = 16.7 - 15.5$
$- 15.9 + 16.4 = 1.7$

$rp_{11} = \bar{y}_{11..} - \bar{y}_{1...} - \bar{y}_{.1..} + \bar{y}_{...,} = 14.3 - 15.5$
$- 16.9 + 16.4 = -1.7$

Here we note that the sum of any set of main effects is zero (within rounding errors) and that the sum of any set of interaction effects for any given level of a specified factor is also zero (within rounding errors). The form of the calculations is the same as for the randomized complete block design since the blocks and the treatments correspond to the two factors and the interaction corresponds to what was called errors. Thus, the interaction may be thought of as discrepance from the additive main effects situations, that is, $\bar{y}_{ij..} - \bar{y}_{....} - r_i - p_j = ij^{th}$ mean minus the overall mean minus the ith rubber effect minus the jth process effect equals the ij[th] interaction effect.

As an example of a 2 × 3 factorial treatment design, suppose that we are studying time spent watching television as compared to time spent in reading plus time spent on civic activities (for example, cub scouts, girl scouts, Red Cross, church visitations, committee for shut-ins, Gray-Ladies at hospital, etc.) for three broad salary categories of married women with a family of four in the 20 to 40 age bracket. Suppose that our population has been stratified into six strata as described above and that a simple random sample of women is selected from each stratum. Suppose that the following data are obtained:

	Total income for a family of size four				
	t_0	t_1	t_2		
Activity	<$6,000	$6-12,000	$12-18,000	Sum	Mean = $\bar{y}_{i..}$
	(hours spent per day)				
Watching TV $= a_1$	$\bar{y}_{10.}=8.25$	$\bar{y}_{11.}=3.50$	$\bar{y}_{12.}=1.75$	13.50	4.50
Reading and civic organizations $= a_2$	$\bar{y}_{20.}=0.25$	$\bar{y}_{21.}=4.50$	$\bar{y}_{22.}=6.25$	11.00	3.67
Sum	8.50	8.00	8.00	24.50	-
Mean = $\bar{y}_{.j.}$	4.25	4.00	4.00	-	$4.08 = \bar{y}_{...}$

activity main effects: $a_1 = \bar{y}_{1..} - \bar{y}_{...} = 4.50 - 4.08 = 0.42$ hrs.

$\quad\quad\quad\quad\quad\quad\quad a_2 = \bar{y}_{2..} - \bar{y}_{...} = 3.67 - 4.08 = -0.41$ hrs.

income main effects: $t_0 = \bar{y}_{.0.} - \bar{y}_{...} = 4.25 - 4.08 = 0.17$ hrs.

$\quad\quad\quad\quad\quad\quad\quad t_1 = \bar{y}_{.1.} - \bar{y}_{...} = 4.00 - 4.08 = -0.08$ hrs.

$\quad\quad\quad\quad\quad\quad\quad t_2 = \bar{y}_{.2.} - \bar{y}_{...} = 4.00 - 4.08 = -0.08$ hrs.

interaction effects: $at_{10} = \bar{y}_{10.} - \bar{y}_{1..} - \bar{y}_{.0.} + \bar{y}_{...} = 8.25 - 4.50$

$\quad\quad\quad\quad\quad\quad\quad\quad\quad\quad\quad - 4.25 + 4.08 = 3.58$ hrs.

$\quad\quad\quad\quad\quad at_{11} = \bar{y}_{11.} - \bar{y}_{1..} - \bar{y}_{.1.} + \bar{y}_{...} = 3.50 - 4.50$

$\quad\quad\quad\quad\quad\quad\quad\quad\quad\quad\quad - 4.00 + 4.08 = -0.92$ hrs.

$\quad\quad\quad\quad\quad at_{12} = \bar{y}_{12.} - \bar{y}_{1..} - \bar{y}_{.2.} + \bar{y}_{...} = 1.75 - 4.50$

$\quad\quad\quad\quad\quad\quad\quad\quad\quad\quad\quad - 4.00 + 4.08 = -2.67$ hrs.

$\quad\quad\quad\quad\quad at_{20} = \bar{y}_{20.} - \bar{y}_{2..} - \bar{y}_{.0.} + \bar{y}_{...} = 0.25 - 3.67$

$\quad\quad\quad\quad\quad\quad\quad\quad\quad\quad\quad - 4.25 + 4.08 = -3.59$ hrs.

$$at_{21} = \bar{y}_{21.} - \bar{y}_{2..} - \bar{y}_{.1.} + \bar{y}_{...} = 4.50 - 3.67$$
$$- 4.00 + 4.08 = 0.91 \text{ hrs.}$$
$$at_{22} = \bar{y}_{22.} - \bar{y}_{2..} - \bar{y}_{.2.} + \bar{y}_{...} = 6.25 - 3.67$$
$$- 4.00 + 4.08 = 2.66 \text{ hrs.}$$

The sums of the various effects add to zero within rounding errors, that is, if the means had been carried out to more decimals the sum would be much closer to zero. The above means may be presented graphically as shown in Figure 8.6. In this hypothetical example, we note that the main effects for income category are very small; the activity main effects are not much larger, and the interaction effects are relatively large.

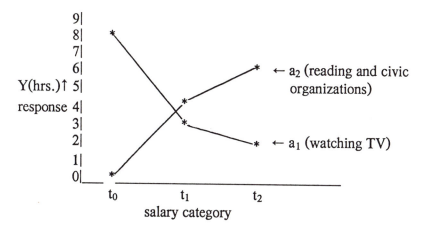

Figure 8.6. Graphical representation of salary versus hours of activity.

As a third example, consider the following data in tons per acre representing the total weight of sugarcane harvested from an experimental unit. Three sugarcane varieties (v_0, v_1, and v_2) were used in all combinations with three levels of nitrogen (150 lbs./A = n_0, 210 lbs./A = n_1, and 270 lbs./A = n_2) resulting in a 3 × 3 factorial treatment design. The treatment means in tons per acre for a sugarcane experiment are presented below:

Variety	Nitrogen			Mean
	n_0	n_1	n_2	
v_0	66.5	69.0	76.0	70.5
v_1	61.4	62.6	70.4	64.8
v_2	68.6	64.5	57.9	63.7
Mean	65.5	65.4	68.1	66.3

Graphically these results are represented in Figure 8.7.

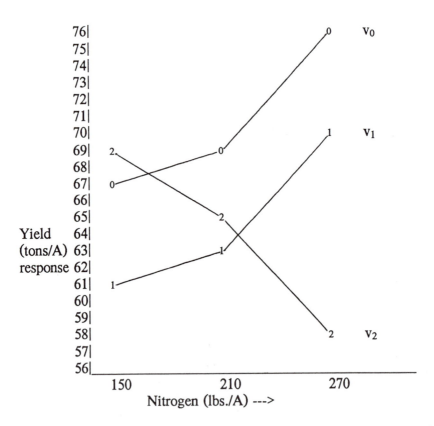

Figure 8.7. Graphical representation of relationship of nitrogen to yield of sugarcane.

A pronounced variety by nitrogen level interaction is present in these data. Varieties v_0 and v_1 respond as might be expected from the effect of additional amounts of nitrogen fertilizer. However, variety v_2 does not behave as one would expect. Does additional nitrogen actually decrease yield? Or is the relationship due to other causes? The investigator found that variety v_2 was susceptible to a disease known as "red hot", so named because of the color of the internodal material of plants affected with the disease. The incidence of the disease also increases with increased vigor and lushness of the plants. The addition of nitrogen caused the young plants to grow fast, and the incidence of the disease increased, resulting in more dead plants at harvest time for the plots receiving the higher amounts of nitrogen. The increased number of dead plants decreased yields. For this variety, life is just a vicious circle! If it tries to produce a lot of sugar, it is highly infected with the "red hot" disease!

As the last example, let us consider the 2^3 factorial treatment design. The illustration used here is taken from Example VII-2 of Federer (1955). The following three factors each at two levels in all combinations were utilized in an engineering education study on the use, preparation and effectiveness of stereophotographic aids. The data and treatments here are for the non-stereophotographs. The treatments are:

Photo and question set = factor b	Class = factor c			
	Freshman = c_0		Senior = c_1	
	Exposure = factor a		Exposure = factor a	
	Left = a_0	Right = a_1	Left = a_0	Right = a_1
b_0	$a_0b_0c_0$	$a_1b_0c_0$	$a_0b_0c_1$	$a_1b_0c_1$
b_1	$a_0b_1c_0$	$a_1b_1c_0$	$a_0b_1c_1$	$a_1b_1c_1$

In order to obtain a valid comparison of non-stereo with stereo pictures, an exposure was made a short distance to the left = a_0 and a short distance to the right = a_1 of the spot where the stereo picture was made. There were two sets of pictures with a different question set associated with each group of pictures; b_0 represented set I photos and the associated set of 27 questions and b_1 represented set II photos and the associated set of 18 questions. The freshman group of boys is represented by c_0 and the senior group of boys by c_1. Six freshmen boys out of 24 were randomly allocated to each of the $a_ib_jc_0$ combinations and six senior students out of 24 seniors were randomly allotted to each of the $a_ib_jc_1$ treatments. Each of the 48 boys was given a question and photo set which was one of the combinations $a_ib_jc_h$. The number of correct (R) and incorrect (W) answers is given below:

Treatment

000		100		010		110		001		101		001		111	
R	W	R	W	R	W	R	W	R	W	R	W	R	W	R	W
10	17	10	17	2	16	10	8	6	21	4	23	6	12	9	9
10	17	6	21	1	17	0	18	5	22	8	19	11	7	5	13
6	21	10	17	5	13	4	14	7	20	9	18	5	13	11	7
12	15	5	22	10	8	4	14	12	15	11	16	7	11	8	10
8	19	2	25	8	10	6	12	11	16	6	21	2	16	8	10
5	22	7	20	3	15	6	12	8	19	5	22	6	12	9	9
51/162		40/162		29/108		30/108		49/162		43/162		37/108		50/108	
.315		.247		.269		.278		.302		.265		.343		.463	

In the investigation more correct answers were obtained by *every* student on the stereophotos than on the non-stereophotos. The differences among category means for all other factors were relatively small, indicating that these factors had little if any effect on the scores of the different boys in the investigation.

8.7 FRACTIONAL REPLICATION IN FACTORIALS

A **fractional replicate** of a factorial treatment design is merely a subset of the complete factorial. For example, if the four treatments 000, 110, 101, and 011 were the only treatments included in an experiment, this would constitute a one-half replicate of a 2^3 factorial. The treatments not included form another one-half replicate and are 100, 010, 001, and 111. The two one-half replicates constitute the complete 2^3 factorial.

A 5 / 9 fraction of a 3^2 factorial that could be used would be the five combinations 00, 01, 12, 10, and 21 and a 2 / 3 fraction could be the six combinations 00, 01, 12, 10, 21, and 22. An optimal main effect plan for a 1 / 9 fraction of a 3^4 would consist of the nine combinations 0000, 0111, 0222, 1012, 1120, 1201, 2021, 2102, and 2210.

For a second type of fractional replicate, suppose that the following numbers of treatments are selected from a 3^2 factorial

	a_0	a_1	a_2
b_0	1	0	1
b_1	0	n	0
b_2	1	0	1

The above fractional replicate of a 3^2 factorial is one of a class of treatment designs known as a **simple response surface treatment design** where the 1 represents one response for that combination, the zero represents no observations recorded, and n means n observations taken for a_1b_1. Likewise, the following fractional replicate of a 5^2 factorial is a simple response surface design (plus sign +) plus four star points (asterisk *) with the blanks indicating no observation:

	a_0	a_1	a_2	a_3	a_4
b_0			*		
b_1		+		+	
b_2	*		n		*
b_3		+		+	
b_4			*		

Thus, there are n center points, four response surface points (pluses), and four star points (asterisks) in the above fractional replicate.

There are numerous types of fractional replicates, and an even more numerous list of published papers on the subject. The methods of constructing fractional replicates, the actual construction, the analyses for fractional replicates, and properties of various fractional replicates have been and are being actively investigated by statisticians. There are many unsolved problems associated with fractional replication.

Fractional replicate treatment designs are useful in specific instances. The fractional replicate is sometimes useful when the investigator wishes to study a large number of factors with each factor at several levels. For example, large factorials such as the 2^{10} with 1024 treatments and 3^7 with 2187 treatments may involve too many treatments for the experimenter to study. He may decide that the number of factors and/or the number of levels cannot be reduced. The alternative is to use a fractional replicate; for instance, a 1 / 8 replicate of a 2^{10} results in $2^7 = 128$ treatments and a 1 / 9 replicate of a 3^7 results in $3^5 = 243$ treatments.

In instances such as the above, fractional replication can be useful, but in cases where the treatments occur more than once in an experiment, it is inefficient to duplicate a fractional replicate. For example, a duplication of a star-point response surface design for n = 4 would be:

	a_0	a_1	a_2	a_3	a_4
b_0			2		
b_1		2		2	
b_2	2		2n=8		2
b_3		2		2	
b_4			2		

If the form of the response were unknown, a much more efficient treatment design with the same number of observations would be the following:

	a_0	a_1	a_2	a_3	a_4
b_0	1	1	1	1	1
b_1	1	1	1	1	1
b_2	1	1	0	1	1
b_3	1	1	1	1	1
b_4	1	1	1	1	1

A general rule to follow in fractional replication *is to take an additional fraction instead of repeating a fraction.* In this manner, the treatment designs nearer to a complete factorial are achieved and a wider coverage of levels of treatment combinations is possible.

8.8 GENETIC TREATMENT DESIGNS

In inheritance studies, the investigator must select the treatment (parents and crosses) included in the experiment. In studying simple Mendelian dominance in plants, the two parents, P_1 and P_2, and the cross $F_1 = P_1 \times P_2$ are sufficient. If, in addition, it is desired to know whether or not one, two, or three pairs of independent alleles are involved, then it would be necessary to include other treatments like, for example, the $F_2 = F_1 \times F_1$ progeny. The ratio of segregation in the F_2 individuals will indicate the number and nature of allelic pairs.

As the inheritance pattern becomes more complex, the greater will be the number of treatments required to ascertain the pattern. Also, as the number of generations of a cross increases, the number of possible crosses increases rapidly. For example, in the kth generation the following are possibilities for k = 0,1,2,3:

k^{th} generation	Crosses possible in generation k	No. = n_k
k=0	P_1, P_2	2
k=1	P_1, P_2, $F_1=P_1 \times P_2$	3
k=2	P_1, P_2, F_1, $F_2=F_1 \times F_2$, $B_1=F_1 \times P_1$, $B_2=F_1 \times P_2$	6
k=3	P_1, P_2, F_1, F_2, B_1, B_2, $P_1 \times F_2$, $P_1 \times B_1$, $P_1 \times B_2$, $P_2 \times F_2$, $P_2 \times B_1$, $P_2 \times B_2$, $F_1 \times F_2$, $F_1 \times B_1$, $F_1 \times B_2$, $F_2 \times F_2$, $F_2 \times B_1$, $F_2 \times B_2$, $B_1 \times B_1$, $B_1 \times B_2$, $B_2 \times B_2$	21
k+1st		$n_k(n_k + 1)/2$

Because the number of possible crosses quickly becomes too large to consider using all crosses, an investigator must necessarily select a sub-set of these; the correct sub-set depends upon the type of inheritance and genetics under study.

In breeding studies on variety evaluation, the genetic design includes selection of the proper controls. Several types of controls or standard varieties may be necessary to evaluate the new varieties being tested. Careful thought is required in obtaining adequate controls for screening new varieties.

An important type of genetic treatment design is the one known as the diallel crossing system for describing the crosses of k lines, for instance, strains of mice. Some possibilities for k = 5 are:

	Design I male lines					Design II male lines					Design III male lines					Design IV male lines				
Female lines	1	2	3	4	5	1	2	3	4	5	1	2	3	4	5	1	2	3	4	5
1	x	x	x	x	x	x	x	x	x	x	-	x	x	x	x	-	x	x	x	x
2	x	x	x	x	x	-	x	x	x	x	x	-	x	x	x	-	-	x	x	x
3	x	x	x	x	x	-	-	x	x	x	x	x	-	x	x	-	-	-	x	x
4	x	x	x	x	x	-	-	-	x	x	x	x	x	-	x	-	-	-	-	x
5	x	x	x	x	x	-	-	-	-	x	x	x	x	x	-	-	-	-	-	-

In the above, x denotes the cross and - denotes no cross. All possible combinations are given in design I, which is the complete $k^2 = 5^2$ factorial. The other three designs are fractional replicates of the complete factorial. Design II represents "selfs" (a line crossed with itself) plus all possible crosses. Design III represents all possible crosses of the k lines plus all possible reciprocal crosses; for instance, if line 1 is the female and is crossed with line 2 as the male, then the reciprocal cross would be line 1 as the male crossed with line 2 as the female. Design IV gives the genetic treatment design of all possible crosses.

The genetic treatment design listed above as III has been used in measuring the amount of communication between individuals in psychological research. Obviously a person does not communicate with or talk to himself when he is a member of a group of individuals. This omits the "selfs". It is interesting that a number of genetic phenomena could be utilized to describe psychological phenomena associated with communication between individuals.

In classification and importance of job studies, it has been found that an individual thinks more highly of his own position or function than do people in other positions. This means that the "selfs", or the self-ratings, should be omitted in an analysis of the data, resulting in design III. For example, suppose that we draw a sample from each of the following university groups and have them rate the relative importance of all groups other than their own:

Category	U	G	P	S	M	A
Undergraduate = U	-	x	x	x	x	x
Graduate student = G	x	-	x	x	x	x
Professor = P	x	x	-	x	x	x
Secretary = S	x	x	x	-	x	x
Maintenance personnel = M	x	x	x	x	-	x
Administrator = A	x	x	x	x	x	-

This would produce design III. This design has also been used to evaluate sequence of pairs of courses and educational experiences.

Design I has been used to study competition among varieties of wheat; the "selfs" represent a line in competition with plants of its own kind. Here again many of the genetic phenomena had direct counterparts in competition terms. If successive doses of drugs are given alone or in combination, the above design is directly useful. Design IV and genetic interpretations have been used in cockfighting experiments; it has also been used frequently in paired-comparisons experiments in the social sciences, and in round-robin tournaments. Design II has been utilized to compare mixtures of beans to determine whether higher yields can be obtained from mixtures than from a variety planted alone (the "selfs").

8.9. BIOASSAY TREATMENT DESIGNS

A **biological assay**, or **bioassay**, is an experimental procedure for identifying the constitution or for estimating the potency of materials by means of their reaction on living material. Examples of use of assays are:
   a. identification of blood groups by serological tests,
   b. estimation of potencies of vitamins from their effects on the growth of microorganisms,
   c. comparison of insecticides by toxicity tests,
   d. others.

An **analytical assay** is a procedure of estimating the potency of a **test preparation** (e.g., a natural source of a vitamin) relative to a **standard preparation** containing the same active material (e.g., a pure chemical form). This type of assay is such that X units of the test preparation produce the same average response as RX units of the standard preparation. The value of R is then defined as the **relative potency**. One important type of response is the following:

$$\text{standard preparation}: \quad Y = a + bX$$
$$\text{test preparation}: \quad Y = a + bRX$$

Now the two response equations intersect when $X = 0$ and the ratio of the slopes of the two lines, that is, $bR / b = R$, gives the relative potency. An experiment designed to estimate R in this way is denoted as a **slope ratio assay**. An example of this type of bioassay is reproduced below in Figure 8.8 (from Finney, 1955, page 125, with the permission of the University of Chicago Press).

An assay which probably has wider applicability than the slope ratio assay is the **parallel line assay**. In this type of bioassay, the relative potency R is measured as the horizontal difference between the following two parallel lines:

$$\text{standard preparation} \quad : Y = a + b \log X$$
$$\text{test preparation} \quad : Y = (a + b \log R) + b \log X$$

An example illustrating this type of assay is taken from page 127 of Finney (1955) and presented in Figure 8.9 (with the permission of the University of Chicago Press).

The problem of selecting the doses, X, to use, the subjects to be utilized in an experiment, and the number of doses represents the selection of the treatment design in bioassays. Many of the concepts and problems involved are discussed by Finney (1955). In addition to the preceding reference, Professor Finney (1978) has written an entire textbook on this important subject. Active research is being pursued on bioassay at the present time.

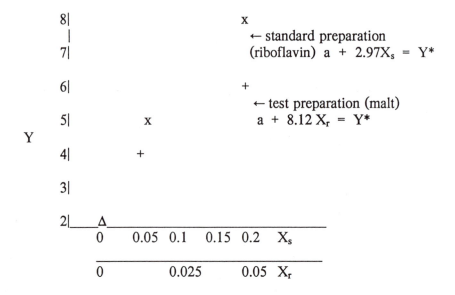

Figure 8.8. Assay of riboflavin in malt, using *L. helveticus* as subject.

Upper horizontal scale (X_s): Dose of riboflavin per tube, in micrograms. Lower horizontal scale (X_r): Dose of malt per tube, in grams. Vertical scale (Y): Titer of N / 10 sodium hydroxide in milliliters. Δ : mean response for 4 tubes without treatment; x: mean responses for 4 tubes on standard preparation; +: mean responses for 4 tubes on test preparation. Two lines intersecting at X = 0 have been fitted by standard statistical techniques. The standard line rises by 2.97 ml per 0.1 mg

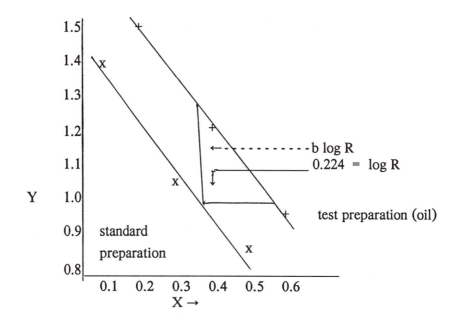

Figure 8.9.   Assay of vitamin D in an oil by chick method.   Horizontal scale (X): log daily dose per chick, in units vitamin D or milligrams oil.

riboflavin, the test line by 8.12 ml per 0.1 gm malt. Hence, the malt is estimated to contain 8.12 / 2.97, or 2.73 mg riboflavin per gram of malt.

Vertical scale (Y): log tarsal-metatarsal distance, in 0.01 mm; x: mean responses for 28 chicks on standard preparation; +: mean responses for 28 chicks on test preparation.   Two parallel lines have been fitted by standard statistical techniques. Measurement shows that the X values of the test line would have to be reduced by 0.224 in order to superimpose it on the standard line. Hence the oil is estimated to contain 0.597 units vitamin D per milligram (since log R = - 0.224 = 0.776 - 1 = log 5.97 - log 10 = log 0.597). Therefore, one gram of oil is equivalent to 0.597 units of vitamin D.

8.10 CLINICAL AND EPIDEMIOLOGICAL TRIALS

A **clinical trial** is a scientifically designed and executed experiment on humans or animals for the purpose of studying health oriented processes, procedures, or materials. Medical, nutritional, psychological, psychiatrical, and even some educational studies are involved. Such practices as bleeding, purging, starvation, portacaval shunts for esophageal varices, and gastric freezing for duodenal ulcers were quickly stopped when subjected to scientific inquiry and scientifically conducted experiments. These practices for the most part were deterimental to patients' health. Statistically designed experiments as described in the previous and present chapters have been and are the foundations for this scientific approach. It has been found that *time to occurrence* of the beginning and endpoints and not just whether or not there has been an effect, has proven to be extremely valuable measurement in clinical trials. This is an illustration of the use of measurement and measuring instruments in experimentation. Also, it has been found that **doubly blind** which means that neither the subject (patient or individual) nor the technician applying the treatment knows the identity of the treatment being applied, studies are essential in most clinical trials. Occasionally, a **singly blind** study, where only the subject does not know the identity of the treatment, is necessary in order to ensure safety for the subject. All of the principles and concepts of experiment design and of treatment design apply in planning and setting up clinical trials.   A special feature of clinical trials is that ethical and moral considerations are (or should be) involved in any clinical trial. These involve selection of subjects, treatment of subjects, and the responsiblity of the investigator to properly design the study and to correctly analyze the results of the study. It is considered immoral to conduct a study involving subjects and have the study turn out to be useless because of poor design or improper analyses. The ethical requirements must be considered prior to conducting all clinical trials.

Epidemiology is the study of the distribution and determinants of health related processes in populations. Disease is one of the processes often studied. Epidemiology has evolved from the study of classical infectious diseases to encompass the "plagues" of industrialized societies, including vascular disease, cancer, and environment and occupation-related diseases. The goal of epidemiology is to elucidate and quantify the relationship between disease and potential risk factors or exposures, in order to prevent future illness. Its emphasis is on the host and the environment; it seeks to give insights about the effect of agents on hosts in different environments and on the importance of various contributing factors for a disease. These concerns contrast with other types of medical procedures where emphasis is on the condition of the *individual* host.

An **epidemiological trial** essentially involves running a clinical trial on a *population.* For example, all the salt sold in one area of a country could be treated with selenium whereas a control area would have salt untreated with selenium and the incidence of skin cancer could be the response variable. Such a study has been set up in China. In a clinical trial, a sample of the population is used and inferences are made from the sample to the population. In an epidemiological trial, the whole population is used.

New York State has collected data for leukemia cases located by Census Tracts over the last several years. These data are a good source to study location and incidence of the disease. The New York State Health Department would like to know if the leukemia cases cluster in certain areas of the State. Statistical methods for detecting clusters are available and the probability that a cluster has not occurred at random can be ascertained. A continuous monitoring of new cases and their location is desired. A surveillance procedure needs to be developed for this epidemiological study of clustering of leukemia cases. Then, if it is decided that the probability that a cluster of leukemia cases exists is high enough, an investigation will be launched into probable causes. This may involve conducting clinical trials, perhaps on animals first, to pinpoint the causes. From the results of the clinical trial, an epidemiological trial may be instituted to possibly reduce or eliminate the appearance of clusters of leukemia cases if in fact they exist.

8.11   SEQUENTIAL SELECTION OF TREATMENTS

In many types of experimentation, it is possible to perform observations in sequence; each succeeding treatment is selected from the results obtained from the investigation from the start until the present response. The object here is to reach some desired goal such as the dosage giving 50% kill (LD_{50} equals mean lethal dose giving 50% kill), the combination of ingredients giving maximum loaf volume, the combination of ingredients making the softest (or hardest) alloy of materials, the combination of fertilizers giving maximum plant response, and so forth. In such investigations, it is assumed that there is *one* combination or dosage which produces the desired results. For example, in seeking an optimum we may liken this to the following example. We proceed as if there were one hill in an area with all land sloping away from the summit. It is so foggy that we can observe only the point at which we are located. We take a reading with a level and proceed in the direction indicated, hoping to reach the summit. If there is more than one hill in the area, or if there are plateaus, we are in trouble and may never find our way to the summit with this procedure. So it is with the above statistical procedures designed for a particular situation. If, for example, the 50% kill is attained with a variety of dosages, there will be no unique dosage giving this prescribed value and there will also be a variety of doses giving other percentage kills.

One of the earlier sequential procedures described is the Dixon-Mood "up-and-down" method, so called because it was first used to determine the heights from which explosives could be dropped before they exploded. If the sample explosive did not explode when dropped at from five feet, say, the explosive was dropped from six feet. If it exploded, the next sample was tried from five feet. If this exploded the next sample was dropped from four feet. This process was continued, and it tended to concentrate the observations around the height at which explosives would explode when dropped.

The same procedure could be utilized to estimate the dosage for any specified percentage kill, for example, 50% kill. Samples of $N = 100$, say, insects would be used. A dosage would be tried and the percentage of insects killed would be observed. If less than 50% of the insects were killed, the dosage would be increased. If more, the dosage would be decreased. The process is continued until the percentage kill oscillates around the desired level.

Another procedure which achieves much the same effect as the above but which goes further in selecting the next treatment level for examination is the Robbins-Munro method. The first observation or treatment is selected from previous knowledge to be as near the desired point as possible. If there is no previous experience with the treatment, then a start is made at any point. The result of the first observation is denoted as Y_1 and of the first treatment (level) as X_1. The result of the second observation is Y_2 observed at level X_2, the third observation is Y_3 observed at X_3, etc.. The resulting levels of X after the first level of X are to be determined as follows:

$$X_2 = X_1 + 2.5 \, (\partial - Y_1)$$
$$X_3 = X_2 + 2.5 \, (\partial - Y_2)$$

$$X_4 = X_3 + 2.5 (\partial - Y_3)$$
...
$$X_{n+1} = X_n + 2.5 (\partial - Y_n)$$

where ∂ is the desired value, say 50% kill and 2.5 is a constant that improves the operation of the procedure. The last value of X, that is X_{n+1}, is the estimated value of the dose giving the desired percentage kill. Note that the above is one possible form of the Robbins-Munro procedure

Professors J. Kiefer and J. Wolfowitz of Cornell University (now deceased) have devised a procedure for obtaining the value of the treatment giving maximum (or minimum) response for situations involving mixtures of compounds such as proportions of two metals in an alloy, proportions of baking powder and soda in a sour milk cake, and so forth. The procedure is almost as simple to apply as the Robbins-Munro procedure. The steps follow:

1. Select two sequences of numbers such as
$$a_i = \frac{1}{i} \text{ and } c_i = (\frac{1}{i})^{1/3}.$$

2. Select a starting point of X, say Z_1.

3. Let $X_1 = Z_1 - c_1$ and $X_2 = Z_1 + c_1$.

4. Obtain responses Y_1 and Y_2 at X_1 and X_2, respectively.

5. Compute $Z_2 = Z_1 + k (Y_2 - Y_1)$

6. Let $X_3 = Z_2 - c_2$ and $X_4 = Z_2 + c_2$ and obtain the corresponding responses Y_3 and Y_4.

7. Let
$$Z_3 = Z_2 + k (1/2)(Y_4 - Y_3)/(1/2)^{1/3}$$

8. The above process continues until n pairs of observations or responses have been obtained. The value of X (say X_{max}) giving the maximum response is then computed as

$$Z_{n+1} = Z_n + k a_n(Y_{2n} - Y_{2n-1})/c_n = X_{max}$$

The value for k may be taken as equal to one, but other values, say 2.5, may make the operation of the procedure more desirable. Note that the value of a_i / c_i becomes smaller and smaller as n increases. This means that Z_n approaches Z_{n+1} in value as n increases, and since the c_i are getting smaller X_{2n} and X_{2n-1} are getting closer together as n increases.

8.12 PRODUCT QUALITY AND IMPROVEMENT

In today's highly competitive manufacturing and business world, high quality products, processes, and services and customer satisfaction are a necessity for continued operation and remaining solvent. Of course, these are not usually requirements for endowed institutions and municipal, state, or federal funded organizations but are when economics rules. The word *quality* means different things to different people. In Quality Control, the word has a restricted practical meaning for each of the control systems in which the technique is used. **Quality** is defined in terms of measurement of the object or process measured. The organization that has low quality products and/or services and uses inefficient procedures will not stay in operation very long. The man, W. Edwards Deming, who is said to have remade "made in Japan", was responsible for the Japanese instituting highly efficient procedures to produce high quality products (See Mann, 1989). The adoption of these procedures by the Japanese has forced some and is forcing more manufacturers and businesses in the United States and elsewhere to adopt similar procedures. As an example, the Ford Motor Company's advertisements state, "Quality is Job One" and General Motors advertises, "We are Putting Quality on the Road". In order to be successful in a truly competitive world, it is essential that efficient procedures be installed and that high quality products or services be produced.

Quality Control is the science related to the development and application of procedures for assessing and for maintaining the quality of an on-going process or operation and for controlling the procedure. Statistical theory and procedures are relied upon heavily for developing and applying quality control procedures. The object of process control is to ensure that the specified statistical tolerances are met and to detect and correct faults as soon as they occur. It allows for variability within certain limits, but the process is kept under control. The ideas of statistical control are employed in keeping track of sample results as they occur through time. This is done through the use of two basic control charts, i.e., for averages and for variability. A **control chart for averages** consists of plotting the average response for n samples, $n = 1, 2, 3, ...,$ against the time when taken. The average of n sample values taken at time one, the average of n samples taken at time two, the average of n samples taken at time three, and so forth are the values used to construct this type of control chart. For example, $n = 5$ light bulbs from the lot of bulbs produced on day one could be tested for number of times the bulb could be turned on and off before it burned out. Then on the second day, a sample of five light bulbs would be tested. The process of sampling and testing would be done continuously through time and the results plotted on a chart or graph. As long as the averages stayed within the prescribed limits, the process would be considered to be under control. To illustrate the construction of such a chart, suppose that twelve inch rulers are being manufactured, that it is desired to have the length of rulers be no more than 12.005 inches long or no shorter than 11.995 inches. The desired length is 12.000 inches but the process is such that not all rulers will be of the desired length. There will be variation in the lengths. Suppose that the results obtained are those given in Figure 8.10. The desired control level is 12.000 inches, the upper

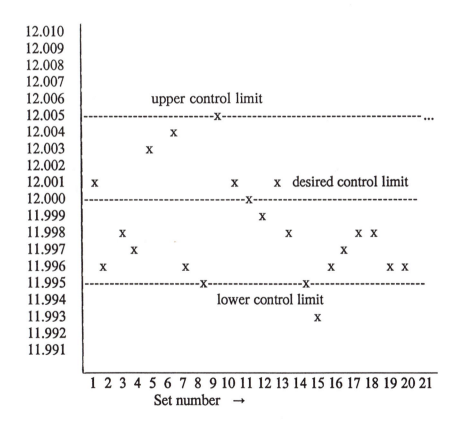

Figure 8.10. Control chart for sets average length of ruler in inches for five rulers taken over time.

control limit is 12.005 inches, and the lower control limit is 11.995 inches. We note that one sample average fell below the lower limit and that the last several averages were hovering just above the lower limit. This may indicate the the manufacturing process is getting out of control on the shorter side. This could be ordinary sampling variation but these results indicate that a close watch on the process should be maintained and perhaps the equipment requires adjustment. A **control chart for ranges** uses the range in values (largest value in the n samples minus the smallest value) for the n samples of a set and the value of the range is plotted against set number in a manner similar to that for averages except that only an upper limit is used. No variation would, of course, be the ultimate in controlling a process but this is often an impossibility.

A control chart for ranges uses the range in values for the n samples of a set and the value of the range is plotted against set number in a manner similar to that for averages except that only an upper limit is used. Suppose that a range of less than 0.015 inches is desired but that a correction will not be made unless the range exceeds 0.025 inches. For the example given in Figure 8.10, suppose the results are as given in Figure 8.11. When the sampling was started, the range was near the warning upper limit. However, as sampling continued, the range was increasing, indicating that the process of manufacturing rulers was getting out of control. When the range exceeded the action upper limit, a correction was made, and the range returned to acceptable values, i.e., less than 0.015.

The basic idea for using control charts is to ascertain that acceptable limits in mean values and ranges are met, and if not, corrective measures are taken to keep the process under control. The use of control charts is a systematic way of controlling quality through means of frequent sampling through time. Control charts may be used effectively in a large variety of applications. Some that come to mind are:

 a. number of pin-feathers left on fowls,
 b. time of arrival of an employee for work,
 c. number of complaints concerning a salesperson or mechanic,
 d. thickness of paper, facial tissue, toilet tissue, etc.,
 e. level of filling milk, soda, beer, wine, etc. containers,
 f. weight of medical pills and vitamins,
 g. porosity in couplings for arteries and veins,
 h. number of raisins in a package of Raisin Bran,
 i. number of recalls on a model of an automobile,

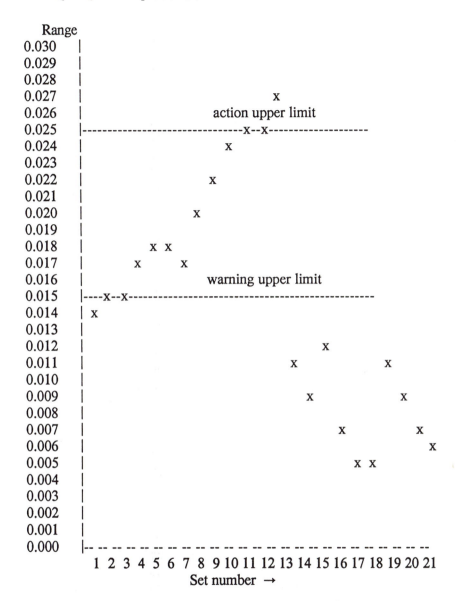

Figure 8.11. Ranges of samples of size n = 5 rulers in inches versus set number.

    j. diameter of nuts and bolts,
    k. thread distance for nuts and bolts,
    m. number of accounting errors,
    n. number of "poor" decisions by a manager,
    o. number of defects in dresses and other articles of clothing,
       and so forth.

Quite often a purchaser or user of a lot of material may wish to sample the material prior to deciding whether or not to purchase or to use it. There usually are three possible decisions to make in situations of this nature. The buyer may decide to accept the lot, to reject the lot, or to continue sampling prior to reaching one of the two previous decisions. This type of sampling of material is known as **acceptance sampling**. It may also be used by a manufacturer in determining whether to discard a lot or to keep it. Also, in some cases, a census of all the material manufactured is used and each item is tested with some being accepted and the others being rejected. One manufacturer of couplings for human veins rejected 98% to 99% of all the couplings made. It would appear that he should make better couplings to begin with and one should have little confidence in the couplings that were accepted. The three factors involved in acceptance sampling are the number of samples n, the size of the lot N, and the number of defectives d. Tables have been constructed allowing a specified percentage of defectives. There are many forms for acceptance sampling. One such form is given in Figure 8.12 where the sampling of the lot continues until it is decided to accept or to reject the lot. For the lot associated with the + symbols, this lot would be rejected. For the lot associated with the o symbols, the lot would be accepted. Note that as the figure is drawn, the sampling for the lot associated with the x symbols, could continue until all the material has been tested. It might be desirable in many situations to discontinue sampling and make a decision to accept or to reject after a specified percentage of the population, say 10%, has been sampled.

The basic idea behind **product improvement** is to produce material that has high quality and that is satisfactory to the customer. Experiments will need to be devised to determine which factors are accounting for poor quality and/or for variation in the material. Since there may be several factors that need to be considered, fractionally replicated factorial designs and response surface designs will be used. These treatment designs are very popular among organizations involved in product improvement. The fraction is selected to be as small as possible and this means that only main effects will be considered to be important in the first stage of an investigation. Then, at some later stage of refinement interactions may be considered to be important. These designs are also very useful in the development of new products. Quite often each of the combinations will not be replicated and hence only a completely randomized design with one replicate on each treatment will be used.

Instead of replicating a particular fraction, it is recommended that a different fraction be used. This has the advantage in that additional effects can be investigated if they are present. Suppose that a manufacturer thought that seven different factors might be affecting the quality of his product. He decides to use the following fractions for the first experiment and for the second experiment should it be needed:

first fraction	second fraction
0 0 0 1 1 1 0	1 1 1 0 0 0 1
0 0 1 1 0 0 1	1 1 0 0 1 1 0
0 1 0 0 1 0 1	1 0 1 1 0 1 0
0 1 1 0 0 1 0	1 0 0 1 1 0 1
1 0 0 0 0 1 1	0 1 1 1 1 0 0
1 0 1 0 1 0 0	0 1 0 1 0 1 1
1 1 0 1 0 0 0	0 0 1 0 1 1 1
1 1 1 1 1 1 1	0 0 0 0 0 0 0

The zeros in the first fraction have been replaced by ones and the ones by zeros to obtain the second fraction. Such a fraction is known as a **fold-over design**. Both fractions used together allow estimation of the main effects of the seven factors as well as sums of two-factor interaction effects among the factors.

8.13 EXPERIMENT DESIGNS FOR VARIOUS TREATMENT DESIGNS

The nature of the treatment design has little or nothing to do with the experiment design used. For example, suppose that the experimenter uses a 2^3 factorial treatment design. Depending only upon the nature of the experiment and the nature of the heterogeneity present, he may use any one of the following experiment designs:

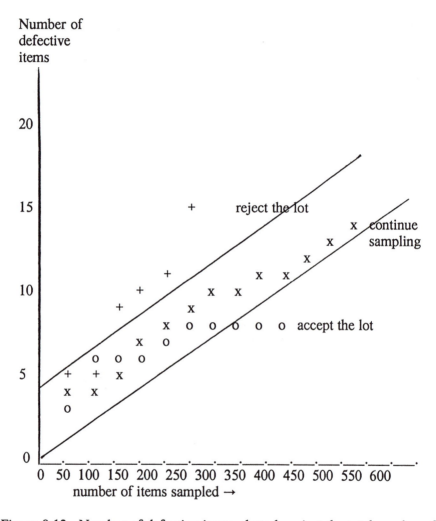

Figure 8.12. Number of defective items plotted against the total number of items sampled with accept and reject lines drawn in.

1.   a completely randomized design with r replicates,
2.   a randomized complete blocks design with r replicates,
3.   an 8 × 8 latin square design with r = 8 replicates,
4.   a 4 × 6 latin rectangle design with r = 3 replicates,
5.   a 4 × 8 latin rectangle design with r = 4 replicates,
6.   an 8 × 7 Youden square design with r = 7 replicates,
7.   an 8 × 9 Youden square design with r = 9 replicates,
8.   a bib design with k = 4, b = 14, r = 7, and λ = 3,
9.   a bib design with k = 2, b = 28, r = 7, and λ = 1, etc..

Thus, we see that there are many experiment designs available for any given treatment design.

8.14 REFERENCES AND SUGGESTED READING

Box, G. E. P. and N. R. Draper (1969). *Evolutionary Operation.* ASQC Quality Press, West Allis, Wisconsin, 237 pp..

Cornell, J. A. (19900. *Experiments With Mixtures--Designs, Models, and the Analysis of Mixture Data, second edition.* John Wiley & Sons, Inc., Somerset , New Jersey, 632 pp..

Cox, D. R. (1958). *Planning of Experiments.* John Wiley and Sons, Inc., New York, vii + 308 pp..
   (Chapters 6 and 7 are, for the most part, at the level of this book. Dr. Cox has covered some of the points discussed in the above; these two chapters will serve to deepen the student's insight into problem of treatment design.)

Devore, J. and R. Peck (1986). *Statistics. The Exploration and Analysis of Data.* West Publishing Company, St. Paul, New York, Los Angeles, San Francisco, xv + 699 pp..
   (An elementary account of statistical procedures at a less difficult level than Snedecor and Cochran.)

Dowdy, S. and S. Weardon (1983). *Statistics for Research.* John Wiley & Sons, New York, Chichester, Brisbane, Toronto, Sinapore,xvi + 537 pp..

Federer, W. T. (1955). *Experimental Design - Theory and Application.*

The Macmillan Company, New York, xix + 544 + 47 pp.. (Republished by Oxford & IBH Publishing Company, New Delhi, India, in 1974.)
(Chapters VII and VIII treat the various aspects of factorial experiments and are more advanced and more detailed than the preceding references. Additional material on factorial experiments may be found in Chapters IX and X.)

Finney, D. J. (1955). *Experimental Design and its Statistical Basis.* The University of Chicago Press, Chicago and London, xi + 169 pp..

Finney, D. J. (1978). *Statistical Method in Biological Assay, Third edition.* Oxford University Press, New York, xii + 508 pp..

Jellinek, E. M. (1946). Clinical tests on comparative effectiveness of analgesic drugs. *Biometrics* 2, 87-91.

Kempthorne, 0. (1952). *The Design and Analysis of Experiments.* John Wiley and Sons, Inc., New York, xix + 631 pp..
(A theoretical and mathematical treatment of treatment design is given in chapters 13-21.)

Mann, N. R. (1989). *The Keys to Excellence. The Story of the Deming Philosophy.* Prestwick Books, Los Angeles, xviii + 196 pp..
(A popularized account of the works and approach of W. Edwards Deming for product improvement.)

Nelson, W. (1990). *Accelerated Testing: Statistical Models, Test Plans, and Data Analysis.* John Wiley & Sons, Inc., Somerset, New Jersey, 601 pp..

Ryan, T. P. (1989). *Statistical Methods for Quality Improvement.* John Wiley & Sons, Inc., Somerset, New Jersey, 446 pp..

Snedecor, G. W. (1946). *Statistical Methods, 4th edition.* Iowa State University Press, Ames, Iowa, xvi + 485 pp..
(Numerous statistical and computational techniques are presented in this text. In order to note several other areas of application, a perusal of the many diverse examples in chapters 6, 7, 13 and 14 and in later editions may be of interest to the reader.)

Yates, F. (1937). Design and analysis of factorial experiments. *Imperial Bureau of Soil Science*, Technical Communication No. 35.
(A classical treatise on the subject of factorial experiments.)

Youden, W. J., Connor, W. S., and Severo, N. C. (1959). Measurements made by matching with known standards. *Technometrics* 1:101-109.

8.15 PROBLEMS

Problem 8.1. (This problem is taken from examples 14.4 and 14.5 of Snedecor's book, 1946, 4th edition, reprinted by permission from *Statistical Methods*, by G.W. Snecedor, copyright by The Iowa State University Press.) In a study of the baking quality of straight grade flour, Q, after being heated to 170°F. for various lengths of time T, in hours, the following data were obtained:

T (hrs.)	Q	log Q	log (Q - 14)	log T
0.25	93			
0.50	71			
0.75	63			
1.00	54			
1.50	43			
2.00	38			
3.00	29			
4.00	26			
6.00	22			
8.00	20			

Obtain log T, log Q, and log(Q - 14) from logarithmic tables or a computer. Then prepare the following graphs:

1. Plot Q as the ordinate against T as the abscissa.
2. Plot log Q against log T and plot the line through the points for log Q = 1.7116 - 0.4678 log T.
3. Plot $1 / Q$ against T.
4. Plot $Y = 1 / (Q - 14)$ against T.
5. (Optional) If you have access to a calculating machine, plot Q against T using the following equation:

$$Q = \frac{1}{.02T + .0055} + 14 .$$

Which of these models do you think best fits the data?

Problem 8.2. In an investigation of the relationship of the number of hours of preparation (T) for an examination and the percentage passing (Y) the following data were obtained:

Number of hours of preparation = T	0	10	20	30	40	60
Percentage passing = Y	66	85	91	95	97	97
log (100 - Y) =						

1. Plot Y, ordinate, against T, abscissa.
2. Plot log (100 - Y) as the ordinate against T. (*Note*: This is a form of the Mitscherlich equation.)

Note the change in relationship obtained in the two plots.

Problem 8.3. In a study of failure rates (F) of motorboat motors after T hours of continuous operation, the following data are available:

Number of hours of operation = T	2.5	5	10	40	120	240	480	720
Percentage of failures = F	55	60	63	66	69	73	74	75
log T								

1. Plot F, ordinate, against, T, abscissa.
2. Plot F against log T, abscissa.

Note the change in the relationship on the log scale.

Note: In the above three problems, note that the type of the investigation is an absolute experiment in the sense that it is a single phenomenon experiment. The interest is on the response to a single factor rather than in comparing the individual levels *per se*.

Problem 8.4. Suppose that Miss I. M. Fashion, "beauty" researcher, wishes to observe the effect of three kinds of oil (olive oil, lanolin, and saffron oil) with two types of pigments (plant pigment and laboratory pigment) on "presentability" of eye shadow after 12 hours, as measured by her procedure, for four different colorings (magnolia pink, blue-bonnet blue, avocado green, and ebony black). She wishes to use the various levels of these three factors in all combinations. Write out the combinations for her.

Problem 8.5.   Given the following effects from a 2×3 factorial experiment reconstruct the yield observations Y_{ij}.

	Two-factor Interaction Effects
$\bar{y}_{..} = 10$	$a_0 b_0 : -3$
$\bar{y}_{0.} - \bar{y}_{..} = -7 = a_0$	$a_0 b_1 : +1$
$\bar{y}_{1.} - \bar{y}_{..} = +7 = a_1$	$a_0 b_2 : +2$
$\bar{y}_{.0} - \bar{y}_{..} = +13 = b_0$	$a_1 b_0 :\ \ 3$
$\bar{y}_{.1} - \bar{y}_{..} = -12 = b_1$	$a_1 b_1 : -1$
$\bar{y}_{.2} - \bar{y}_{..} = -1 = b_2$	$a_1 b_2 : -2$

Graph the responses, Y_{ij}, using the three levels of the second factor as the abscissa and Y as the ordinate.

Problem 8.6. The following represents the actual layout of an experiment and actual yields of wheat. The letters represent the treatments with n referring to nitrogen, p to phosphorous and k to potassium fertilizers. If a letter is present this means that element in the fertilizer was present. What appears to be the experiment design and the randomization procedure? What is the treatment design?

n	p	np	k	nk	(1)	npk	pk
19	12	18	16	11	12	19	19

n	nk	pk	npk	p	k	np	(1)
13	7	17	17	20	12	19	16

nk	np	n	p	(1)	npk	pk	k
11	18	10	18	10	17	18	14

pk	k	npk	(1)	n	np	p	nk
18	13	14	12	11	15	17	16

np	(1)	nk	n	pk	p	k	npk
18	13	13	11	17	16	11	17

k	pk	(1)	np	npk	n	nk	p
15	18	13	17	16	10	9	21

npk	p	k	pk	np	nk	(1)	n
19	19	11	17	18	9	10	15

(1)	npk	p	nk	k	pk	n	np
18	20	21	16	17	18	14	23

Problem 8.7. Sometimes researchers unwittingly obtain an apparent relation which turns out to be spurious. For example, suppose that one has a measure of variation, say s, and a mean $\bar{y}$ for a series of values of $\bar{y}$ and s. Suppose that one plots relative variation $= s / \bar{y}$ against $\bar{y}$; if s stays relatively constant compared to $\bar{y}$, then one is essentially plotting $1 / \bar{y}$ against $\bar{y}$ which gives a beautiful curve. Another example of this appears when one plots new recruits relative to parent stock in the population against parent stock in the population. If the number of new recruits is relatively constant, then one is essentially plotting the reciprocal of numbers of parent stock against numbers of parent stock. Construct a set of ten numbers, d_i, whose sum is zero, make the range in these ten numbers small, say $-2 \geq d_i \geq 2$, select a set of ten values of y_i whose range is 20, say 1 to 20, add the d_i randomly to the y_i, add a constant, say 4, to each of the $y_i + d_i$ values, and then plot $1 / (y_i + d_i + 4)$ against $y_i + d_i + 4$. Observe the relatively regular curve which, of course, merely indicates that one is plotting a set of values against a function of these same values.

Problem 8.8. Obtain thickness measurements for pages 1 to 10, 11 to 30, 31 to 60, 61 to 100, 101 to 150, and 151 to 210 of this book. Plot the thickness against number of pages. Describe your measuring instrument. What functional relation exists between number of pages and thickness of pages? How would you estimate the thickness of a single page? Repeat the experiment but measure the thickness of page 1, pages 2 to 3, 4 to 6, 7 to 10, 11 to 15, and 16 to 21 each 10 times around the edge of the page. Plot the 10 measurements against number of pages. In both cases you measured 210 pages; which procedure do you believe to be the more accurate for determining the thickness of a page and why? Does the functional relation change? Why or why not? Which procedure gives greater variation in measurements?

Problem 8.9 The following data are loaf volumes in a baking experiment involving two varieties of yeast cells and three baking temperatures:

Yeast variety	350° F	Baking temperature 375° F	400° F	$Y_{.j}$
v_0	10	20	24	54
v_1	6	16	8	30
$Y_{i.}$	16	36	32	84

Show how to compute the arithmetic mean, yeast variety effects, temperature effects, and interaction effects. Prepare a graph showing interaction effects. Describe three experiment designs for this treatment design and explain when you would use each design.

Problem 8.10. Three factor interaction effects may be computed from the following formula

$$Y_{hij} + \bar{y}_{h..} + \bar{y}_{.i.} + \bar{y}_{..j} - \bar{y}_{hi.} - \bar{y}_{h.j} - \bar{y}_{.ij} - \bar{y}_{...}$$

Show how to compute the eight interaction effects for the 2^3 factorial on stereophoto versus non-stereophoto example described in Section 8.6.

Problem 8.11. For the example discussed in the previous problem, prepare graphical plots for each of the three two factor interactions and of the three factor interaction. Note that a two factor interaction compares the difference of the main effect for factor two, say B, at each of the levels of factor one, say A. The three two factor interaction effects for the 2^3 factorial may be represented as follows:

$$a_1 (b_1 - b_0) - a_0 (b_1 - b_0) = \text{A by B interaction}$$
$$a_1 (c_1 - c_0) - a_0 (c_1 - c_0) = \text{A by C interaction}$$
$$b_1 (c_1 - c_0) - b_0 (c_1 - c_0) = \text{B by C interaction}$$

A three factor interaction compares the two factor interaction effects at the different levels of the third factor. For a 2^3, factorial this is

$$a_1 [b_1 (c_1 - c_0) - b_0 (c_1 - c_0)] - a_0 [b_1 (c_1 - c_0) - b_0 (c_1 - c_0)] =$$
$$a_1 b_1 c_1 - a_1 b_1 c_0 - a_1 b_0 c_1 + a_1 b_0 c_0 - a_0 b_1 c_1 + a_0 b_1 c_0 + a_0 b_0 c_1 - a_0 b_0 c_0$$

where responses are inserted for the above treatment combinations, i.e., Y_{111} is inserted for $a_1 b_1 c_1$ = 111 in the notation of this chapter.

Problem 8.12. We know for the parameter $\emptyset$ that all responses Y_i lie in the interval $\emptyset \leq Y_i \leq \emptyset + 10$. The procedure is to sample sequentially until we find that the lowest two values differ by one. To illustrate, we also know that the largest value we observe that this value minus 10 sets a lower bound. Also, if any two observations are ten units apart, we know that we have the upper and lower limits and we know $\emptyset$; let the first observation we observe be 41; then, 41 - 10 = 31 $\leq \emptyset \leq$ 41. Suppose the second observation is 38; then, 31 $\leq \emptyset \leq$ 38. Let the third observation be 42; then, 41 $\leq \emptyset$ + 10 $\leq$ 42 and 31 $\leq \emptyset \leq$ 32. We stop sampling because the criterion has been satisfied. Suppose that the observations had been 41, 39, 35, 38, 36, 41, 34, 33, 32, 35, 40, and 33 taken sequentially in the order given. Using the above procedure, when could the sampling have stopped?

Problem 8.13. For the four genetic treatment designs described in Section 8.8, discuss how these designs could be used for situations other than those listed in the text.

Problem 8.14. Substitute sweeteners for sugar in food products are in common usage. Show how to design a bioassay experiment demonstrating how much of the sweetener is equivalent to one gram of sugar.

Problem 8.15. In an educational experiment, students were given tests to determine retention and comprehension for the following eight treatment combinations:

One presentation				Two presentations			
Visual		Auditory		Visual		Auditory	
Immediate	Delayed	Immediate	Delayed	Immediate	Delayed	Immediate	Delayed
66.1	41.7	56.4	24.7	77.0	70.3	66.3	55.6

Show how to compute the main effects and interactions for the above data. Graph your results.

Problem 8.16. Suppose that four bridge players wish to play three games on Monday night for each of three weeks. They want to have each of the other three people as their partner in one game each night. Set up the design using a latin square of order three. Suppose that you would also like to have each player play in each of the four positions S, E, N, and W. Show how to design an experiment meeting these conditions.

Problem 8.17. For a study on reaction time while under the influence of alcohol, age is thought to be another variable that could affect reaction time. Suppose that the following data came from a randomized complete block experiment design with r = 5 replicates and with 50 subjects making up an experimental unit. The treatment design was a 3 age groups × 3 levels of alcohol consumption. The average reaction times of 50 subjects replicated five times are:

age group	amount of alcohol none	1 oz.	2 oz.	sum
20-39	0.42	0.47	0.65	1.54
40-59	0.51	0.62	0.66	1.79
60 or over	0.57	0.73	0.79	2.09
sum	1.50	1.82	2.10	5.42

Show how to estimate effects and plot the results graphically. (Problem taken from Dowdy and Wearden, 1983).

Problem 8.18. In learning about the shrinkage of cotton knit undershirts when washed and dried at military base laundries, the U. S, Army Quartermaster Corps took a random sample of four brands of shirts from several hundred brands available for purchase (from Dowdy and Wearden, 1983). Two shirts were randomly selected such that two of each brand were washed at three different temperatures of drying and two different temperatures of washing, i.e., cold water and hot water. The response variable was shrinkage of length in centimeters and the results are given below for each shirt (48 shirts in all):

Cold water wash   Hot water wash
Drying temperature   Drying temperature

Brand	210° F	218° F	226° F	210° F	218° F	226° F
A	1.9 2.1	3.3 3.7	7.5 7.9	3.4 3.6	8.0 7.6	7.5 7.7
B	2.2 2.4	4.8 5.0	9.8 9.2	4.6 4.4	9.3 9.5	10.1 9.7
C	2.8 3.2	6.6 6.5	13.2 13.0	6.3 5.7	12.9 13.3	13.1 13.3
D	3.1 3.7	4.8 4.5	11.2 10.8	5.6 5.0	10.9 10.7	11.4 11.7

Graph the results for cold water and also for hot water. Explain how to estimate the various effects. What are the population sizes for each of the three factors? Note that when the levels of a factor are selected at random the effect is denoted as a **random effect** and when a census of levels of a factor are used the effects are denoted as **fixed effects**. Which of the three factors in the above experiment could be considered as a random effect and which as a fixed effect?

Problem 8.19. Suppose an experimenter is interested in conducting an experiment where the response variables are weight loss and amount of sleep required. He wishes to use four distances of jogging, zero, one kilometer, two kilometers, and four kilometers, two types of aerobic exercise, none and one hour, and four diets, high protein, high fat, high carbohydrate, and normal. Write out the combinations for the experimenter.

Problem 8.20. Design an experiment design and a treatment design for a dietician who wishes to study tenderness of beef roasts and pork roasts using four roasting times (30, 50, 70, 90 minutes per pound) and five tenderizers (none, vinegar, papain, wine, lemon juice). What combinations should the dietician use? How should the roasts be selected? What type of blocking and replication would you suggest?

Problem 8.21. Suppose that an experimenter used three cooking oils (M, C, W) and four brands of flour (0, 1, 2, 3,) in all possible combinations. The 12 treatments were used in an experiment designed as a randomized complete block design with four blocks (replicates). The response was volume in cubic centimeters of 24 donuts. Suppose that the following means for the 12 treatment combinations were obtained:

Brands of flour

Oil	0	1	2	3	$\bar{y}_{h..}$	$\bar{y}_{h..} - \bar{y}_{...}$

C	$\bar{y}_{C0.}=12$ $\bar{y}_{C1.}=15$ $\bar{y}_{C2.}=21$ $\bar{y}_{C3.}=12$	15	-5
M	$\bar{y}_{M0.}=14$ $\bar{y}_{M1.}=18$ $\bar{y}_{M2.}=25$ $\bar{y}_{M3.}=15$	18	-2
W	$\bar{y}_{W0.}=19$ $\bar{y}_{W1.}=24$ $\bar{y}_{W2.}=23$ $\bar{y}_{W3.}=42$	27	7
$\bar{y}_{.i.}$	15 19 23 23	$20 = \bar{y}_{...}$	
$\bar{y}_{.i.} - \bar{y}_{...}$	-5 -1 3 3		

	Estimated interaction effects				
	0	1	2	3	sum
C	2	1	3	?	0
M	1	1	?	?	0
W	?	?	?	?	?
sum	0	0	0	0	0

Show how to obtain the values where the question marks appear and graph the two-way interaction.

Problem 8.22. List five possible experiment designs for the treatment design given in Problem 8.21 and describe when each would be used.

Problem 8.23. The tolerance of cats for tinctures of strophanthus A and B and ouabain was obtained for 23 cats, seven each on A and B and nine on ouabain. The mean values obtained were:

	Strophanthus A (0.01 c.c. / kg.)	Strophanthus B (0.01 c.c. / kg.)	Ouabain (0.01 c.c. / kg.)
Mean tolerance	1.68	1.99	6.12

Suppose that strophanthus A is the standard preparation and that the other two are test preparations, how is the relative potency of the test preparations estimated? (Problem from Finney, 1978, page 18.)

Problem 8.24. The following data are for increase in height + length of comb of capons (Y) in millimeters when injected with testosterone propionate from two sources where the means are for five capons:

	Standard preparation				Test preparation		
Dose	20 ug.	40 ug.	80 ug.		20 ug.	40 ug.	80 ug.
Mean	6.0	10.4	15.2		5.8	11.4	16.8

Plot mean response against log dose and draw an eye-fitted line of "best fit" for both the standard and test preparations. Estimate the slope of each line where slope is the increase in Y for a unit of log dose. Note for ease of computation use log 20 = 1, log 40 = 2, and log 80 = 3 since the doses are equally spaced on a log scale. Then, show how to compute relative potency of the test to the standard preparations.

Problem 8.25. The following data are from Finney (1978), page 150, and represent responses (Y) and are for an assay of nicotinic acid in a meat extract where the test preparation was a solution containing 0.2 mg. per ml. Duplicate assay tubes were prepared for each of the five doses of standard nicotinic acid and three doses of the test preparation. The means are presented below:

Dose of standard / tube (ug.)						Dose of test / tube (ml.)		
0.05	0.10	0.15	0.20	0.25		1.0	1.5	2.0
3.25	4.85	6.05	7.85	9.45		4.85	6.40	7.70

The mean response for zero dose per tube was 1.45. This is a slope ratio assay experiment. Plot the data and estimate slopes using "best fit" by eye method. Compute the relative potency of test to standard.

Problem 8.26. For the data of Problem 8.21, consider that C is a standard treatment and the other two are test treatments. Using the parallel line method, estimate the relative potency of the other two oils using graphical methods.

Problem 8.27. Manufacturer number one has an operator continuously adjusting a machine for making 12-inch rulers after every measurement. If the measurement is below 12 inches, the machine is adjusted to make longer rulers and if the measurement if above 12 inches the machine is adjusted to make the rulers shorter. Manufacturer number two uses a different procedure, i.e., the one described for Figure 8.10. Which of the manufacturers will have the most uniform lengths of rulers? Explain how you reached your answer and how you justified it.

CHAPTER 9. SUMMARIZATION OF DATA: GRAPHS, CHARTS, AND FIGURES

9.1 INTRODUCTION

The previous chapters have been concerned with procedures for obtaining meaningful, informative, and reliable data. This chapter and Chapters 10 and 12 describe some methods for summarizing data from a survey or an experiment. In the present chapter, we describe different kinds of graphs, charts, diagrams, and figures useful in summarizing some relevant facts from a mass of data. The data obtained for the 1967 class survey are to be utilized to illustrate some statistical procedures. Since data collection often results in large quantities of numbers, some form of summarization is necessary. However, prior to summarizing any data set, the investigator should carefully scrutinize the design of the investigation and the method of conducting it. If the design, conduct, and/or analysis of the investigation were questionable, more than likely the data set is not worth summarizing. The person summarizing a data set should have confidence in how the investigation was carried out as it would be a waste of one's time to summarize a data set which contains little or no information about the goals of the investigation.

The consequences of analyzing a data set which was obtained by inappropriate methods or was obtained with unreliable procedures are more than just wasting the time of the person doing the summarization The publication of results from the summarization of a poor data set can be misleading to readers and other investigators in the field of inquiry. Once the results appear in printed form, the reader is often unable to distinguish between a data set which contains reliable information and one which does not. In printed form in the same publication, both publications appear of equal value. It should be realized that data sets vary in value from being absolutely worthless to being excellent. However, it may often be difficult to evaluate data sets and, hence, one should be skeptical and cautious. A person should have full confidence in the reliability of a data set before any summarizations are performed.

9. 2 PARTIAL SUMMARIZATION OF A CLASS SURVEY

The statement that "A picture is worth a thousand words" (It was attributed to an unknown Chinese of bygone days, or so Bob Hope said at the Oscar awarding ceremony on April 18, 1966.) is often heard. This statement applies when presenting the numerical results from an experiment, survey, or any other type of investigation involving data. The individual data may be too numerous to present, and/or the meaning may not be clear in the individual data whereas it is in summary form such as in a picture, graph, table, or chart. Summarization of individual observations is essential in order to glean the information from the data and to interpret the results from the survey, the experiment, or other investigation.

In the class survey on heights, ages, weights, eye and hair color, and class standing, we obtained the data given in Table 9.1. The first items to note are that this represents a lot of numbers and that data for some individuals are incomplete. Of the 101 individuals registered for credit, 85 individuals were in class on the day of the survey, and an additional 13 who were not in class reported height, weight, age, hair and eye color, and class standing data for a total of 98 individuals. Of the 93 who submitted measurements on themselves, four failed to report a weight measurement and one failed to report his version of his eye and hair color. Also, in scanning the data of Table 9.1, student number 5 was measured as 1800 millimeters tall by team A and as 1706 millimeters tall by team B. It would appear from "own" height that team A made an error in measuring or in recording and that the height should have read 1700 millimeters instead of 1800. Also, it is possible that this was a typographical error.

As may be observed from Table 9.1, a large amount of data can often be accumulated quickly. Such a volume of data is not easily assimilated, and it is highly desirable to use some method of condensing the data into summary form. One method is to group into classes and present the number of individuals in each class. A **class** is a category broader than the original data that is determined by the nature of the data or by the interests of the experimenter. A class could be for nominal data such as blond, brown, black, or red, or for ordinal data such as 62 inches, 63 inches, and so forth. A **class interval** is used with ordinal data and is the interval from the lower value to the upper value of a class such as, e.g., 61.5 inches to 62.5 inches. A **class center** is the mid-point of a class interval. A **bar chart or graph** is a figure obtained by plotting frequency, percent, or areas (amount) against the class mid-points where the classes may be nominal or ordinal values. A **histogram** is a bar chart or graph where the areas of classes are plotted against the class centers and when the class centers are contiguous and are ordinal values. When the class intervals are of equal size for all classes, then the frequency or percent of individuals in a class may be plotted against class center values. If the class intervals are not of equal sizes, it is essential that areas of classes be used in order not to present a misleading and/or distorted visualization from the graph. Bar graphs have proven to be a very effective and useful method of summarizing data. Some forms of bar charts and graphs and tabular presentations are used below to summarize the data in Table 9.1.

One method of summarizing some of the information obtained is given in Figures 9.1 and 9.2 and Tables 9.2, 9.3, and 9.4. In Figures 9.1 and 9.2, we have plotted the frequency of heights that occurred at any given measurement. Such frequency distributions are called **histograms**. For example, when measuring in feet, there were nine individuals in the five-foot class and 76 in the six-foot class. Obviously, we had mostly six-footers in the class! This most frequent class is called the **modal class** and the class center, six feet, is called the **mode**. The class intervals are from 4'6" to 5'6" and from 5'6" to 6'6" for the two classes. When heights are recorded in inches as reported on the class questionnaire, the range is from 62 inches to 75 inches. In 1966, the range was from 61 to 78 inches. One or more students were represented in all 14 classes, but women appeared only in the first six classes. The modal classes here are the 68 and 71 inch classes and the modes are 68 and 71 inches. When the frequency distribution of heights in 2.5 centimeters class intervals for recording instrument and recorder A is constructed, we note that the range in class centers is from 160.0 to 192.5 centimeters whereas the range in millimeters is from 1600 to 1920. The modal class is 175.0 with the mode equal to 175.0. For recording instrument and recorder B, the range is from 1613 millimeters to 1919 millimeters with a modal class of 172.5 centimeters and with a mode of 172.5 centimeters.

It should be noted here that measuring instrument A was biased. The metersticks were screwed to the pine board three millimeters above the bottom of the stick. Thus, before any comparisons are made between the heights obtained by the two recorders using the two instruments, three millimeters must be subtracted from each of the heights obtained by recorder A. Perhaps the measuring instrument should have been biased by three centimeters as the measurers did not notice that they were using a biased measuring instrument in 1967 although they did in later years. If the differences in measurements taken by the two teams and listed as A - B

Table 9.1. Measurements on 1967 class.

own Stu't[a] in	mm(A)	mm(B)	height ft. (A)	ft. (B)	yd. (A)	yd. (B)	weight own	(B)	age own (yr.,mo.)	eye color own(A)		hair color own(A)		yr*	A-B	Hts. (mm) (A+B)/2 (in)
1. 71	1810	1819	6	6	2	2	148	152	20, 1	Bl	Bl	Br	Br	So.	-9	1814(71)
2. 65							120		32, 6	Br		Bk		Gr.		
3. 72	1800	1832	6	6	2	2	175	174	20, 7	Bl	Bl	Blo	Blo	Jr.	-32	1816(71)
4. 70	1800	1791	6	6	2	2	173	182	18, 5	Br	Br	Bk	Br	So.	9	1796(71)
5. 67	1800	1706	6	6	2	2	152	159	21, 11	Bl	Bl	Blo	Br	Sr.	94	1753(69)
6. 69	1750	1738	6	6	2	2	160	163	22, 6	Br	Br	Bk	Br	Jr.	12	1744(69)
7. 64	1660	1658	5	5	2	2	131	143	21, 2	Bl	Bl	Br	Br	Sr.	2	1659(65)
8. 72	1820	1820	6	6	2	2	180	184	20, 0	Bl	Bl	Blo	Blo	Jr.	0	1820(72)
9.* 66	1680	1680	6	6	2	2	136	136	18, 2	Bl	Bl	Blo	Blo	Fr.	0	1680(66)
10. 69	1760	1768	6	6	2	2	150	151	18, 10	Br	Br	Br	Br	Fr.	-8	1764(69)
11.* 63	1640	1617	5	5	2	2	140	141	19, 6	Br	Br	Br	Br	So.	23	1628(64)
12.* 67	1730	1712	6	6	2	2	123	123	19, 11	Bl	Bl	Br	Br	Sr.	18	1721(68)
13. 70	1760	1770	6	6	2	2	150	154	19, 0	Gy	Bl	Br	Br	So.	-10	1765(69)
14. 71							171		18, 2	Br		Br		Fr.		
15. 75	1900	1890	6	6	2	2	170	169	22, 6	Bl	Bl	Bk	Br	Sr.	10	1895(75)
16.	1780	1784	6	6	2	2		144			Br		Br		-4	1782(70)
17. 72	1840	1848	6	6	2	2	160	160	21, 10	Bl	Bl	Br	Br	Sr.	-8	1844(73)
18. 68	1740	1737	6	6	2	2	160	169	32, 0	Br	Br	Bk	Bk	Gr.	3	1738(68)
19.* 65	1670	1681	5	5	2	2	145	150	20, 0	Bl	Bl	Br	Br	So.	-11	1676(66)
20. 71	1790	1779	6	6	2	2		181	21, 11	Gy	Br	Br	Br	Sr.	11	1784(70)
21. 72	1840	1839	6	6	2	2	153	158	19, 9	Br	Br	Br	Br	Fr.	1	1840(72)
22. 73	1890	1874	6	6	2	2	177	185	18, 6	Gr	Bl	Br	Br	Fr.	16	1882(74)
23. 72	1840	1833	6	6	2	2	224	220	32, 2	Bl	Bl	Br	Br	Oth	7	1836(72)
24.	1700	1703	6	6	2	2		158			Gy		Br		-3	1702(67)
25. 75	1900	1914	6	6	2	2	195	201	19, 5	Br	Br	Br	Br	So.	-14	1907(75)
26. 72	1840	1839	6	6	2	2	148	143	19, 8	Br	Br	Br	Br	So.	1	1840(72)
27.* 66	1690	1703	6	6	2	2	125	121	19, 5	Br	Br	Br	Br	So.	-13	1696(67)
28. 71	1800	1792	6	6	2	2	148	149	18, 2	Bl	Bl	Br	Br	Fr.	8	1796(71)
29. 70	1703	1765	6	6	2	2	130	129	19, 2	Gr	Br	Br	Br	So.	-62	1734(68)
30. 65	1603	1640	5	5	2	2	142	141	19, 4	Bl	Bl	Br	Br	So.	-37	1622(64)
31. 71	1820	1800	6	6	2	2	168	170	19, 6	Gr	Br	Br	Br	So.	20	1810(71)
32. 69	1790	1765	6	6	2	2	180	186	21, 3	Bl	Bl	Br	Br	Sr.	25	1778(70)

Table 9.1. continued

Stu'tᵃin	own	height mm(A)	mm(B)	ft. (A)	ft. (B)	yd. (A)	yd. (B)	weight own	(B)	age own (yr.,mo.)	eye color own(A)	(A)	hair color own(A)	(A)	yr*	A-B	Hts. (mm) (A+B)/2 (in)
33.	72							175		20, 0	Br		Br		So.		
34.	72	1840	1836	6	6	2	2	140	145	21, 7	Bl	Bl	Br	Br	Jr.	4	1838(72)
35.*	65	1660	1651	5	5	2	2	125	127	18, 9	Br	Br	Bl	Bl	Fr.	9	1656(65)
36.	71							145		21, 1	Br		Br		Sr.		
37.	69	1781	1783	6	6	2	2	145	149	19, 1	Bl	Br	Br	Br	So.-	2	1782(70)
38.	71	1820	1828	6	6	2	2	146	148	21, 6	Br	Br	Br	Br	Sr. -	8	1824(72)
39.	67	1680	1708	6	6	2	2	125	120	19, 5	Bl	Bl	Br	Br	So.	-28	1694(67)
40.*	66	1700	1706	6	6	2	2	128	129	19, 6	Br	Br	Br	Br	So.-	6	1703(67)
41.	74	1910	1919	6	6	2	2		187	19, 7	Bl	Bl	Br	Br	So.-	9	1914(75)
42.	74	1920	1895	6	6	2	2	175	184	19,11	Bl	Bl	Br	Br	So.	25	1908(75)
43.	73	1850	1853	6	6	2	2	168	169	21, 0	Br	Bl	Blo	Br	Sr. -	3	1852(73)
44.		1800	1782	6	6	2	2		166			Bl	Blo			18		1791(71)
45.	74	1880	1872	6	6	2	2	151	159	19, 4	Bl	Bl	Br	Br	So.	8	1876(74)
46.	71	1782	1818	6	6	2	2	130	135	17, 0	Bl	Bl	Br	Br	Fr.	-36	1800(71)
47.	68	1720	1721	6	6	2	2	155	159	22, 0	Gr	Bl	Br	Br	Jr. -	1	1720(68)
48.*	63							110		21, 5	Br		Br		Sr.		
49.	68	1710	1708	6	6	2	2		199	38, 8	Br	Br	Bk	Bk	Gr.	2	1709(67)
50.	66	1690	1683	6	6	2	2	165	173	19, 8	Gr	Gr	Br	Br	So.	7	1686(66)
51.	72	1840	1814	6	6	2	2	178	182	20, 9	Br	Br	Bk	Bk	Sr.	26	1827(72)
52.	68	1740	1724	6	6	2	2	156	158	20,11	Gr	Gy	Bk	Br	So.	16	1732(68)
53.	72	1783	1812	6	6	2	2	151	152	20, 2	Gr	Br	Br	Br	So.	-29	1798(71)
54.	74	1920	1885	6	6	2	2	213	217	19, 7	Bl	Bl	Br	Br	So.	35	1902(75)
55.	69	1750	1770	6	6	2	2	174	181	20, 3	Gy	Gy	Br	Br	Jr.	-20	1760(69)
56.	70	1800	1804	6	6	2	2	155	161	25, 5	Bl	Bl	Br	Br	Sr. -	4	1802(71)
57.	68	1740	1750	6	6	2	2	165	168	24, 7	Gy	Bl	Br	Br	Gr.	-10	1745(69)
58.	68	1720	1721	6	6	2	2	115	116	20, 2	Br	Br	Br	Br	Jr. -	1	1720(68)
59.	74	1901	1890	6	6	2	2	163	168	20, 2	Bl	Bl	Br	Br	Jr.	11	1896(75)
60.	69	1740	1750	6	6	2	2	157	163	22, 0	Bl	Bl	Br	Br	Sr.	-10	1745(69)
61.*	67	1740	1725	6	6	2	2	140	149	21, 9	Bl	Gy	Br	Br	Sr.	15	1732(68)
62.	71	1830	1821	6	6	2	2	135	132	19, 4	Bl	Bl	Br	Br	Sr.	15	1732(68)
63.	69	1740	1751	6	6	2	2	145	150	20, 2	Br	Br	Br	Br	Sr.	-11	1746(69)
64.	63	1620	1619	5	5	2	2	126	128	23, 1	Bk	Br	Bk	Bk	Sr.	1	1620(64)
65.		1790	1781	6	6	2	2		166			Br	Bk			9		1786(70)
66.	68	1730	1736	6	6	2	2	145	146	17, 9	Br	Br	Br	Br	Fr. -	6	1733(68)
67.	67	1680	1703	6	6	2	2	165	176	20, 0	Br	Br	Br	Br	So.	-23	1692(67)

Table 9.1. continued

Stu't^a (own)	in (own)	mm(A)	mm(B)	ft.(A)	ft.(B)	yd.(A)	yd.(B)	weight own(A)	(B)	age own (yr.,mo.)	eye own	eye (A)	hair own	hair (A)	yr*	A-B	Hts. (mm) (A+B)/2 (in)
68.	70	1800	1797	6	6	2	2	126	131	19, 0	Bl	Bl	Br	Br	So.	3	1798(71)
69.	66	1700	1715	6	6	2	2	148	154	19, 3	Gr	Br	Br	Br	So.	-15	1708(67)
70.	68	1730	1748	6	6	2	2	145	150	20, 0	Bl	Bl	Br	Br	So.	-18	1739(68)
71.	67	1710	1708	6	6	2	2	153	153	20, 0	Br	Br	Br	Br	Sr.	2	1709(67)
72.*	63								110	21, 9	Br		Br		Sr.		
73.*	67	1701	1713	6	6	2	2		147	19, 8	Br	Br	Br	Br	So.	-12	1707(67)
74.	74								172	20, 2	Bl		Blo		So.		
75.	68								148	21,11	Bl		Br		Sr.		
76.	69	1750	1748	6	6	2	2	135	142	20, 2	Gr	Br	Br	Br	So.	2	1749(69)
77.	71	1820	1795	6	6	2	2	172	179	19, 3	Gy	Br	Br	Br	So.	25	1808(71)
78.	68								184	20, 5					So.		
79.*	63	1620	1637	5	5	2	2	110	112	17,10	Bl	Bl	Br	Br	Fr.	-17	1628(64)
80.*	65	1680	1660	6	6	2	2	125	125	18, 0	Gr	Br	Br	Br	Fr.	20	1670(66)
81.*	65	1600	1613	5	5	2	2	95	95	20, 2	Br	Br	Br	Br	Sr.	-13	1606(63)
82.	74	1900	1870	6	6	2	2	175	176	20, 1	Br	Br	Br	Br	Jr.	30	1885(74)
83.	68	1750	1744	6	6	2	2	215	218	23, 4	Gr	Br	Br	Br	Jr.	6	1747(69)
84.	72	1890	1893	6	6	2	2	240	252	21,10	Br	Br	Br	Br	Sr.	- 3	1892(74)
85.	65	1640	1641	5	5	2	2	142	155	17, 3	Br	Br	Br	Bk	Fr.	- 1	1640(65)
86.	68	1720	1733	6	6	2	2	150	151	20, 1	Gy	Br	Br	Br	Jr.	-13	1726(68)
87.	71								170	20, 0	Br		Br		So.		
88.		1800	1801	6	6	2	2		144			Bl		Br		- 1	1800(71)
89.	71	1820	1794	6	6	2	2	143	141	20,11	Bl	Bl	Br	Br	Jr.	26	1807(71)
90.	68	1680	1716	6	6	2	2	165	172	25, 5	Br	Br	Rd	Br	Jr.	-36	1698(67)
91.	68	1740	1737	6	6	2	2	160	170	20, 7	Gr	Bl	Br	Blo	Jr.	3	1738(68)
92.	72	1840	1840	6	6	2	2	184	188	22, 8	Bl	Bl	Br	Br	Sr.	0	1840(72)
93.	71								166	21, 0	Gr		Bk		Jr.		
94.	69	1720	1739	6	6	2	2	160	160	20, 0	Bl	Bl	Bk	Bk	Jr.	-19	1730(68)
95.	71								185	20, 6	Gr		Bk		Jr.		
96.	66	1710	1713	6	6	2	2	149	153	28, 0	Gr	Bl	Br	Br	Oth	- 3	1712(67)
97.	73	1870	1856	6	6	2	2	213	215	20, 8	Gy	Br	Br	Br	Jr.	14	1863(73)
98.	72								184	18, 3	Gr		Br		Fr.		
No. 93	85	85	85	85	85	85	89	85		93	92	85	92	85	93		

Asterisk indicates female student as determined by name.   Year in school indicated by Fr. = Freshman, So. = Sophomore, Jr. = Junior, Sr. = Senior, Gr. = Graduate, Oth = Other.

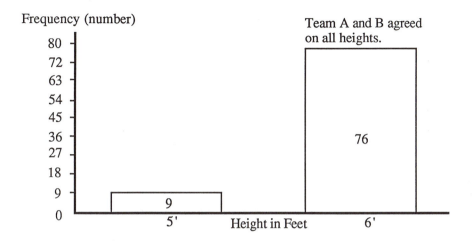

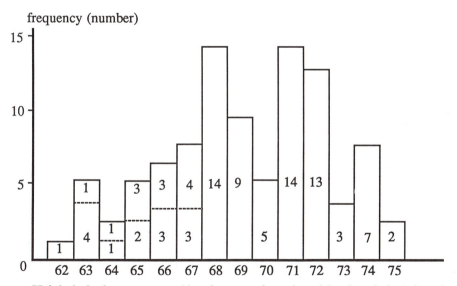

Height in inches as returned by class questionnaire. (Numbers below dotted line refer to number of women out of 14.)

Figure 9.1 Heights in feet as measured by teams A and B and in inches as recorded by the students themselves, Spring class, 1967.

Height in cm. as measured by team A (number below dashed line refer to number of women.

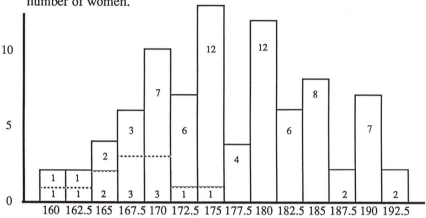

Height in cm. as measured by team B (number below dashed line refer to number of women.

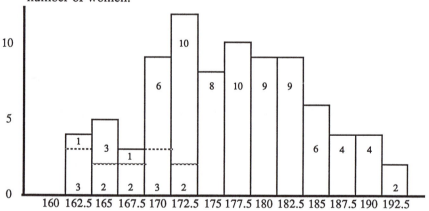

Figure 9.2 Height in cm. as measured by teams A and B, Spring Class, 1967.

Table 9.2. Frequency distribution of eye and hair color, Spring, 1967.

Eye color	black	brown	blond	red	Total	Proportion
		For own Hair color				
black	1	0	0	0	1	1/92
brown	6	25	2	1	34	17/46
green	2	13	0	0	15	15/92
blue	2	28	5	0	35	35/92
grey	0	7	0	0	7	7/92
Total	11	73	7	1	92	- -
Proportion	11/92	73/92	7/92	1/92	-	

Eye color	black	brown	blond	red	Total	Proportion
		For A Hair color				
black	0	0	0	0	0	0
brown	6	33	1	0	40	8/17
green	0	1	0	0	1	1/85
blue	1	34	5	0	40	8/17
grey	0	4	0	0	4	4/85
Total	7	72	6	0	85	- -
Proportion	7/85	72/85	6/85	0	-	

in Table 9.1, we note that the differences range from -62 mm to +94 mm, or from -6 to +9 cm. There are 41 differences with a minus sign, three ties, and 41 with a plus sign. After subtracting three millimeters from the A measurements, there are 51 differences with a minus sign, three ties, and 31 with a plus sign. Since we expect $85 / 2 = 42.5$ pluses on the basis of a hypothesis of no difference, i.e., the null hypothesis, in the measurements A and B, we note that this is a fairly poor fit with expectation. (Note, for

Table 9.3. Frequency distribution of age and class standing, Spring, 1967.

Age class*	Freshman	Sophomore	Junior	Senior	Graduate	Other	Total
16,7 to 17,6	2						2
17,7 to 18,6	8	1					9
18,7 to 19,6	2	15					17
19,7 to 20,6	1	16	8	4			29
20,7 to 21,6		1	5	7			13
21,7 to 22,6			3	9			12
22,7 to 23,6			1	2			3
23,7 to 24,6							0
24,7 to 25,6			1	1	1		3
> 25,6					3	2	5
Total	13	33	18	23	4	2	93

* 16,7 means 16 years and 7 months of age; 17,6 means 17 years and 6 months of age.

those who know about the chi-square statistic (See Chapter 14.), X^2 (one degree of freedom) = $[(52.5 - 42.5)^2 + (32.5 - 42.5)^2] / 42.5 = 200 / 42.5 = 4.7$, which is a rather large value of chi-square. One and a half of the ties were put into the minus class and one and a half into the plus class. Hence, we probably would conclude that team B obtained larger heights on the average than did team A (the short team).)

Instead of constructing a frequency distribution as a figure or graph, we may construct a table or array. A **two-way array or table** is a tabular array of the frequencies of individuals appearing in the combination ij for the ith level of one classification and the jth level of a second classification. The number and nature of the combinations is very much akin to a two-factor factorial. and the two classifications correspond to the two factors. This method of summarization is popular with many investigators. We may wish to study the two-way array of hair color and eye color as in Table 9.2 and to note the differences between own and A classifications.

The most frequent combination is brown hair and blue eyes; the next most frequent is brown hair and brown eyes. Many combinations had zero frequency in our survey. These classes *may* also have zero frequency in the entire universe of individuals, as for instance, black eyes with naturally occurring red hair. Another manner of summarizing data is to present the results as proportions, as in the last row and last column of Table 9.2. Of the 92 individuals returning a questionnaire 1 / 92 had red hair, 7 / 92 had blond hair, 73 / 92 had brown hair, and 11 / 92 had black hair. Thus, approximately 3 / 4 of the class had brown hair. We have no idea of how representative this is of the student body at Cornell University because we have no idea of how many individuals were allotted to this class with respect to the color of their hair. *Perhaps* this class could be considered as a representative sample of students at Cornell with respect to hair color or eye color, the sample could be considered as a simple random on the basis that eye or hair color of a student who signed up for this class, is a random event. If so, then about one-fourth (25 / 92) of Cornell students in the Spring of 1967 reported that they had brown eyes and brown hair, and, as determined by Team A, about one-third (33 / 85) had brown eyes and brown hair.

In Table 9.3, we have given the frequencies of age and class standing. The most frequent group in the two-way array is of 20 year olds, plus or minus six months, in the sophomore class with 16 individuals. The most frequent class in the class standing classification is the sophomore group. The most frequent age class is the 20 year plus or minus six months group with 29 individuals. In this table, the class with zero frequencies was left blank to contrast with the previous table, Table 9.2, where zeros were included. Either form is acceptable, but Table 9.3 may be easier to read.

From the two-way array of weights and heights in Table 9.4, we may note that all heights and weights of the 13 female students (There were 15, but one presented no data, and one did not report her own weight) in this class fall in the upper left-hand corner of the table and that there appears to be a relationship of higher weights with the taller people. The upper right-hand and the lower left-hand portions of the table are devoid of frequencies in this sample. We know that individuals of the "5 × 5" and of the "beanpole" categories do exist in the population, even though infrequent. The relatively heavier concentration of individuals along the diagonal, sloping downward, of the table indicates a positive relationship between height and weight. In general, taller people do weigh more than shorter people. There is simply more of them!

Other methods of summarizing these data will be considered later.

Before proceeding to another example, we should note that the size of the class interval was selected arbitrarily. This was also true of the class center. In general, we may note that if the class interval is too broad, all observations fall into one class. This was true of the height measurement made in yards. If the class interval is too fine, for instance one millimeter, only zero, one, or a few individuals will fall in the class interval. To obtain some idea of the modal class or classes and of the shape of the frequency distribution, we need more than 93 observations as recorded here. 1000 to 5000 observations with 20 to 30 classes should suffice to indicate the form of the distribution and the modal class for most ordinal data. Also, such a number gives a good idea of the range of the sample observations. The smaller the number of observations, the broader the class intervals will need to be in order to obtain observations. Measurements of height in feet resulted in nine in the 5-foot class and 76 in the 6-foot class; those in inches, however, produced some observations in all classes from 62" to 75". Had we grouped by 2-inch intervals starting with 61-62, there would have been more individuals in each class.

9. 3   A MORE EXTENSIVE EXAMPLE OF A FREQUENCY DISTRIBUTION

In order to obtain an idea of the frequency distribution of heights of men and women, data originally reported by Karl Pearson and A. Lee in volume 2 of *Biometrika,* 1903, and referring to heights of English people about 1900, are reproduced in Table 9.5. The class interval in daughters' and fathers' heights was one inch. The class center was on the half-inch mark. If a height was reported as 62 inches, one-half was put in the 61.5 inch class and one-half in the 62.5 inch class. Likewise, if a father 63 inches tall had a daughter 52 inches tall one-fourth of this observation was put into each of the four classes: (62.5, 52.5), (62.5, 53.5), (63.5, 52.5), and (63.5, 53.5). This splitting of an observation accounts for the fractions of observations found in the table. The most frequent class in the two-way classification is the one with 67.5" for fathers and 63.5" for daughters. The modal class for fathers' heights is 68.5" and for daughters is the 63.5" class. None of the 1376 observations are to be found in the upper right-hand corner or the lower left-hand corner of the table. Very tall daughters did not have very short fathers and *vice versa.*

One item of interest is the symmetrical bell-shaped form of the frequency distribution represented by a histogram of the frequencies for

Table 9.4. Frequency distribution of weights and heights of class, Spring, 1967.*

Own

Height class (in.)	\	\	\	\	Weight class (pounds)	\	\	\	\	\	\	\	\	\	Total
	91-100	101-110	111-120	121-130	131-140	141-150	151-160	161-170	171-180	181-190	191-200	201-210	211-220	221-230	231-240	
62	g															1
63		3g		1	g											5
64				g	1											2
65			1	g		g,2										5
66			2g	g		2		1								6
67			g,1	g			2	1								6
68				1			4	4	2		1		1			13
69						1	3	3		2						9
70				2			1	1		1						5
71				1		1	5		3	2	1					13
72						1	1	3		4	2			1	1	13
73								1	1				1			3
74							1	1	3				1			6
75								1			1					2
Total	1	3	2	10	7	19	14	10	13	4	1		3	1	1	89

Team B

Height class (in.)	\	\	\	\	Weight class (pounds)	\	\	\	\	\	\	\	\	\	Total
	91-100	101-110	111-120	121-130	131-140	141-150	151-160	161-170	171-180	181-190	191-200	201-210	211-220	221-230	251-260	
64	g		g	1			g									4
65				2g			2	1								5
66						g		g		1						3
67			1	3g			g	4		1		1				11
68				1			g,1	5	3	1						12
69				1			3		2		1		1			8
70							2	2	2		2					8
71					1		3	1	2	1	2					10
72					2		3	2		1	2		1			11
73								1	1				1			3
74								1	2	1	1		1			6
75											1		1		1	3
76											1					1
Total	0	3	2	7	4	18	17	12	6	10	1	1	4		1	85

(Note: In Table 9.4, g refers girl student and xg refers to x girl students, x = 1, 2, 3. A number alone refers to number of boy students in a given class.)

Table 9.5. Heights (times 10) in inches of English fathers and daughters about 1900.

Fathers	\ Daughters (frequency times 10)*																					Total
	525	535	545	555	565	575	585	595	605	615	625	635	645	655	665	675	685	695	705	715	725	
585				2	2	2	5	7														20
595				2	2	7	10	7	5	10												45
605					5	5	20			17	22	2	2									75
615				2	15	7				25	20	20	20	25	5	5						145
625	2	2			12	45	7		60	80	97	45	60	17	10	5		2	2			450
635	2	2			5	10	10	47	62	115	120	82	32	5	15			2	2			515
645					15	17	50	125	130	227	110	92		110	32	10	2	2				925
655				10	10	15	12	62	182	237	260	272	230	122	72	57	2	2				1550
665					5	25	50	117	202	237	330	357	187	92	87	70	15	2				1780
675					5		27	35	110	202	282	372	285	197	160	40	30	2				1750
685						5	5	35	90	165	247	315	330	300	262	142	55	10	17	5	10	1995
695						5	2	20	47		102	142	262	342	265	267	132	42	25	2		1660
705							17	25	42		137	162	245	222	205	120	57	65	45	5		1350
715								5	12	30	47	77	117	150	185	112	52	22	7	5		825
725									12	12	7	15	55	47	77	45	37	27	12	15		365
735											5	7		10	37	42	37	25	20	7	7	200
745													2	2	20	2	7	15	10	2	2	65
755															10	10	5		20			45
Total	5	5	0	10	45	145	155	485	990	1415	1905	2120	1985	1595	1425	775	360	195	95	40	10	13760

* Frequencies in table have been multiplied by 10; 5 is to be appended to all last digit numbers 2 and 7 and 0 is to be appended to all last digit numbers 0 and 5, i.e., 2 is 0.25, 37 is 3.75, 5 is 0.50, and 10 is 1.00. This convention was used in order to fit this table on a single page without going to a smaller fontsize.

heights of fathers and of the daughters. Where the measurements follow a bell-shaped frequency distribution such as this, we say that the observations follow the form for a **normal frequency distribution**. Many measurements follow a normal frequency curve, but many do not. For example, weights of adult humans would tend to follow a frequency distribution which is not bell-shaped, since there would be too many weights in the right-hand portion of the frequency distribution.

9. 4. GRAPHS AND CHARTS

There are many, many ways in which to draw graphs, in which to present the basic results in a meaningful manner, and in which to disillusion the reader. The reader is referred to Huff (1954, Chapters 3, 5, 6, 7 and 9), Moroney (1956, Chapter 3), Schmid (1956), Bevan (1968, Chapter 5), and Campbell (1974) for additional and interesting reading on graphs and charts.

Too illustrate various ways of graphically representing a set of data, we shall consider an example given by Bevan (1968) wherein he states that "in an exclusive interview with the author", Yogi Bear stated that of $100 he received for television and film work during a given month, he gave the reservation manager $25 for board and lodging, spent $45 on extra honey, gave $10 to his friend Bubu, bribed someone with $15, and did not remember how he spent the remaining $5. We may represent these expenditures in an **ideograph** or **ideogram form** as follows:

Twenty $5 bills spent by Yogi Bear

$5 $5 $5 $5 $5	$5 $5 $5 $5 $5 $5 $5 $5 $5	$5 $5	$5 $5 $5	$5
five $5 bills were spent on board and lodging	nine $5 bills were spent on extra honey	two $5 bills were given to his friend Bubu	three $5 bills were spent on bribing some-one	one $5 bill could not be accounted for

Another form for presenting the above data would be in the form of a **pictogram** or a **pictograph** using money bags of varying sizes (varying only one dimension such as height), replicas of $5 bills, or pictures of various activities with the amount of money indicated. Still another type of

presentation of the disappearance of Yogi's $100 could be in the form of a
pie chart or **pie graph** as shown in Figure 9.3.

Horizontal and vertical bar charts may be utilized to show the
expenditures as shown in Figure 9.4, 9.5, and 9.6. A **horizontal bar
graph** is a bar graph for the bars run in a horizontal position whereas a
vertical bar graph is a bar graph in which the bars are drawn vertically
as in Figures 9.1 and 9.2. The above results for Yogi Bear may also be
presented in a vertical or horizontal line chart or graph as shown in
Figures 9.7 and 9.8, respectively.

For the example considered in this section, there is no scale for the
expenditure items as there is for dollars spent, i.e., the scale is nominal. In

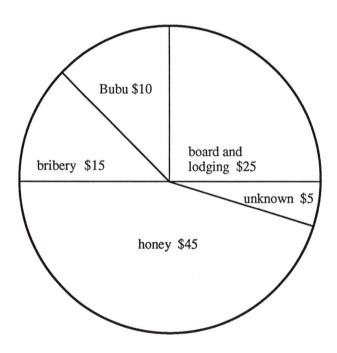

Figure 9.3. Pie graph.

Figures 9.1 and 9.2, both axes had known scales; the vertical one or the ordinate represented number or frequency of occurrence, and the abscissa or horizontal scale represented units of length measurements. The graphs in these pictures were denoted as frequency histograms. Thus, we may note that a histogram is a bar chart which measures the relative area, proportion, or frequency of occurrence in a given class interval which is measured on the abscissa in scaled units of measurement. This means that the spacing between bars in a histogram is represented by ordinal data and, therefore, cannot be for nominal data, as it was for presenting data similar to Yogi Bear's expenditures.

Instead of bars on a graph, it may be desirable to use a single line to represent the frequency or amount of response for each or the classes. Such graphs are denoted as **horizontal or vertical line graphs**; the direction of the lines determine whether it is vertical or horizontal. Yogi Bear's expenditures are presented as line graphs in Figures 9.7 and 9.8.

If the frequency at the successive midpoints of class intervals are connected by straight lines and if the ordinate and abscissa are scaled units of measurement, the resulting figure is denoted as a **frequency polygon**. An example of four frequency polygons in one graph is given in Figure 9.9.

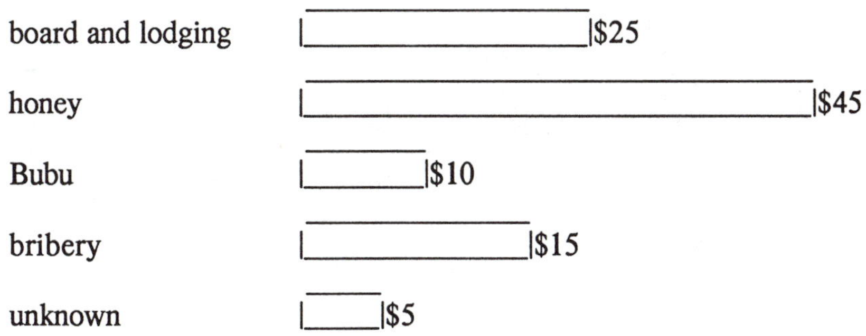

board and lodging |_____|$25

honey |_____|$45

Bubu |_____|$10

bribery |_____|$15

unknown |_____|$5

Figure 9.4. Horizontal bar graph.

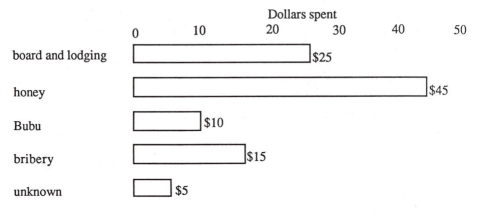

Figure 9.5   Horizontal bar graph.

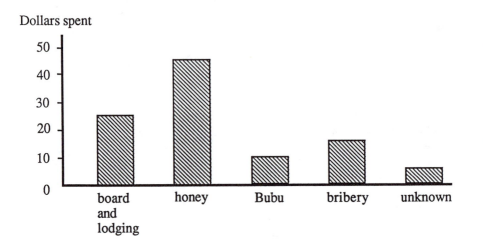

Figure 9.6   Vertical bar graph.

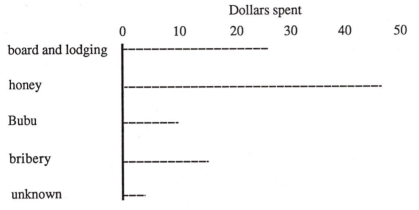

Figure 9.7   Horizontal line graph.

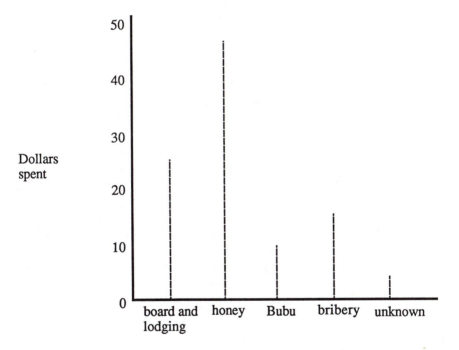

Figure 9.8   Vertical line graph.

The number of deaths from motor vehicle accidents is given in terms of 10,000,000 born alive, to take into account the differences in proportions of various categories in the population. From the graph one *can* conclude that more men are killed in motor vehicle accidents than women. However, from these data *per se,* it would be unwise to conclude that one should always go driving with a woman! It may be that women simply do not drive under the same conditions that men do and perhaps at the same time of day. Likewise, we should not conclude from these data in themselves that middle-aged white drivers are safer drivers than middle-aged non-whites. Cause and effect must be considered in light of all the facts and not just of one such set, as is done by persons who wish to "prove" (really to substantiate) their contention. For example, suppose that one were very much opposed to lowering the voting age to 18. One could use Figure 9.9 or, even better, a pooled estimate for all male and female data, to contend that immaturity is amply illustrated in the 15 - 20 and 20 - 25 age classes by the increased number of deaths, that instead of lowering the voting age to include more immature people, we really should raise it, and that the data in Figure 9.3 "prove" one's point! (Perhaps the data substantiate the fact that women are more mature than men but do not prove it!) To determine whether or not the contentions are true would require considerable study and more precise definitions.

Instead of utilizing a frequency polygon, a researcher may wish to use cumulative frequencies in the form of an **ogive** or a **cumulative frequency** curve or graph. To illustrate, let us accumulate deaths over age classes for non-whites as in Figure 9.10. At each age, the number of deaths due to motor vehicle accidents per 10 million born alive has been accumulated at each 5-year age interval. Likewise, one could reverse the cumulative frequency curves and have "number-yet-to-die" instead of "number-died-to-age-class". Both forms are sometimes used in "intelligence quotients" = (mental age / actual age) × 100 charts where it is desired to project "more than" or "less than" a given I.Q. score (See Moroney, 1956, page 25, for an illustration.).

Note that a smooth curve was drawn through the midpoints of the age classes. Such a curve can be misleading because it would appear that these results are arrived at from a much larger sample of data than is actually the case. It is recommended that for samples the cumulative frequency graph be used, instead of the ogive which should be reserved for the population. Also, one could connect the class midpoints with straight lines instead of

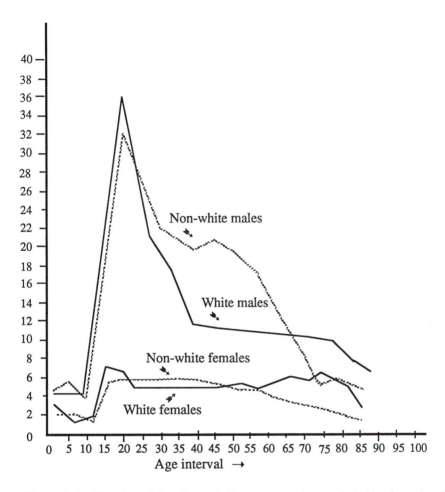

Figure 9.9   Number dying in each five year age interval (midpoints given) from motor vehicle accidents, 1959-61. (Source: U.S. Life Tables by Causes of Death: 1959-61, Volume 1, No. 6, U.S. Dept. of Health, Education and Welfare, May 1968.)

curved ones if such a representation were desirable. With appropriate
explanation, the reader should not be deceived.

Many other forms of graphs and charts are available to convey the
pertinent facts from an experiment or survey. It is essential that a result be
portrayed in its true form. Sometimes three-dimensional figures in the
form of volumes, as for instance a loaf of bread or a pound of butter, are
useful in portraying facts from a given set of data. Colors may be utilized

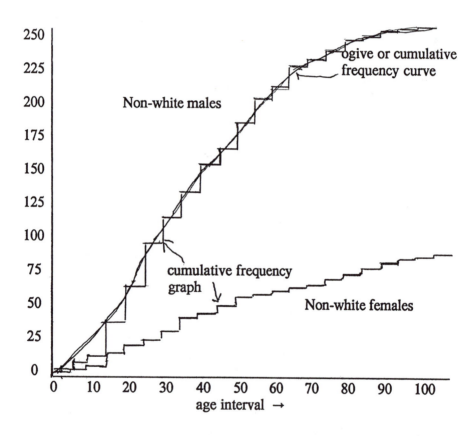

Figure 9.10. Cumulative frequency of number of deaths due to motor vehicle
accidents for non-white males and females over age classes. (Source: U.S. Life
Tables by Causes of Death: 1959-61, volume 1, no. 6, U.S. Dept. of Health,
Education, and Welfare, May, 1968.)

to liven up a drab but important set of figures. All such methods are useful in summarizing facts from a set of data, and they are also useful in deceiving the reader about the facts, as we shall see in Section 9.6.

9.5 STEM AND LEAF DIAGRAMS

Professor John W. Tukey of Princeton University proposed a simple and straight-forward graphical method for obtaining a histogram without first ordering the data. In addition, considerably more information about the data set is immediately available. This method involves construction of a stem and leaf diagram as a first step. The construction of a **stem and leaf diagram or graph** consists first of determining the **stems** or classes which may be the thousands, hundreds, or tens digit of the data, secondly of writing down the stems in the first column of the diagram, thirdly of drawing a line after the column of stems, and lastly copying the last digit or digits of the data after the appropriate stem in the order the data are tabulated. No re-ordering of the data is necessary as it is with a bar chart. The resulting stem and leaf graph is a bar chart and when the stems are ordinal data, it is a histogram. More detail on this type of graph may be found in Mosteller and Tukey (1977) and Koopmans (1987) as well as in other references.

To illustrate the method of constructing a stem and leaf graph, consider the following data representing the 102 examination (prelim) grades for the first test given to students in a statistics class using this text in Spring 1986:

```
96  79  82  87  75  86  84  86  86  94
82  95  81  88  71  86  84  87  94  70
81  77  97  88  73  86  83  94  83  69
82  77  82  96  75  86  92  88  83  90
82  78  81  87  96  94  84  86  91  70
81  78  77  86  92  96  83  90  84  68
81  77  79  92  68  86  95  86  84  90
82  77  89  87  69  91  83  97  83  47
82  91  49  90  73  88  83  88  83  56
90  91  64  88  75  87  83  87  83  61
65  64
```

For the first stem and leaf graph for the above data set, use the tens digit to form the stems. The units digits will form the leaves. As an illustration, we first construct a skeletal or partial stem and leaf graph using only the ten grades in the first row reading from left to right. The selected stems are 9, 8, 7, 6, 5, and 4 and the ten resulting leaves are 6, 9, 2, 7, 5, 6, 4, 6, 6, and 4; the graph for these ten grades is:

Stems	Leaves
9	6 4
8	2 7 6 4 6 6
7	9 5
6	
5	
4	

The stem and leaf graph or diagram for the 102 grades is given in Figure 9.11. The above procedure is used to construct this graph.

Although the graph in Figure 9.11 is a horizontal histogram, it could be turned 90 degrees to obtain a vertical histogram. Additional information is easily obtained from this graph or may be appended to it. Two columns have been added to the graph. These are a column of frequencies for each of the stems and a column of cumulative frequencies. The latter item makes it un-necessary to obtain a count of the data before making the graph as the final cumulative frequency gives the total number of grades, i.e., 102. Also, the range is immediately obvious, being from 97 down to 47 or a spread of 50 points. Almost one-fourth, 24 / 102, of the grades were in the 90s, and approximately one-half, 50 / 102, were in the 80s.

The selection of stems using the tens digit may be sufficient for the purpose of giving grades of A for the 90s, B for the 80s, C for the 70s, D for the 60s, and F for the rest of the grades. However, having 50 grades in one class might be too coarse a grouping. It may be that a finer partitioning of the 102 grades is desirable. In such a case, stems could be selected to be five grade points, i.e., 46 - 50, 51 - 55, 56 - 60, 61 - 65, 66 - 70, 71 - 75, 76 - 80, 81 - 85, 86 - 90, 91 - 95, and 96 - 100, or any other set of stems meeting the needs of the investigator. The stem and leaf graph of the grades for the preceding stems is given in Figure 9.12. This selection of stems appears to portray the distribution of grades in a better manner than Figure 9.11. The modal class is the 86 - 90 class with 28 grades in the class. The 51st grade had a value of 84 and the 52nd grade value was also 84. Thus, the median is 84. The upper quartile was from

Stems	Leaves	Frequency	Cumulative frequency
9	|6454746206412602501710001	24	24
8	|2764662186471863322683217461634166 4		
	|297332838387373	50	74
7	|95107375808779735	7	91
6	|98894154	8	99
5	|6		100
4	|79	2	102

Figure 9.11. Stem and leaf diagram with frequencies and cumulative frequencies of 102 prelim grades for Spring 1986.

Stems	Leaves	Frequency	Cumulative frequency
96-100	|676667	6	6
91-95	|4544241225111	13	19
86-90	|76668678668076606609708808 77	28	47
81-85	|242141332232141341423323333	27	74
76-80	|977887797	9	83
71-75	|513535	6	89
66-70	|090889	6	95
61-65	|4154	4	99
56-60	|6	1	100
51-55	|	0	100
46-50	|79	2	102

Figure 9.12. Stem and leaf diagram of 102 grades by five point stems.

89 to 97 while the lower quartile was from 47 to 78. The middle 50%, the inter-quartile range, was from 78 to 88. A five grade point interval still may be too coarse a grouping if the instructor gives grades of A+, A, A-, B+, B, B-, C+, C, C-, D+, D, D-, and F. The instructor may wish to construct a stem and leaf graph with stems of two or three grade points.

9.6 SOME COMMENTS ON GRAPHS, CHARTS, AND PICTURES

Huff's (1954) and Campbell's (1974) delightful little books contain a wealth of examples of misuses of statistics and deceptive tricks. These books are highly recommended reading in connection with this chapter, as are the other references listed at the end of this chapter. As may be observed from these references, there are many ways of deceiving the un-suspecting reader. The scales of the ordinate and abscissa can be changed relative to each other to increase or to decrease the slope of a curve. The scales can be chopped off and the picture of the remaining segment enlarged to make the slope of a curve appear steep when in fact it was almost flat. Class intervals can be selected to produce a histogram that may present a biased picture. In the class survey for heights measured in yards, the class interval was from 4'6" to 7'6"; all 85 students fell into this class, resulting in no variation in heights as presented. Class intervals can always be selected so as to exclude zero frequencies when in fact they should be indicated as being present.

An interesting example from Huff (1954), pages 118 - 119, relates to the use of percentages and the selection of a base which can make prices appear to go up or down as the person pleases. The data given are:

Item	price last year	this year	ratio this year / last year	last year / this year
milk	20¢	10¢	50%	200%
bread	5¢	10¢	200%	50%
		arithmetic average	125%	125%

Now let us present the data in graphical form; in Figure 9.13, last year is represented by the base period (denominator) for the left graph and this year is the base for the graph on the right. In viewing either of the graphs, the reader could easily be misled. The pertinent fact is that this year it costs 5¢ less for one loaf of bread plus one quart of milk than it did last year! The use of percentages did nothing to clarify the issue. Also, the selection of items and prices to be utilized in constructing percentages can have a considerable effect; in the above, an arithmetic average of prices was used, whereas a geometric average, $\{50\%(200\%)\}^{1/2} = 100\%$ (See Chapter 10), would show that prices did not change, that is, one doubled in price and the other item was one-half the price, and it does not matter which year is used as the base.

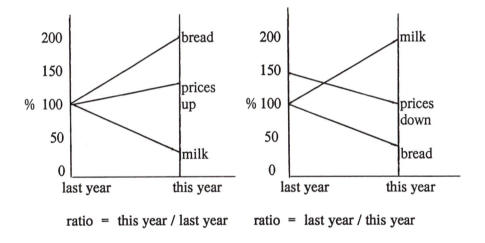

Figure 9.13. Two ways of depicting the same data.

An example illustrating misuse of a picture is Figure 9.14, depicting consumption of cheese from various sources in the United Kingdom in 1953. This appeared in "Season at a Glance" in the New Zealand Dairy Board Annual Report (p. 13, 30th edition). The practice of presenting pictures such as this leads to the misconception that foreign cheese is only about 1 / 16 that of Commonwealth cheese, whereas in fact it is about 1 / 4; the height as well as the diameter of the cylinder was changed in this graph. A picture such as Figure 9.15 would be more appropriate, since here only the heights are changed. Another method of presenting the same idea is to use one cylinder of cheese cut into segments as shown in Figure 9.16, or

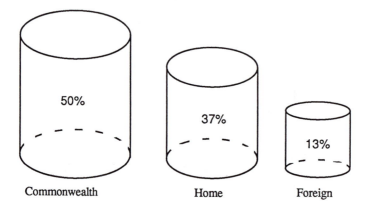

Figure 9.14   Biased cylinder graph.

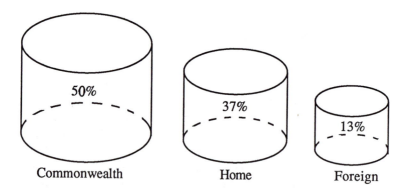

Figure 9.15   Cylinder graph.

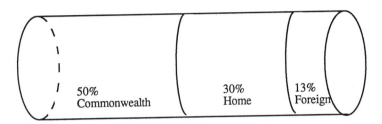

Figure 9.16 Cylinder graph.

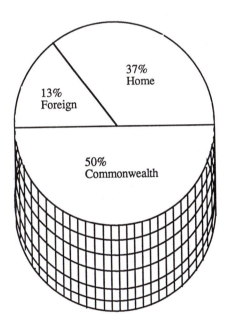

Figure 9.17 Cylinder pie graph.

into wedges as in Figure 9.17.

A procedure that may create misconceptions is a change from actual numbers to proportions or percentages. For example, suppose that politician K wishes to create the impression that he is doing better than politician E relative to number of people unemployed in his area. Suppose that the number in the labor force under the administration of E was 60 million, and under K it was 70 million, and that his aides prepared the two bar graphs shown in Figures 9.18. Politician K might be tempted to use the right-hand bar graph. In fact, he might consider chopping off the graph below 4.2%, to indicate that he is doing twice as well as E had. On the other hand, E could use the left-hand bar graph to show that there were 20,000 fewer individuals unemployed under him than under K. In situations like this, one should present both bar graphs and explain that the number in the labor force had changed from 60 to 70 million workers, that the number of unemployed had increased by 20,000, and that the percentage had decreased from 5.8% to 5% of the total labor force.

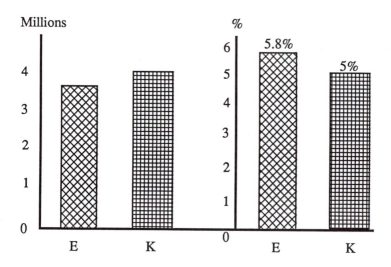

Figure 9.18. Bar graphs for number of unemployed in millions and for percentages of unemployed.

The process of reading off intermediate values on a graph between two observed values is known as **interpolation**. This may be all right in some situations, say where the variates on the abscissa and ordinate are continuous, that is, take on all possible values, but may be non-sensical in others. For example, consider the following set of data:

Number of children per family	Number of families observed
0	31
1	25
2	35
3	20
5	16
6	12
7	5
8	1
Total	145

The above results are presented in the following two forms in Figure 9.19, one a histogram and the second as a linegraph or frequency polygon. Interpolation for non-integer values of number of children per family would be meaningless. It would be non-sensical to say that ten families had 4.4 children (as seemingly indicated by the dotted lines)!

The process of extending a graph beyond the data points is known as extrapolation. There are many dangers associated with extrapolation beyond the range of the data. If the investigator had the correct form of the response function or model, then it is possible to use extrapolation to advantage. However, in the majority of situations, the investigator does not know the correct form of the model and he uses an approximation to the true model. Too often investigators forget that the model or response function they are using is only an approximation, sometimes very rough, of the true model.

Figure 9.20 illustrates the decline in average annual death rate decade by decade from 1890 to 1930 (solid line). The estimated annual death rate (dashed line) for the 1930-1940 decade is under $0.05 / 1000 = 5 / 100,000$. Since we have gone this far, why not also estimate the annual death rate for the 1940-1950 decade? This gives the unrealistic result that there is a return from death, that is, the death rate is about $-2 / 100,000$. By the 1950-1960 decade, the dead are returning by hordes, an example of resurrection from the dead! Of course, this is nonsense. We would suspect that the annual death rate would flatten out and remain nearly constant

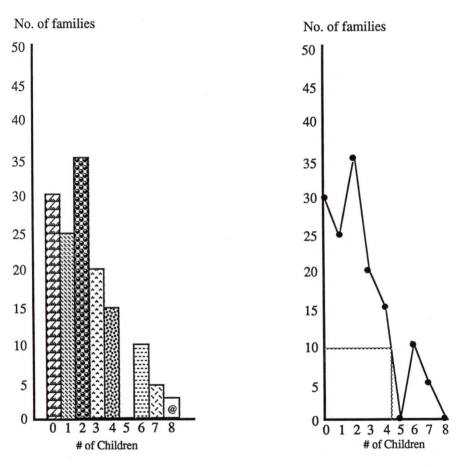

Figure 9.19   Histogram on the left and linegraph on the right for number of
families observed as ordinate and number of children per family as abscissa.

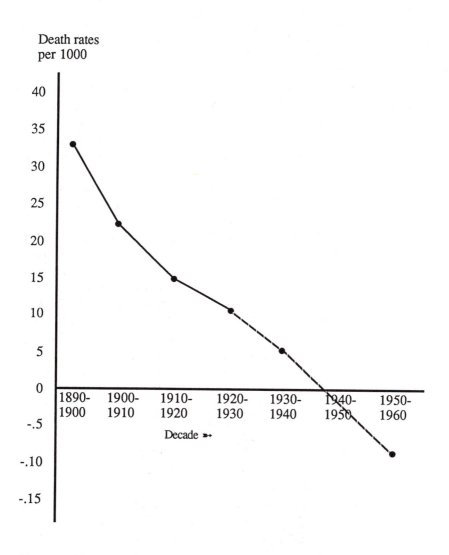

Figure 9.20. Extrapolation of linegraph.

somewhere above the zero point. Here we see one of the dangers of extrapolation when the wrong response function is used. The straight line fitting to these data is only a fairly good approximation *within the range of the data*. It is worthless for predicting death rates for later decades for which we have no data.

For certain goals, it is necessary to project into the future. Governments and insurance companies need estimates of future population growth, of birth rates, of population sizes, of life expectancies, and so forth in order to make appropriate plans to conduct their operations. Demographers are concerned with making these projections and need to make them in order for a Society to function smoothly. They do the best they can with the data and methods that are available. Despite this, the user of such predictions should realize that most are subject to many external influences and may be wide of the mark.

As a second example of the danger of extrapolation, suppose that the percentage of seasonally unemployed workers for factory A runs between 1.5% and 6%. Suppose that the percentage unemployed by month is graphed as in Figure 9.21. (solid line). Now suppose that only the data from the first six months were available to an investigator and that he is a "straight-line advocate". He fits a straight line to the data points by standard linear regression techniques (the dashed line) and gleefully predicts that there will be no unemployment by mid-August!

Suppose, on the other hand, that only the last six months' unemployed percentages were available to our "straight-line advocate". After drawing his standard linear regression (the dashed line) through the data points, he would woefully predict that over 7% would be unemployed in January! As plainly illustrated above, extrapolation can quickly lead to misleading and non-sensical results. Therefore, the experimenter must be wary of any extrapolations, especially when he does not know the form of the response function over all values of the abscissa.

As a third example, we note the study by Professor U. Bronfenbrenner, Cornell University, on the response by girls and by boys to different degrees of severity of punishment. There is no doubt but that a straight line fit for these data would be unsuitable. Despite this, numerous investigators did use a straight line fit to the data and reached the wrong conclusions. The "straight-line advocates" stated that girls should not be

Percentage unemployed
in Factory A

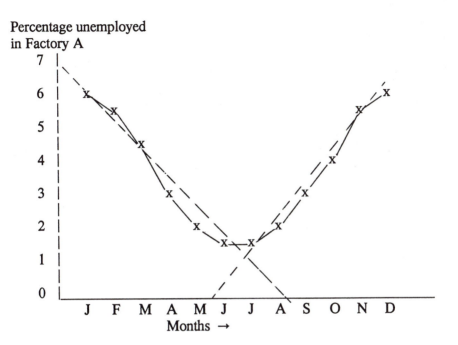

Figure 9.21. Straight fitting to selected parts of a curved function for percentage of workers unemployed in factory A for a 12 month period.

punished and that boys should be punished severely! The fact was that the maximum response to severity of punishment was reached much faster by girls than by boys and that the response functions for both groups were curved and not straight lines. This should not have been a surprise finding for a person acquainted with human nature. The only explanation for many investigators reaching the wrong conclusion appears to be that they had learned about straight line fitting and nothing else. They then used the procedures without critcally evaluating it.

From various sources, data are available on the number of passengers per car on highways for each year. For example, in 1940 a survey showed that the average number of passengers per car was 3.2, in 1950 the figure was 2.1, and in 1960 the average was only 1.4. If we plot these three pairs of figures as in Figure 9.22 and connect the adjacent points with straight lines, a sharp decline is to be noted. Suppose that our "straight-line advocate" draws a "best fitting" line through all three points and extends it to the year axis, the abscissa. We note that in 1964, cars should have been going down the highways with no one in them! Even if the "straight-line advocate" is playing it "safe" and only uses a line with a smaller decrease, the dashed line on the right, it will still indicate that cars will contain no passengers by 1982. Probably the true situation is that this is a curved function which approaches one passenger per car as a limit. Although some of these examples may be amusing, they are intended to illustrate the dangers of extrapolation beyond the data available and the consequences of using the wrong model.

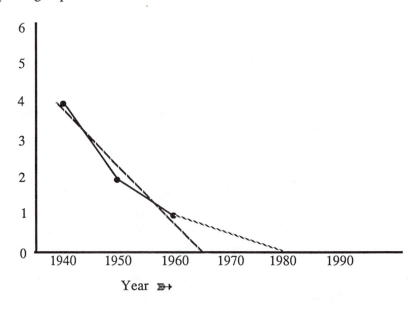

Figure 9.22 Number of passengers per car on the highway by year.

In presenting a graph for a set of data, one should present a scatter diagram or two-way representation of the data (See Table 9.4, for e.g..) of the points as given in Figure 9.23. If a line is drawn through the points in the scatter diagram, the method used to plot the points on the line should be given. Frequently a person will draw an "eye-fitted" line through the points, and no explanation is given about the method of determining which line was selected. This can be misleading, especially if the points are omitted. For example, suppose that one has the scatter diagram in Figure 9.24. Any straight line using linear regression methods will fit equally well, as there is no relationship of the Y_i and X_i values. The eye-fitter then can draw any line to suit his purpose. If he omits the points in the scatter diagram, the reader could be misled into thinking that the Y_i values are influenced by the value of the X_i, when in fact they were not. It is common practice to omit the original data and only present straight lines or curved lines to represent the data. Omission of the original data may be detrimental for the reader in being able to comprehend fully the nature of the data. Also, a graph may be presented as a smoothed curve when it should have been presented as a line polygon as in Figures 9.19 and 20.

Be wary of any graph or figure for which the scale has been omitted. Figure 9.25 was adapted from Moroney (1956), page 29. Here we note that the two halves of the figure may, and quite probably do, have different scales. The figure would indicate that the return to "normal state" is much faster using "Snibbo" than was the increase in "Inter-Pocula Index". Also, note that no scales are given and that it would appear that the subjects were more than half intoxicated when the study began on the left-hand figure and that they were completely normal at the end of the study on the right-hand portion of the figure. Obviously, the advertiser wishes to sell "Snibbo"!

We have covered some of the honest and some of the dis-honest ways of presenting results. There are others, but the general idea should be clear from what has been presented thus far. Charts, graphs, and figures can lead or mislead equally well. Be certain to read figures carefully and determine what they mean rather than what they appear to mean. Factual and misleading graphs and charts may be found in all types of literature, often side by side. They are of daily occurrence in newspapers and magazines. Frequently there is no attempt to mislead, since an uncritical writer may not even realize that he has "stretched" one scale, changed scales in the middle of a graph, omitted a scale, or otherwise misled the reader. In any event, the reader must be wary and critical.

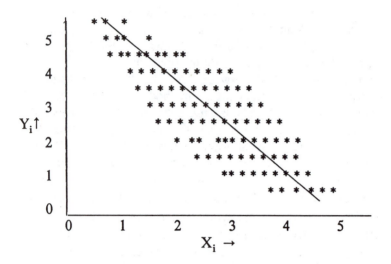

Figure 9.23. Scatter diagram of Y_i values plotted against X_i values.

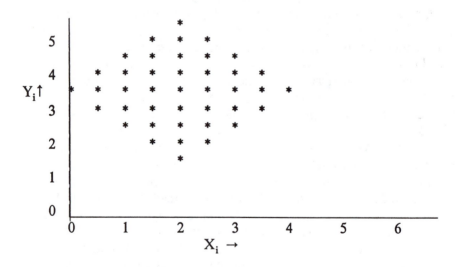

Figure 9.24. Scatter diagram when there is no relationship between the Y_i and X_i values.

"Trust your doctor - he knows."

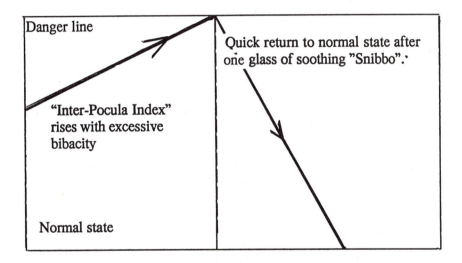

After consumption of alcohol your Inter-Pocula Index rises to what may prove a dangerous level, with serious risk to muscular atony. In such cases the taking of a therapeutic nostrum has untold effect as a sedative and restorative. There is no finer acting nostrum than quick-acting soothing Snibbo.

Figure 9.25. Scaleless graph.

9.7 REFERENCES AND SUGGESTED READING

Bevan, J. M. (1968). *Introduction to Statistics.* Philosophical Library, Inc., New York, vii + 220 pp., chapter five.

Campbell, S. K. (1974). *Flaws and Fallacies in Statistical Thinking.* Prentice Hall, Inc., Englewood Cliffs, New Jersey, viii + 200 pp..

Gelfand, I. M., E. G. Glagoleva, and E. E. Schnol (1969). *Functions and Graphs*, Volume 2. The M.I.T. Press, Cambridge, MA., and London, England, v + 105 pp..

Huff, D. (1954). *How to Lie with Statistics.* W. W. Norton and Company, Inc., New York, 142 pp..

Koopmans, L. (1987). *Introduction to Contemporary Statistical Methods.* Duxbury Press, North Scituate, Mass., 683 pp..

Moroney, M. J. (1956). *Facts from Figures, 3rd edition.* Penguin Books, Baltimore, Md., xiii + 472 pp..

Mosteller, F. and J. W. Tukey (1977). *Data Analysis and Regression.* Addison-Wesley Publishing Company, Reading, Mass., Menlo Park, Cal., London, Amsterdam, Sydney, Don Mills, Ontario, xviii + 588 pp..

Schmid, C. F. (1956). What price pictorial charts? *Estadística* 14:12-25.

9. 8. PROBLEMS

Problem 9.1. Find two or more examples from newspapers, magazines, or other published literature of mis-leading graphs, and show how to present the results in order not to mis-lead but to present the facts. Describe how the graphs are or could be mis-leading.

Problem 9.2. Prepare a scatter diagram of the 93 pairs of height and weight measurements as presented by the student from Table 9.1. Use small o to represent the measurements for women and small x to represent the measurements for men.

Problem 9.3. From the data in Table 9.1, complete the following table and comment on the results.

Team	Frequency of last digit of heights in millimeters digit									
	0	1	2	3	4	5	6	7	8	9
A										
B										
expected	8.5	8.5	8.5	8.5	8.5	8.5	8.5	8.5	8.5	8.5

The expected number 8.5 is obtained as 85 / 10 and should be the same for all ten digits if there is no bias in selecting a last digit of a measurement.

Problem 9.4. From the data in Table 9.1, complete the following table:

	Frequency of last digit for weights in pounds digit									
	0	1	2	3	4	5	6	7	8	9
Own										
-observed										
-expected	8.9	8.9	8.9	8.9	8.9	8.9	8.9	8.9	8.9	8.9
Team B										
-observed										
-expected	8.5	8.5	8.5	8.5	8.5	8.5	8.5	8.5	8.5	8.5

The expected numbers are computed as 89 / 10 and 85 / 10. Do people tend to select one digit over another in reporting weights, for instance, zeros, fives, and nines?

Problem 9.5. Eight matchboxes with the number of matches in each box as indicated are arranged as follows:

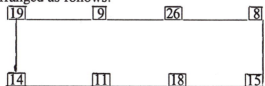

The first box contains 19 matches, the second 9 matches, etc.. It is permissible to move any match from one box to another if the boxes are connected by a line. The matches must be shifted in such a manner as to have an equal number of matches in each box. Show how this can be done to minimize the total number N of matches shifted and show graphically that your solution is a minimum.

Problem 9.6. Construct the graph for the following function using different colors for the different parts of the graph:
$$y = 1/(x - 1) + 1/(x + 1)$$

for all value of x and y between zero and 50.

Problem 9.7. We have illustrated methods of deceiving the reader with various graphical presentations. Words which in themselves are vague or convey mis-interpretations, may be attached to graphs to make them even more mis-leading. A good source for finding such words is in literature circulated by campus or other groups promoting a cause. Such words or phrases as imperialism, working class, middle-class values, liberation groups, exploitation, anarchist, liberal, conservative, sexual liberation, and generation-gap, are often used to mislead. Find ten examples of such words, and illustrate how you believe that the word has been used to mislead.

Problem 9.8. In describing TA training and development efforts in CUE, Cornell Undergraduate Education, volume 3(2), page 7, it is reported that 140 students applied but not all could be accommodated for a Fall 1989 Workshop, and 136 were enrolled in the Spring 1990 Workshop. These graduate students were from eight colleges or schools at Cornell University. The following percentages were reported in the article:

College	Fall 1989	Spring 1990
Architecture, Art, Planning	6.9%	8.1%
Arts and Sciences	29.2%	30.1%
Agriculture and Life Sciences	31.3%	25.1%
Biological Sciences	6.9%	6.6%
Engineering	11.1%	13.2%
Hotel Administration	*	0.7%
Human Ecology	11.1%	11.8%
Industrial and Labor Relations	*	2.2%
Veterinary Medicine	*	2.2%

* Individual percentages were not given for these three; instead the combined total was given as 3.5%.

For the above data, prepare three different graphs for these data using various shadings for the parts of the graphs.

Problem 9.9. For the data in Problem 9.8, use different colors and/or color shadings and show how to present misleading graphs for these data.

Problem 9.10. The grades for Prelims II and III for Spring 1986 were:

Prelim II:

90	81	70	59	42	67	90	84	76	60
52	43	46	52	66	76	84	95	93	84
72	63	54	44	91	85	70	65	53	46
97	88	72	62	54	49	97	85	73	64
55	45	91	88	75	69	59	80	80	89
86	87	82	83	80	82	88	85	76	74
74	77	78	78	72	76	72	72	71	77
72	73	74	71	72	67	64	63	64	63
67	68	67	66	69	64	63	68	69	67
65	58	52	53	56					

Prelim III:

30	42	72	45	56	60	70	80	91	50
60	60	70	56	80	90	37	43	46	54
70	60	80	90	80	91	91	81	73	62
56	47	47	56	62	73	81	91	49	56
62	74	82	91	57	63	74	82	92	58
64	74	82	92	65	65	66	66	66	67
67	69	79	79	78	77	77	76	76	76
76	76	75	75	82	83	84	84	84	86
86	86	87	87	88	88	82	93	93	94
95	88	88	89						

Prepare stem and leaf diagrams for the grades from the two prelims using two different sets of stems. Find the modal class, the range, the upper and lower quartiles, and the median for each prelim. Also, include a column of frequencies for each stem and a column of cumulative frequencies.

Problem 9.11. Prepare a pie graph for the data in Problem 9.10 for the various stems, using the set of stems with the larger interval. Do this for all three prelims. Also, if 90s are A grade, 80s are B grades, 70s are C grade, 60s are D grade, and anything less than 60 is an F grade, prepare bar graphs for all three prelims.

Problem 9.12. Prepare stem and leaf diagrams for hair and eye color for Own measurements as given in Table 9.1.

Problem 9.13. Prepare bar graphs for ages from Table 9.1.

Problem 9.14. The following are for apartment rental prices in dollars as obtained from a copy of *The Ithaca Journal*, 3/30/78:
250, 185, 120, 200, 181, 181, 181,181, 181, 259, 259,259, 259 259,225, 225, 225, 225, 225, 120, 120, 120, 240, 240, 240, 360, 360, 360, 150,165, 195, 180, 185, 185, 185, 150, 270, 187, 150, 170, 180, 250, 200, 185, 250, 230, 250, 185, 240, 250, 125, 156, 202, 145, 210, 290, 350, 291, 190, 180, 256, 449.

Prepare a stem and leaf graph of these apartment rentals, and find the median value and the modal class.

Problem 9.15. Telefund is fund-raising project whereby students and faculty of Kansas State University telephone alumnae. The results of the telephoning for the 1989-1990 year as reported in *Ways and Means*, Number 11, May 1990, by College and year are given below:

	1990	1989
College of Agriculture	$ 90,731	$ 78,046
College of Architecture & Design	32,580	25,635
College of Arts & Science	177,292	151,035
College of Business Administration	72,706	64,068
College of Education	88,110	68,704
College of Engineering	117,499	104,557
College of Human Ecology	60,262	50,806
College of Veterinary Medicine	35,175	32,710
Intercollegiate Athletics	12,755	12,173

Select graphs to depict the above results which you think would be appropriate for the officials of Kansas State University.

CHAPTER 10. SUMMARIZATION OF DATA: NUMERICAL

10.1 MEASURES OF CENTRAL TENDENCY

In the previous chapter, several methods of graphical summarization of data were presented. In the present chapter, numerical procedures for summarizing information about the data set are presented. Such procedures form the basis for the second part of the definition of Statistics, i.e., statistical methodology. Several references to the arithmetic mean, to the mode, and to the median were made in previous chapters. Further discussion of these and other measures of central tendency of a data set is presented in this chapter. In addition, several measures of variation are discussed; some of these are interquartile range, range, variance, standard deviation, variance of a mean of n observations, standard error of a mean of n observations, variance of a difference between two arithmetic means, and standard error of a difference between two arithmetic means.

The definition and calculation of the arithmetic mean, the median, and the mode are described below and are illustrated with two data sets.

Example 10.1. Suppose the following two sets of data are available:
Set 1:
Data: 5, 7, 8, 9, 3, 2, 4, 5, 6, 5
Ordered data: 2, 3, 4, 5, 5, 5, 6, 7, 8, 9
The sum of the ten numbers is 54

Set 2:
Data: 8, 3, 5, 7, 4, 7, 4, 5, 6
Ordered data: 3, 4, 4, 5, 5, 6, 7, 7, 8
The sum of the nine numbers is 49.

The **mode** is defined to be the most frequent (or popular) number in the data set. In set 1, the number 5 occurs three times and no other number occurs more than once, and hence, 5 is the mode In the second set of data, there are three modes, 4, 5, and 7, since each of these numbers occur twice and all other numbers occur once.

The **median** of a data set or sample is that value which lies midway between the $n/2$ and $(n+2)/2$ ordered variates for n an even number and is the value of the $(n+1)/2$ ordered observations when n is odd. The population median is the value which divides the totality of observations in the population such that 50% fall below the median and 50% fall above the median. For the set 1 sample, there are ten observations, and the sample median falls between the 5th and 6th observations when they are arrayed in lineal order. Since both the 5th and 6th observations have the number 5, the median value is 5. If the values had been different numbers, say 5 and 8, the median would have been $(5+8)/2 = 6.5$ which is mid-way between these two numbers. For the second set of data, the number of observations is odd and the median is the value associated with the $(n+1)/2$[th] observation which is the fifth observation. Since the value of the 5th observation is 5, this is the median for these data .

The **arithmetic mean** or **average** is defined to be the sum of the n observations divided by the number of observations. Often the term **mean** is used for the arithmetic mean, and we shall use this same convention here. If other types of means are used, the term mean will be pre-fixed with an adjective to denote the type of mean being used. The averages of the observations for the two sets are:

$$(5 + 7 + 8 + 9 + 3 + 2 + 4 + 5 + 6 + 5) = 54/10 = 5.4$$

and

$$(8 + 3 + 5 + 7 + 4 + 7 + 4 + 5 + 6) = 49/9 = 5.44$$

Example 10.2. Suppose that we have a camp of 99 workers whose salaries are all equal, say $20,000 per year, and that their leader, a political appointee, receives $1,000,000 per year. The mode and the median are both $20,000, but the arithmetic mean is $[99(\$20,000) + \$1,000,000]/100 = \$2,980,000/100 = \$29,800$. Here we note that there is only one individual above the arithmetic average, and there are 99 below. The formula expressing this mean of n observations, with the ith $(i = 1, 2, ... , n)$ observation being defined as Y_i, is

$$(Y_1 + Y_2 + Y_3 + ... + Y_n)/n = \Sigma_{i=1 \text{ to } n} Y_i/n = Y/n = \bar{y}_{..}$$

The symbol ... indicates that all values of Y_i from Y_3 to Y_n are to be included. The symbol $\Sigma_{i=1 \text{ to } n}$ is a summation sign indicating that the sum is over all observations from 1 to n. Likewise in the the symbol Y., the dot indicates a summation over all values of i in Y_i . These two symbols are simply shorthand notations for the left side of the above equation. The arithmetic average may be likened to a center of gravity or to a fulcrum for a scaled board on which the data are placed in their respective positions relative to their values. Thus for the set 1 data, where each datum is represented by a square and the mean centered or balanced on a fulcrum, we can picture the result as shown in Figure 10.1. When the fulcrum is placed at the mean value of 5.4 the board balances and is level. For the data of Example 10.2, the fulcrum would be placed at $29,800 as shown in Figure 10.2. Since there are 99 values of $20,000 the fulcrum is set at $29,800 to balance the board which has one value at $1,000,000. The example illustrates that the three averages arithmetic mean, median, and mode do not yield the same values, and that the proportion of the observations below the arithmetic mean depends upon the symmetry of the relative frequencies of the data. Since these different averages yield different values, it is imperative that the type of average be specified and understood by the user. All averages are useful for specific but different purposes.

If the distribution of values of observations is not symmetrical, that is, it is skewed, the relative locations of the mode, median and mean are shown in Figure 10.3. One half of the area under the curve is to the left of the median (unshaded area) and one half is to the right of the median (shaded area).

A situation wherein the arithmetic average leads to the wrong answer is the following. The speeds on four sides of a square of side 100 miles are 100 miles per hour, 200 miles per hour, 300 miles per hour, and 400 miles per hour (m.p.h.), respectively. The arithmetic average of the four speeds is (100 + 200 + 300 + 400) / 4 = 250 m.p.h.. That using the arithmetic average is not correct is easily seen from the following:

Time to travel first side = 1 hour = 60 minutes
Time to travel second side = 1 / 2 hour = 30 minutes
Time to travel third side = 1 / 3 hour = 20 minutes
Time to travel fourth side = 1 / 4 hour = 15 minutes
Total time = 25 / 12 hours = 125 minutes

Therefore, total distance / total time = 400 miles / 25 / 12 hours = 192 m.p.h., which is the correct result. There are other ways to obtain the correct answer; for instance, one could use a weighted mean as follows:

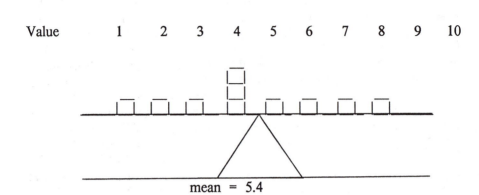

Figure 10.1. Arithmetic mean represented as a center of gravity for the data of Example 10.1, set 1.

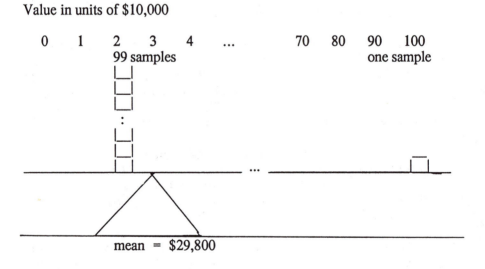

Figure 10.2. Arithmetic mean represented as a center of gravity for data of Example 10.2.

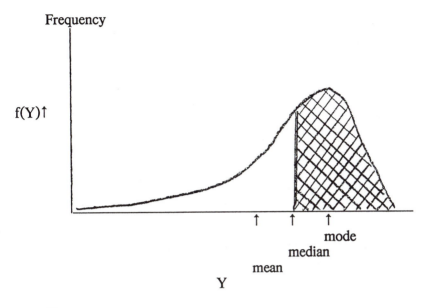

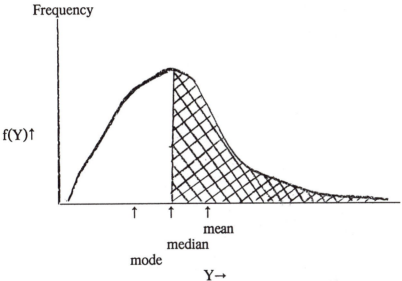

Figure 10.3. Relative locations of mean, median and mode for skewed distributions.

$$\frac{1(100) + \frac{1}{2}(200) + \frac{1}{3}(300) + \frac{1}{4}(400)}{1 + \frac{1}{2} + \frac{1}{3} + \frac{1}{4} = 25/12} = \frac{12(400)}{25} = 192 \text{ m.p.h.}$$

$$\frac{60(100) + 30(200) + 20(300) + 15(400)}{60 + 30 + 20 + 15 = 125} = \frac{2400}{125} = 192 \text{ m.p.h.}$$

In lieu of these, one could use a **harmonic mean or average**, which is defined as the reciprocal of the mean of the reciprocals. The formula for computing the harmonic mean is:

$$\bar{y}_h = 1/\sum_{i=1}^{n} 1/Y_i \, /n = n/\sum_{i=1}^{n} 1/Y_i \, .$$

This type of average is appropriate when dealing with rates or prices. For the example on speeds

$$\bar{y}_h = 4/(\frac{1}{100} + \frac{1}{200} + \frac{1}{300} + \frac{1}{400}) = \frac{4(1200)}{12 + 6 + 4 + 3} = 192 \text{ m.p.h.,}$$

which, we may note, is the correct average. If the times had been constant and the distances variable, then the arithmetic average for miles per hour (m.p.h.) would have been correct; however, with the distances are constant and the times are variable, the harmonic average or mean is the correct one to use.

Another type of average is the **geometric mean or average**, which is defined to be the nth root of the product of n observations. The formula for computing the geometric mean is:

$$\bar{y}_g = (Y_1 \times Y_2 \times Y_3 \times \ldots \times Y_n)^{1/n} = (\textstyle\prod_{i=1 \text{ to } n} Y_i)^{1/n}$$

where the symbol $\prod_{i=1 \text{ to } n}$ indicates a product of n variables. This type of average is appropriate for situations where the Y_i values increase in a geometric or exponential manner. The growth data for chick embryos from six to 12 days in Chapter 8 is an example of data following an exponential increase. Suppose that we had the 6-day weight = Y_6 = 0.029 and the 16-day weight = Y_{16} = 2.812 and wished to estimate the 11-day weight. The geometric mean is:

$$\bar{y}_g = \sqrt{0.029(2.812)} = 0.286$$

whereas the observed Y_{11} was 0.261. The arithmetic mean would be $(0.029 + 2.812) / 2 = 1.420$ which is considerably different from $Y_{11} = 0.261$ or from $\bar{y}_g = 0.286$. Another example wherein a geometric mean would be appropriate is the case of the change in population of a city when the birth and death rates stay constant and when there is no migration in or out of the city.

Another type of average is one called an index number. Moroney (1956), for example, gives an introduction to index numbers through the use and construction of a simple Cost of Living Index. He has little use for the many numbers used by governments to characterize the economic and social status of a nation. However, since these indices are of frequent occurrence and use even though many are antiquated, Moroney does discuss them. The "expenditures" of a "standard family" are computed in some fashion for a given list of items; these are converted to percentages of expenditures. These percentages are compared to some base period and then used as the weights in a Cost of Living Index. Rate of inflation, percentage of the work force unemployed, parity prices, Gross National Product (GNP), and many others are in common use in our Society and other societies. *The Wall Street Journal* contains many indices every issue. One that is very popular is the Dow Jones Industrial Average which some use as a measure of economic health of the United States. Other measures of economic health are the Standard and Poor's, Dow Jones Futures Index, Dow Jones Industrials Index, Transportation Index, and Utilities Index. There are many more which businesses and governmental organizations find useful in everyday work and planning.

A percentage or ratio is an average in the same sense that the preceding ones are. Some of us are acquainted with the batting averages of baseball players. We know that this average is a ratio of the number of hits, say h, to the total times at bat, say b. The batting average is $[h(1) + (b - h)(zero)] / b =$ ratio expressed as a three-decimal number such as 0.341. This means that $h / b = 34.1\%$ of the times at bat $= b$, the batter had a hit $= h$, and $(b - h) / b = 65.9\%$ of the time he did not get a hit or got zero hit. The variability of a proportion or percentage depends upon the sample size. A batting percentage based on three times at bat is much more variable than when based on 100 times at bat.

We utilized proportions to summarize some of the data from the class survey as, for instance, in Table 9.2. The results in Table 9.3 could have been similarly treated. The use of proportions to summarize data is common among investigators.

10.2 MEANS FROM ORTHOGONAL EXPERIMENT, TREATMENT, AND SAMPLE SURVEY DESIGNS

In the completely randomized design with v treatments each repeated r times, or in a sample survey design with a random selection of the clusters and of the r elements in each cluster, data of the following nature are available:

	Treatment or cluster number				
	1	2	3	...	v
	Y_{11}	Y_{21}	Y_{31}	...	Y_{v1}
	Y_{12}	Y_{22}	Y_{32}	...	Y_{v2}
	Y_{13}	Y_{23}	Y_{33}		Y_{v3}
	⋮	⋮	⋮	⋱	⋮
	Y_{1r}	Y_{2r}	Y_{3r}	...	Y_{vr}
Total	$Y_{1.}$	$Y_{2.}$	$Y_{3.}$	...	$Y_{v.}$
Mean	$\bar{y}_{1.}$	$\bar{y}_{2.}$	$\bar{y}_{3.}$	...	$\bar{y}_{v.}$

Grand total = sum of all the observations =

$$\sum_{i=1}^{v} \sum_{j=1}^{r} Y_{ij} = Y_{..}$$

overall mean = $Y_{..}/rv = \bar{y}_{..}$

The treatment or cluster means, $Y_{i.}/r = \bar{y}_{i.}$, and the overall mean $\bar{y}_{..}$ are the means which summarize the information about arithmetic averages for these data when the yield equation is of the form:

Y_{ij} = true population mean + true treatment or cluster effect
+ deviation of observation from true cluster or treatment mean

$$= \mu + \tau_i + \varepsilon_{ij}$$
= estimated population mean + estimated treatment or cluster effect +
estimated random deviation

$$= \bar{y}_{..} + (\bar{y}_{i.} - \bar{y}_{..}) + e_{ij}$$

In a stratified sample survey design with c randomly selected clusters per stratum and with r randomly selected elements per cluster, data of the following nature are available:

	Stratum number				
	1	2	...		s
	cluster number	cluster number	...		cluster number
	1 2 ... c	1 2 ... c	...		1 2 ... c

$Y_{111} Y_{121} \cdots \quad Y_{1c1}$	$Y_{211} Y_{221} \quad Y_{2c1}$	...		$Y_{s11} Y_{s21} \quad Y_{sc1}$
$Y_{112} Y_{122} \cdots \quad Y_{1c2}$	$Y_{212} Y_{222} \cdots Y_{2c2}$	...		$Y_{s12} Y_{s22} \cdots Y_{sc2}$
$\vdots$				
$Y_{11r} Y_{12r} \cdots Y_{1cr}$	$Y_{21r} Y_{22r} \cdots Y_{2cr}$	...		$Y_{s1r} Y_{s2r} \cdots Y_{scr}$

$Y_{11.} Y_{12.} \cdots Y_{1c.}$	$Y_{21.} Y_{22.} \cdots Y_{2c.}$	...		$Y_{s1.} Y_{s2.} \cdots Y_{sc.}$
$\bar{y}_{11.} \bar{y}_{12.} \cdots \bar{y}_{1c.}$	$\bar{y}_{21.} \bar{y}_{22.} \cdots \bar{y}_{2c.}$	...		$\bar{y}_{s1.} \bar{y}_{s2.} \cdots \bar{y}_{c.}$
Stratum total $Y_{1..}$	$Y_{2..}$	...		$Y_{s..}$
Stratum mean $\bar{y}_{1..}$	$\bar{y}_{2..}$	...		$\bar{y}_{s..}$

The ij^{th} cluster mean is equal to the ij^{th} total divided by r, that is, $Y_{ij.} / r = \bar{y}_{ij.}$. The ith stratum mean is equal to $Y_{i..} / rc = \bar{y}_{i..}$ where $Y_{i..}$ is the total for the ith cluster and rc is the number of observations in a stratum total.

For a randomized complete block design with v treatments in r different complete blocks, the data rearranged in a systematic fashion may be represented as:

Treat-ment	Block					Treatment Total	mean
	1	2	3	...	r		
1	Y_{11}	Y_{12}	Y_{13}	...	Y_{1r}	$Y_{1.}$	$\bar{y}_{1.}$
2	Y_{21}	Y_{22}	Y_{23}	...	Y_{2r}	$Y_{2.}$	$\bar{y}_{2.}$
3	Y_{31}	Y_{32}	Y_{33}	...	Y_{3r}	$Y_{3.}$	$\bar{y}_{3.}$
$\vdots$	$\vdots$	$\vdots$	$\vdots$	$\ddots$	$\vdots$	$\vdots$	$\vdots$
v	Y_{v1}	Y_{v2}	Y_{v3}	...	Y_{vr}	$Y_{v.}$	$\bar{y}_{v.}$
Bl. total	$Y_{.1}$	$Y_{.2}$	$Y_{.3}$	...	$Y_{.r}$	$Y_{..}$	-
Bl. mean	$\bar{y}_{.1}$	$\bar{y}_{.2}$	$\bar{y}_{.3}$	...	$\bar{y}_{.r}$	-	$\bar{y}_{..}$

The arithmetic means are computed as the total divided by the number of observations. The block means $\bar{y}_{.j} = Y_{.j} / v$, the treatment means $\bar{y}_{i.} = Y_{i.} / r$, and the overall mean $\bar{y}_{..} = Y_{..} / rv$ are the means required to summarize the information in a randomized complete block design when the yield datum is represented by the linear additive equation

$$Y_{ij} = \mu + \tau_i + \beta_j + \epsilon_{ij}$$

= true overall mean + true treatment effect + true block effect

+ true random error.

$$Y_{ij} = \bar{y}_{..} + t_i + b_j + e_{ij}$$

= estimated mean + estimated treatment effect + estimated

block effect + estimated random error

$$= \bar{y}_{..} + (\bar{y}_{i.} - \bar{y}_{..}) + (\bar{y}_{.j} - \bar{y}_{..}) + (Y_{ij} - \bar{y}_{i.} - \bar{y}_{.j} + \bar{y}_{..}) .$$

The various totals in terms of true population effects, i.e., parameters, are

$$Y_{..} = rv\,\mu + r \sum_{i=1 \text{ to } v} \tau_i + v \sum_{j=1 \text{ to } r} \beta_j + \sum_{i=1 \text{ to } v}\sum_{j=i \text{ to } r} \epsilon_{ij}$$

$$Y_{i.} = r\,(\mu + \tau_i) + \sum_{j=1 \text{ to } r} (\beta_j + \epsilon_{ij})$$

$$Y_{.j} = v\,(\mu + \beta_j) + \sum_{i=1 \text{ to } v} (\tau_i + \epsilon_{ij})$$

Here we note that the difference between any two $Y_{i.}$ or any two $\bar{y}_{i.}$ contains only the difference between two treatment effects and some error terms. The same is true for any two block means $\bar{y}_{.j}$, and, hence, the treatment and the block effects in this experiment design are orthogonal by definition.

Example 10.3. To illustrate the above, consider the following numerical example with $v = 3$ and $r = 4$ for a randomized complete block design. The yields or responses have been nearly always re-arranged from the arrangement used in the actual experiment for convenience of computations. The responses are given below:

Treatment	Block				Treatment Total	Mean
	1	2	3	4		
1	5	13	7	3	28	7
2	9	10	8	9	36	9
3	1	7	6	6	20	5
Block total	15	30	21	18	84	-
Block mean	5	10	7	6	-	7

In the latin square design, there are three different kinds of arithmetic means, the row mean, the column mean, and the treatment mean. These, together with the overall arithmetic mean, constitute all the means necessary to summarize the information from a $k \times k$ latin square design when the yield datum for a single observation is of the following nature:

$$Y_{hij} = \mu + \gamma_h + \tau_i + \rho_j + \varepsilon_{hij}$$

= true overall mean + true column effect + true treatment effect

+ true block or row effect + true random error,

= estimated overall mean + estimated column effect + estimated

treatment effect + estimated row effect + estimated random error

$$= \bar{y}_{...} + c_h + t_i + r_j + e_{hij}$$

$$= \bar{y}_{...} + (\bar{y}_{h..} - \bar{y}_{...}) + (\bar{y}_{.i.} - \bar{y}_{...}) + (\bar{y}_{..j} - \bar{y}_{...}) +$$

$$(Y_{hij} - \bar{y}_{h..} - \bar{y}_{.i.} - \bar{y}_{..j} + 2\bar{y}_{...}),$$

where there is a one-to-one correspondence between the letters in the preceding equation and the terms in parentheses in the last equation.

The various totals in terms of effects are:

$$Y_{...} = k^2 \bar{y}_{...} + k \sum_h c_h + k \sum_i t_i + k \sum_j r_j + \sum_h \sum_i \sum_j e_{hij},$$

$$Y_{h..} = k(\bar{y}_{...} + c_h) + \sum_i t_i + \sum_j r_j + \sum_{i\&j} e_{hij},$$

$$Y_{.i.} = k\left(\overline{y}_{...} + t_i\right) + \sum_h c_h + \sum_j r_j + \sum_{h\&j} e_{hij},$$

and

$$Y_{..j} = k\left(\overline{y}_{...} + r_j\right) + \sum_h c_h + \sum_i t_i + \sum_{h\&i} e_{hij}.$$

The summations are over the subscripts indicated on the summation signs $\sum$. The difference between any two treatment means, say $\overline{y}_{.1.}$ and $\overline{y}_{.2.}$, *contains only the difference* due to the two treatment effects, t_1 and t_2, and some error terms. The same is true for row and column means. The same relations hold when the parameters are inserted in place of the estimated values of the parameters. Therefore, by the definition in Chapter 7, treatments are orthogonal to rows and columns, and rows and columns are orthogonal to each other.

The yields $= Y_{hij}$, totals, and means of rows, of columns, and of treatments in a latin square design may be of the following form. The particular form will depend upon the particular randomization utilized by the experimenter in the investigation.

Row	Column 1	2	3	...	k	Row total	mean
1	Y_{111}	Y_{221}	Y_{331}	...	Y_{k41}	$Y_{..1}$	$\overline{y}_{..1}$
2	Y_{122}	Y_{232}	Y_{342}	...	Y_{k52}	$Y_{..2}$	$\overline{y}_{..2}$
3	Y_{133}	Y_{243}	Y_{353}	...	Y_{k63}	$Y_{..3}$	$\overline{y}_{..3}$
⋮	⋮	⋮	⋮	⋱	⋮	⋮	⋮
k	Y_{1kk}	$Y_{2,k-1,k}$	$Y_{3,k-2,k}$	...	Y_{k1k}	$Y_{..k.}$	$\overline{y}_{..k}$
Col. total	$Y_{1..}$	$Y_{2..}$	$Y_{3..}$	...	$Y_{k..}$	$Y_{...}$	-
Col. mean	$\overline{y}_{1..}$	$\overline{y}_{2..}$	$\overline{y}_{3..}$	...	$\overline{y}_{k..}$	-	$\overline{y}_{...}$

Treatment	Treatment number 1	2	3	...	k	Sum
Total	$Y_{.1.}$	$Y_{.2.}$	$Y_{.3.}$	...	$Y_{.k.}$	$Y_{...}$
Mean	$\overline{y}_{.1.}$	$\overline{y}_{.2}$	$\overline{y}_{.3.}$	...	$\overline{y}_{.k.}$	$\overline{y}_{...}$

Example 10.4. As a small numerical example to illustrate the above general form, suppose that $k = 3$ and that the following data and arrangement of treatments A, B, and C are available from an experiment designed as a latin square:

Row	Column 1	Column 2	Column 3	Row Total	Row Mean
1	B 23	A 17	C 29	69	23
2	A 16	C 25	B 16	57	19
3	C 24	B 18	A12	54	18
Col. total	63	60	57	180	-
Col. mean	21	20	19	-	20

Treatment	Treatment A	B	C	Sum
Total	45	57	78	180
Mean	15	19	26	60 = 3(20)

Another orthogonal design is the simple change-over design. The treatment, row, column, and grand totals are obtained in much the same manner as for the latin square design. The schematic layout for v treatments in vs columns and v rows is, for yield $= Y_{hij}$:

Row	Column 1	2	...	v	v+1	...	vs	Row Total	Mean
1	Y_{111}	Y_{221}	$...Y_{v,v,1}$		$Y_{v+1,11}$	$...Y_{vs,v,1}$		$Y_{..1}$	$\bar{y}_{..1}$
2	Y_{1v2}	Y_{212}	$...Y_{v,v-1,2}$		$Y_{v+1,22}$	$...Y_{vs,v-1,2}$		$Y_{..2}$	$\bar{y}_{..2}$
⋮	⋮	⋮	⋮		⋮	⋮		⋮	⋮
v	Y_{12v}	Y_{23v}	$...Y_{v1v}$		$Y_{v+1,vv}$	$...Y_{vs,1v}$		$Y_{..v}$	$\bar{y}_{..v}$
Col. total	$Y_{1..}$	$Y_{2..}$	$...Y_{v..}$		$Y_{v+1..}$	$...Y_{vs..}$		$Y_{...}$	
Col. mean	$\bar{y}_{1..}$	$\bar{y}_{2..}$	$...\bar{y}_{v..}$		$\bar{y}_{v+1..}$	$...\bar{y}_{vs..}$			$\bar{y}_{...}$

	Treatment					Sum
	1	2	3	...	v	
Treatment total	$Y_{.1.}$	$Y_{.2.}$	$Y_{.3.}$	...	$Y_{.v.}$	$Y_{...}$
Treatment mean	$\bar{y}_{.1.}$	$\bar{y}_{.2.}$	$\bar{y}_{.3.}$	...	$\bar{y}_{.v.}$	$v\bar{y}_{...}$

The yield equation for a single datum is the same as for the latin square design.

Example 10.5. A small numerical example for v = 2 and s = 3 is used to illustrate the computation of the row, column, treatment, and grand or overall means. The example is given below where the letter A or B designates the treatment and the number designates the yield data:

Row	Column						Total	Mean
	1	2	3	4	5	6		
1	A 10	B 6	B 7	A 11	A 10	B 4	48	8
2	B 8	A 12	A 9	B 9	B 6	A 16	60	10
Col. total	18	18	16	20	16	20	108	
Col. mean	9	9	8	10	8	10	-	9

	Treatment (letter)		Sum
	A	B	
Treatment total	68	40	108
Treatment mean	11.33	6.67	18 = 2(9)

From the definition of orthogonality in Chapter 7, we note that if the difference between two treatment means (or other means) does not contain any effects other than those due to the two specific treatments and random error, then the other effects are orthogonal to the treatments, and the experiment design is an orthogonal one. Since this is true, the arithmetic means and linear combinations of these means summarize the information from the data.

As may be noted from Chapter 8, the factorial treatment design is one in which the various effects of factors and the corresponding interactions are orthogonal to each other. The various arithmetic means and linear combinations of means suffice to estimate the various main effects and interactions. This is not true for most fractional replicates of a factorial treatment design nor for the balanced incomplete block design. When non-orthogonality is present in a statistical design, the computations for obtaining the solutions to a set of effects involves the solution of a simultaneous set of equations. With modern software packages such as Gauss, the solutions can be obtained with relative ease. Without such computer aids, the solutions are often very difficult and time-consuming. In making use of the software packages, it is essential that the user completely comprehends the nature and goals of the procedure. It is rather easy to mis-use a complex computational procedure and to obtain solutions for the effects which are incorrect.

10.3 SOLUTIONS FOR EFFECTS FROM A BIB DESIGN

The estimation of treatment and block effects and adjusted treatment means in a balanced incomplete block (bib) design are included here to illustrate the fact that arithmetic means for treatments do not suffice to estimate the differences between treatments in a non-orthogonal design. The bib design was selected because the arithmetic computations remain relatively simple.

A particular systematic layout for a bib design with $v = 4$ treatments in $b = 6$ blocks of size $k = 2$ each and with the various totals and means used in the analysis is given below:

Block number

1	2	3	4	5	6
Y_{11}	Y_{12}	Y_{13}	Y_{24}	Y_{25}	Y_{36}
Y_{21}	Y_{32}	Y_{43}	Y_{34}	Y_{45}	Y_{46}

Block total	$Y_{.1}$	$Y_{.2}$	$Y_{.3}$	$Y_{.4}$	$Y_{.5}$	$Y_{.6}$	$Y_{..}$ = grand total
Block mean	$\bar{y}_{.1}$	$\bar{y}_{.2}$	$\bar{y}_{.3}$	$\bar{y}_{.4}$	$\bar{y}_{.5}$	$\bar{y}_{.6}$	$\bar{y}_{..}$ = grand mean

Treatment

	1	2	3	4	Sum
Treatment totals	$Y_{1.}$	$Y_{2.}$	$Y_{3.}$	$Y_{4.}$	$Y_{..}$

Note that a particular yield is written as

$$Y_{ij} = \mu + \beta_j + \tau_i + \varepsilon_{ij}$$

= true overall mean + true jth block effect + true ith treatment effect

+ a true random error.

Then,

$$Y_{1.}/3 = \bar{y}_{1.} = (3\mu + \beta_1 + \beta_2 + \beta_3 + 3\tau_1 + \varepsilon_{11} + \varepsilon_{12} + \varepsilon_{13})/3,$$

$$Y_{2.}/3 = \bar{y}_{2.} = (3\mu + \beta_1 + \beta_4 + \beta_5 + 3\tau_2 + \varepsilon_{21} + \varepsilon_{24} + \varepsilon_{25})/3,$$

$$Y_{3.}/3 = \bar{y}_{3.} = (3\mu + \beta_2 + \beta_4 + \beta_6 + 3\tau_3 + \varepsilon_{32} + \varepsilon_{34} + \varepsilon_{36})/3,$$

$$Y_{4.}/3 = \bar{y}_{4.} = (3\mu + \beta_3 + \beta_5 + \beta_6 + 3\tau_4 + \varepsilon_{43} + \varepsilon_{45} + \varepsilon_{46})/3,$$

$$Y_{.1}/2 = \bar{y}_{.1} = (2\mu + 2\beta_1 + \tau_1 + \tau_2 + \varepsilon_{11} + \varepsilon_{21})/2,$$

$$Y_{.2}/2 = \bar{y}_{.2} = (2\mu + 2\beta_2 + \tau_1 + \tau_3 + \varepsilon_{12} + \varepsilon_{32})/2,$$

$$Y_{.3}/2 = \bar{y}_{.3} = (2\mu + 2\beta_3 + \tau_1 + \tau_4 + \varepsilon_{13} + \varepsilon_{43})/2,$$

$$Y_{.4}/2 = \bar{y}_{.4} = (2\mu + 2\beta_4 + \tau_2 + \tau_3 + \varepsilon_{24} + \varepsilon_{34})/2,$$

$$Y_{.5}/2 = \bar{y}_{.5} = (2\mu + 2\beta_5 + \tau_2 + \tau_4 + \varepsilon_{25} + \varepsilon_{45})/2,$$

$$Y_{.6}/2 = \bar{y}_{.6} = (2\mu + 2\beta_6 + \tau_3 + \tau_4 + \varepsilon_{36} + \varepsilon_{46})/2, \text{ and}$$

$$Y_{..}/12 = \bar{y}_{..} = [12\mu + 2\sum_{j=1 \text{ to } 6}\beta_j + 3\sum_{i=1 \text{ to } 4}\tau_i +$$
(sum of 12 error terms)] / 12.

From the above means, we note that any difference between treatment means $\bar{y}_{i.}$ contains block effects, and any difference between block means $\bar{y}_{.j}$ contains treatment effects. Hence, by the definition of orthogonality, the treatment and block effects are not orthogonal.

Now how do we estimate a treatment effect $= t_i$, which is an estimate of τ_i? If we set up values denoted as $Q_{i.} = Y_{i.} -$ (sum of means of blocks in which treatment i occurs), and omit error terms from the equations, we obtain simple solutions as follows, where t_i is the solution for τ_i:

$$t_1 = (Y_{1.} - (\bar{y}_{.1} + \bar{y}_{.2} + \bar{y}_{.3}))/2 = Q_{1.}/2$$

$$t_2 = (Y_{2.} - (\bar{y}_{.1} + \bar{y}_{.4} + \bar{y}_{.5}))/2 = Q_{2.}/2,$$

$$t_3 = (Y_{3.} - (\bar{y}_{.2} + \bar{y}_{.4} + \bar{y}_{.6}))/2 = Q_{3.}/2$$

$$t_4 = (Y_{4.} - (\bar{y}_{.3} + \bar{y}_{.5} + \bar{y}_{.6}))/2 = Q_{4.}/2.$$

Likewise, $\bar{y}_{..}$ is a solution for μ in the above response model for Y_{ij}. Therefore, a treatment mean adjusted for block effects is computed as $\bar{y}_{..} + t_i = \bar{y}_{..} + Q_{1.}/2 = \bar{y}_{i.}$ (adjusted). The difference between any two adjusted treatment means or effects is, say, for treatments 1 and 2,

$$\bar{y}_{1.}\text{(adj.)} - \bar{y}_{2.}\text{(adj.)} = \bar{y}_{..} + t_1 - \bar{y}_{..} - t_2 = t_1 - t_2 = (Q_{1.} - Q_{2.})/2.$$

The solution b_j for the β_j effect may be obtained by setting the error terms equal to zero (that is, omitting them) and substituting in the solutions for the t_i in the block totals minus k times the mean. For example,

$$b_1 = (Y_{.1} - 2\bar{y}_{..} - t_1 - t_2)/2 = (Y_{.1} - 2\bar{y}_{..} - Q_{1.}/2 - Q_{2.}/2)/2.$$

In general, the solution for any t_i in a bibd with v treatments in b blocks of size k is given by

$$t_i = Q_{i.}\,k/v\lambda,$$

where λ is the number of times any given pair of treatments occurs together in the b blocks. In the example above, $\lambda = 1$ and $r = 3 =$ the number of times each treatment occurs in the experiment design.

Example 10.6. We shall approach the illustration a little differently this time in order to obtain deeper insight into the nature of the yield data in terms of the effects. Suppose that some one tells us that he obtained the following solutions from an experiment designed as a bib design with $v = 4, r = 3, b = 6, k = 2$, and $\lambda = 1$ as described above:

$\bar{y}_{..} = 10$	$t_3 = 5$	$b_2 = -3$	$b_5 = 1$
$t_1 = -1$	$t_4 = 0$	$b_3 = 1$	$b_6 = -2$
$t_2 = -4$	$b_1 = 3$	$b_4 = 0$	
$e_{11} = -1$	$e_{32} = 0$	$e_{24} = 1$	$e_{45} = 2$
$e_{21} = 1$	$e_{13} = 1$	$e_{34} = -1$	$e_{36} = 1$
$e_{12} = 0$	$e_{43} = -1$	$e_{25} = -2$	$e_{46} = -1$

Note that the yield in a bib design in terms of solutions for effects is

$$Y_{ij} = \bar{y}_{..} + t_i + b_j + e_{ij}$$

which for $i = 1$ and $j = 1$ is

$$Y_{11} = 10 - 1 + 3 - 1 = 11.$$

Continuing this process for all 12 observations, we obtain the following reconstruction of the experimenter's yields:

		Blocks			
1	2	3	4	5	6
$Y_{11} = 11$	$Y_{12} = 6$	$Y_{13} = 11$	$Y_{24} = 7$	$Y_{25} = 5$	$Y_{36} = 9$
$Y_{21} = 10$	$Y_{32} = 7$	$Y_{43} = 15$	$Y_{34} = 9$	$Y_{45} = 18$	$Y_{46} = 12$
$Y_{.j}$ 21	13	26	16	23	21

Grand total $= Y_{..} = 120$

The treatment totals are:

$$Y_{1.} = 28, Y_{2.} = 22, Y_{3.} = 25, \text{ and } Y_{4.} = 45.$$

The process the experimenter went through to give us the above solutions was to compute the various totals above. Then, he computed the $Q_{i.}$ values as follows:

$$Q_{1.} = Y_{1.} - (Y_{.1} + Y_{.2} + Y_{.3})/2 = 28 - (21 + 13 + 26)/2 = -2,$$

$$Q_{2.} = Y_{2.} - (Y_{.1} + Y_{.4} + Y_{.5})/2 = 22 - (21 + 16 + 23)/2 = -8,$$

$$Q_{3.} = Y_{3.} - (Y_{.2} + Y_{.4} + Y_{.6})/2 = 25 - (13 + 16 + 21)/2 = 0,$$

and

$$Q_{4.} = Y_{4.} - (Y_{.3} + Y_{.5} + Y_{.6})/2 = 45 - (26 + 23 + 21)/2 = 10.$$

Then,

$$t_1 = Q_{1.}/2 = -1,$$

$$t_2 = Q_{2.}/2 = -4$$

$$t_3 = Q_{3.}/2 = 0,$$

$$t_4 = Q_{4.}/2 = 5, \text{ and}$$

$$\bar{y}_{..} = 120/12 = 10.$$

From these values we compute the b_j values as:

$$b_1 = [21 - 2(10) - (-1) \quad (-4)]/2 = 3,$$

$$b_2 = [13 - 2(10) - (-1) - 0]/2 = -3,$$

$$b_3 = [26 - 2(10) - (-1) - 5]/2 = 1,$$

$$b_4 = [16 - 2(10) - (-4) - 0]/2 = 0$$

$$b_5 = [23 - 2(10) - (-4) - 5]/2 = 1,$$

and

$$b_6 = [21 - 2(10) - 0 - 5]/2 = -2.$$

Likewise, the e_{ij} may be computed as:

$$e_{ij} = Y_{ij} - \bar{y}_{..} - t_i - b_j,$$

which for $i = 1$ and $j = 1$ is:

$$e_{11} = 11 - 10 \; (-1) - 3 = -1.$$

The remaining e_{ij} are computed similarly.

The treatment means adjusted for block effects are computed as follows:

$$\bar{y}_{1.} \text{(adjusted)} = \bar{y}_{..} + t_1 = 10 - 1 = 9,$$

$$\bar{y}_{2.} \text{(adjusted)} = \bar{y}_{..} + t_2 = 10 - 4 = 6,$$

$$\bar{y}_{3.} \text{(adjusted)} = \bar{y}_{..} + t_3 = 10 + 0 = 10,$$

and

$$\bar{y}_{4.} \text{(adjusted)} = \bar{y}_{..} + t_4 = 10 + 5 = 15.$$

The formulae gave us the same values we started with. This should increase our confidence in the formulae and give us some insight into the structure of the yields, of the effects, and of the computations.

10.4 BOX PLOTS AND ASSOCIATED STATISTICS

In order to better comprehend the results from an investigation, it is usually better to combine graphical and numerical methods of statistical analysis of a data set. A procedure devised by Professor John W. Tukey and described in Mosteller and Tukey (1977), Koopmans (1987), and Seigal (1988) was partially discussed in Section 9.5 and is further described below. The first part of the procedure is to prepare a stem and leaf diagram as in Figure 9.12. Then within each stem, order the leaves numerically as shown in Figure 10.4 for the data of Figure 9.12. By ordering the data within each stem, the data are ordered from lowest to highest and it is simple matter to find the percentiles, the median, and the range. The various ten percentiles of these data are 47-69, 69-77, 77-81, 81-83, 83-84, 84-86, 86-88, 88-90, 90-94, and 94-97. The grades 93, 85, 80, 76, 74, and 72 did not appear and there are no obvious breaks in the grades which might be used if grades of A, B, C, D, and F were to be given. An asterisk could be placed after the appropriate number in the above graph to denote the percentiles. Rather than using percentiles, the investigator may wish to present the data in a box plot. A **box plot** is a graphical display of the median, interquartile range, and points computed from the previous statistics as in Figure 10.5. The **interquartile range** represents the middle 50% of a data or of a population set. For the above set of grades, we note that the median is 84, and

the upper quartile level = UQ = 89,
the lower quartile level = LQ = 78,

and the

interquartile range = IQR = 89 - 78 = 11.

We next compute a **scale factor** equal to 1.5 IQR, which is equal to 1.5 (11) = 16.5 for the above set of grades. The number 1.5 is a constant used for all cases. We next compute **fences** or boundaries based on the

Stems	Leaves	Frequency	Cumulative frequency
96-100	|666677	6	6
91-95	|1111222444455	13	19
86-90	|6666666666777777888888900000	28	47
81-85	|111112222222233333333344444	27	74
76-80	|777778899	9	83
71-75	|133555	6	89
66-70	|889900	6	95
61-65	|1445	4	99
56-60	|6	1	100
51-55	|	0	100
46-50	|79	2	102

Figure 10.4 Ordered stem and leaf diagram of 102 grades by five point stems.

scale factor which indicate points to be expected based on the IQR. Fences are useful to indicates data points that appear to be very different from the main body of the data. Two kinds of fences, **minor fences** which indicate points beyond which observations may be considered as being somewhat or mildly different, i.e., **minor outliers**, and **major fences** which indicate points beyond which the data are considered to be quite different from the main body of the observations, i.e., **major outliers**. The upper and lower minor fences are computed as

and
$$fu = UQ + 1.5\,IQR = 89 + 16.5 = 105.5$$

$$fl = LQ - 1.5\,IQR = 78 - 16.5 = 61.5.$$

The upper and lower major fences are computed as

and
$$FU = fu + 1.5\,IQR = 105.5 + 16.5 = 122$$

$$FL = fl - 1.5\,IQR = 61.5 - 16.5 = 45.$$

For this data set, we notice that there are no minor or major outliers for the higher grades. However, for the lower grades, the grades of 47, 49, 56, and 61 may be considered to be minor outliers. For this data set, there are no major outliers although 47 and 49 are not far from FL = 45. This procedure is useful for quickly detecting observations which may be outliers due to a mistake, a change in the process, or introduction of other factors affecting the value of an observation. It is based on the nature of the middle 50% of the data.

Note that the box plot could be superimposed on the stem and leaf ordered plot in Figure 10.4. The lines would be drawn in at the appropriate values for the box and for the values of the fences. In addition, asterisks could be included to denote the various percentiles.

If outliers are considered to be present in a data set, the experimenter may wish to use a mean excluding the outliers. A **trimmed mean** is an arithmetic mean of data after a specified lower and upper percent, say α%, of the data have been omitted, trimmed, and the inner $1 - \alpha$% of the data are used to compute the mean. For the above example, suppose that we trim the values 47 and 49 and 97 and 97 from the data before computing the arithmetic mean resulting in a mean of $8,123 / 98 = 82.99$ whereas the untrimmed mean was $8,423 / 102 = 82.58$. In this case the trimmed mean and the untrimmed mean are essentially equal. The median value is unchanged by this trimming. Further details on trimmed means and many other procedures may be found in Mosteller and Tukey (1977).

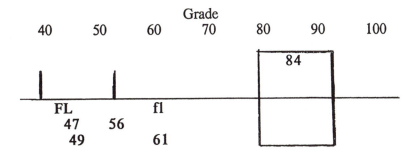

Figure 10.5. Box plot for 102 grades with major and minor lower fences, minor lower outliers, and median value 84. (The upper minor and major fences exceed 100% and are not included.)

10.5 MEASURES OF VARIATION

As we have observed thus far, not all values fall at the average, regardless of what kind of average is used. There will be a scattering of values on both sides of an average. In addition to obtaining an average for a set of data, it is often of interest to have a statistic for summarizing information about scatter. One of the simplest measures of scatter is the **range**, which is simply the difference between the largest observed value and the smallest observed value in the sample, experiment, or population. This measure of variation has already been discussed previously. For the class survey data, the range in heights (2nd column of Table 9.1) was 75" - 62" = 13", and the range in weights (9th column in Table 9.1) was 240 lbs. - 95 lbs. = 145 lbs.

Associated with the range is the idea of percentiles as discussed in the previous section. For example, ten percentiles or 25 percentiles (or quartiles) are frequently used to describe the scatter of grades on an examination or in a course. We often hear the term "in the upper 10%" or "in the upper 25%" of the class. Associated with this idea would be presentation of data from a frequency distribution as percents. We essentially did this for the data of Table 9.2 by presenting the proportion of individuals in the various hair and eye color classes. The percentage of individuals in the eye color classes black, brown, green, blue, and grey was

2%, 37%, 15%, 37%, and 9%, respectively. If there is a gradation of color from black to grey, we could speak of the lower two percentile, the lower 39%, and so forth .

Another measure of variation is **absolute error** or **absolute mean deviation**, which is the average of the absolute deviations from the mean, or

$$\Sigma_{i=1 \text{ to } n} \mid Y_i - \bar{y}_. \mid / \, n,$$

where the vertical pair of lines means that the sign of the deviation is ignored. For example, suppose that $Y_1 = 1, Y_2 = 3, Y_3 = 5, Y_4 = 7$, and $Y_5 = 9$ are the five sample values. Then $\Sigma_{i=1 \text{ to } 5} Y_i = 25 =$ sum of the $n = 5$ observations, and $\bar{y}_. = 25 / 5 = 5 =$ arithmetic mean. The deviations from the mean are $Y_1 - \bar{y}_. = 1 - 5 = -4, Y_2 - \bar{y}_. = 3 - 5 = -2, Y_3 - \bar{y}_. = 5 - 5 = 0, Y_4 - \bar{y}_. = 7 - 5 = 2$, and $Y_5 - \bar{y}_. = 9 - 5 = 4$. The absolute mean deviation is

$$\Sigma_{i=1 \text{ to } 5} \mid Y_i - \bar{y}_. \mid / \, 5 = \{ \mid -4 \mid + \mid -2 \mid + \mid 0 \mid + \mid 2 \mid + \mid 4 \mid \} / \, 5$$
$$= 12 / 5 = 2.4.$$

This measure of variation has not been used to any extent for statistical summarization of data. The statistical properties of this measure have not been developed to the stage where the absolute error could be used routinely in practice

Another statistic that is useful in summarizing information about scatter is the **variance**, which for a sample of size n is defined to be the sum of the squares of the deviations from the mean divided by n - 1. This is denoted symbolically as:

$$s^2 = \Sigma_{i=1 \text{ to } n} (Y_i - \bar{y}_.)^2 / (n - 1) .$$

This statistic is an estimate of the population variance usually denoted by σ^2. The sample variance is an average of the squared distances from the arithmetic mean and refers to the scatter of single observations. The number n - 1 is used as the denominator rather than n in order to obtain an unbiased estimate of the population variance σ^2. Since there are n observations and since one statistic $\bar{y}_.$ was estimated from the data, only n - 1 deviations $Y_i - \bar{y}_.$ are free to vary as their sum must equal zero. The nth deviation is determined from the first n - 1 deviations. For this case, n - 1 is denoted as the number of deviations free to vary and has been called the number of **degrees of freedom**. For the above example, the estimated variance with 5 - 1 = 4 degrees of freedom is computed as follows:

$$s^2 = \{(-4)^2 + (-2)^2 + 0^2 + 2^2 + 4^2\} / (5 - 1) = 40 / 4 = 10.$$

In referring to the scatter of means from a sample of size n, the **variance of a sample mean of n observations** is defined to be the variance divided by n, the number of observations in the sample. Symbolically this is denoted by

$$s^2_{\bar{y}} = \Sigma_{i=1 \text{ to } n} (Y_i - \bar{y})^2 / n(n - 1) = s^2 / n.$$

The number of degrees of freedom for this statistic is determined by the number of degrees of freedom associated with the variance. Here we note that as the sample size increases, the variation among sample means decreases. With small samples, the scatter among means from sample to sample may be large. However, with relatively larger samples, this variation among sample means is much smaller and the investigator can be assured that on the average his sample mean will be relatively close to the population mean.

The **standard deviation of a single observation**, or simply the **standard deviation**, is the square root of the sample variance, or

$$s = \sqrt{\Sigma_{i=1 \text{ to } n} (Y_i - \bar{y})^2 / (n - 1)} = \sqrt{s^2}.$$

The **standard deviation of a mean of n observations**, or more commonly the **standard error of a mean**, is equal to

$$s_{\bar{y}} = \sqrt{\Sigma_{i=1 \text{ to } n} (Y_i - \bar{y})^2 / n(n - 1)} = \sqrt{s^2 / n}.$$

Also, since we have talked about differences between two independent sample means such as might occur in a completely randomized experiment design or in a stratified random sample survey design, we need to define another term, the **standard error of a difference** between two treatment (or stratum) means. This is equal to the square root of the sum of the variances of a mean for the two treatments, say treatments 1 and 2, and may be denoted symbolically as

$$s_{\bar{y}_1 - \bar{y}_2} = \{ \Sigma_{j=1 \text{ to } n_1} (Y_{1j} - \bar{y}_{1.})^2 / n_1(n_1 - 1)$$

$$+ \Sigma_{j=1 \text{ to } n_2} (Y_{2j} - \bar{y}_{2.}) / n_2(n_2 - 1)\}^{1/2}$$

where n_1 observations are used to estimate the mean and variance for treatment (or stratum) one and n_2 observations are used to estimate the mean and variance for treatment (or stratum) two. The **variance of a difference between two means** is simply the square of the standard error of a difference of two means.   If the deviations ($Y_{1j} - \bar{y}_{1.}$) and ($Y_{2j} - \bar{y}_{2.}$) have the same degree of variation or variability, that is, they both come from a population with a population variance of σ^2, then the variance of a difference between the two means becomes:

$$s^2_{\bar{y}_{1.} - \bar{y}_{2.}} = \{[\Sigma_{j=1 \text{ to } n_1} (Y_{1j} - \bar{y}_{1.})^2 + \Sigma_{j=1 \text{ to } n_2} (Y_{2j} - \bar{y}_{2.})^2] /$$

$$[(n_1 - 1) + (n_2 - 1)]\} \times \{1 / n_1 + 1 / n_2\},$$

which for $r = n_1 = n_2$ becomes:

$$s^2_{\bar{y}_{1.} - \bar{y}_{2.}} = \Sigma_{i=1 \text{ to } 2} \Sigma_{j=1 \text{ to } r} (Y_{ij} - \bar{y}_{i.})^2 / r (r - 1).$$

If all treatments in a completely randomized design of v treatments with r replicates on each treatment have the same degree of scatter, then the estimated variance of a single observation is equal to

$$s_e^2 = \Sigma_{i=1 \text{ to } v} \Sigma_{j=1 \text{ to } r} (Y_{ij} - \bar{y}_{i.})^2 / v (r - 1)$$

$$= \Sigma_{i=1 \text{ to } v} \Sigma_{j=1 \text{ to } r} e^2_{ij} / v (r - 1).$$

The subscript e is placed on s^2 to denote that s_e^2 , the estimated error variance, refers to the scatter of the e_{ij} = estimated random errors = non-assignable errors. The standard error of a single mean of r replicates in a completely randomized design is simply s_e^2 / r , and the standard error variance of a difference between two treatment means, say 1 and 2, each with r replicates, is $2s_e^2 / r$; the standard error of a difference of two means is $\sqrt{2s_e^2/4} = s_e\sqrt{2}/r$.

With the above formulation, we can proceed to obtain the error variance, s_e^2, and related quantities for other orthogonal designs such as the randomized complete block design, the simple change-over design, and the latin square design. We simply obtain the sum of squares of the estimated random errors $= e_{ij}$ or e_{hij} and divide by the appropriate constants, which are denoted as degrees of freedom, to obtain the estimated error variance. **Degrees of freedom** for a survey or an experiment design are the total number of observations minus the number of independent constants estimated. For the various designs, these and other quantities are computed as follows:

Completely randomized design (v treatments and r replicates):

degrees of freedom for error variance $= v(r - 1)$

error deviation $= Y_{ij} - \bar{y}_{i.} = e_{ij}$

error variance $= s_e^2 = \Sigma_{i=1 \text{ to } v}\, \Sigma_{j=1 \text{ to } r}\, (Y_{ij} - \bar{y}_{i.})^2 / v(r - 1) =$

$$\Sigma_{i=1 \text{ to } v}\, \Sigma_{j=1 \text{ to } r}\, e^2_{ij} / \text{(degrees of freedom)}$$

variance of a difference between two means $= s^2_{\bar{y}_{i.} - \bar{y}_{i'.}} = 2s_e^2/r$,

where $e_{ij} = Y_{ij} - \bar{y}_{i.} = Y_{ij} -$ estimated treatment mean $=$ estimated random error and $i \ne i'$.

Randomized complete block design (v treatments and r replicates):

degrees of freedom for error variance $= (v - 1)(r - 1)$

error deviation $= Y_{ij} - \bar{y}_{i.} - \bar{y}_{.j} + \bar{y}_{..} = e_{ij}$

error variance $= s_e^2 = \Sigma_{i=1 \text{ to } v}\, \Sigma_{j=1 \text{ to } r}\, (Y_{ij} - \bar{y}_{i.} - \bar{y}_{.j} + \bar{y}_{..})^2 / (r$

$$- 1)(v - 1) = \Sigma_{i=1 \text{ to } v}\, \Sigma_{j=1 \text{ to } r}\, e^2_{ij} / (r - 1)(v - 1)$$

variance of a difference between two treatment means $= 2s_e^2/r$

where $e_{ij} = Y_{ij} - \bar{y}_{i.} - \bar{y}_{.j} + \bar{y}_{..} = Y_{ij}$ - estimated treatment mean - estimated block mean + estimated grand mean = estimated random error.

<u>Latin square design (k treatments, k rows (replicates), and k columns):</u>

degrees of freedom = $(k - 1)(k - 2)$

error deviation = $Y_{hij} - \bar{y}_{h..} - \bar{y}_{.i.} - \bar{y}_{.j} + 2\bar{y}_{...} = e_{hij}$

error variance = $s_e^2 = \sum_{h, i \text{ \& } j=1 \text{ to } k} e^2_{hij} / (k - 1)(k - 2)$

variance of a difference between two treatment means = $2s_e^2 / k$,

where $e_{hij} = Y_{hij}$ - estimated column mean - estimated treatment mean - estimated row mean + 2 (estimated grand mean) = estimated random error.

<u>Simple change-over design (v treatments, vs columns (replicates), and v rows):</u>

degrees of freedom = $(v - 1)(vs - 2)$

error deviation = $Y_{hij} - \bar{y}_{h..} - \bar{y}_{.i.} - \bar{y}_{.j} + 2\bar{y}_{...} = e_{hij}$

error variance = $s_e^2 = \sum_{h=1 \text{ to } vs} \sum_{j=1 \text{ to } v} e^2_{hij} / (v - 1)(vs - 2)$

variance of a difference between two treatment means = $2s_e^2 / vs$

where $e_{hij} = Y_{hij}$ - estimated column mean - estimated treatment mean - estimated row mean + 2(estimated grand mean) = estimated random error.

The computation of the sum of squares of the deviations is similar for all designs, the only difference involved is in the computation of the estimated error deviations and in the degrees of freedom. There is a mathematical reason for the latter, but we shall not go into that in this text.

We proceed in the same manner for a bib design with v treatments in b

incomplete blocks of size k each. Here we cannot use arithmetic means but
have to use the effects or the adjusted means instead. Since

$$Y_{ij} = \mu + \tau_i + \beta_j + e_{ij} = ..\bar{y}.. + t_i + b_j + e_{ij},$$

then

$$e_{ij} = Y_{ij} - \bar{y}.. - t_i - b_{j.} .$$

The error variance for a bib design is:

$$s^2_e = \sum_{\text{over all bk obs.}} e^2_{ij} / (bk - v - b + 1),$$

where the degrees of freedom for the error variance $= bk - v - b + 1$.
Note that $bk = rv$ for a bib design. Since the design is not orthogonal, the
variance of a difference between two means or effects adjusted for block
effects is different from orthogonal designs. The variance of a difference
between two adjusted means, say $\bar{y}_{i.}$ and $\bar{y}_{i'.}$, or two treatment effects, say
t_i and $t_{i'}$, from a bib design is:

$$s^2_{\bar{y}_{i.} - \bar{y}_{i'.}} = s^2_{t_i - t_{i'}} = 2 k s^2_e / v \lambda = [2 s^2_e \{1 + (v - k) / v(k - 1)] / r.$$

Let us now apply these formulae to the four numerical examples above.
In Example 10.1, the e_{ij} from the randomized complete block design of v
$= 3$ treatments in $r = 4$ replicates are:

Treatment	Block (e_{ij} values) 1	2	3	4	Sum	$\sum_i \sum_j e^2_{ij} = 38$
1	0	3	0	-3	0	
2	2	-2	-1	1	0	$(r - 1)(v - 1) =$
3	-2	-1	1	2	0	$(4 - 1)(3 - 1) = 6$
Sum	0	0	0	0	0	$s^2_e = 38 / 6 = 19 / 3.$

$$s^2_{\bar{y}_{i.} - \bar{y}_{i'.}} = 2 (19) / 3 (4) = 19 / 6, \text{ and } s^2_{\bar{y}_{i.}} = 19 / 3(4) = 19 / 12.$$

For the latin square design as given in Example 10.2 with $k = 3$, the
various computations are:

Row	Column (e_{hij} values)			Sum	
	1	2	3		
1	B 0	A -1	C 1	0	$e_{2A1} + e_{1A2} + e_{3A3} = 0$
2	A 1	C 0	B -1	0	$e_{1B1} + e_{3B2} + e_{2B3} = 0$
3	C -1	B 1	A 0	0	$e_{3C1} + e_{2C2} + e_{1C3} = 0$
Sum	0	0	0	0	$s^2_e = 6 / (3 - 1)(3 - 2) = 3.$

$$s^2_{\bar{y}_{i.} - \bar{y}_{i'.}} = 2 (19) / 3 (4) = 19 / 6, \text{ and } s^2_{\bar{y}_{i.}} = 19 / 3 (4) = 19 / 12.$$

For the simple change-over design in Example 10 3, the error deviations are:

Row	Column (e_{hij} values)						Sum
	1	2	3	4	5	6	
1	-1/3	1/3	7/3	-1/3	2/3	-7/3	0
2	1/3	-1/3	-7/3	1/3	-2/3	7/3	0
Sum	0	0	0	0	0	0	0

$$\sum_{h=1 \text{ to } 6} \sum_{j=1 \text{ to } 2} e^2_{hij} / (2 - 1)(6 - 2) = 240 / 9(4) = 20 / 3 = s^2_e .$$

$$s^2_{\bar{y}_{.A.} - \bar{y}_{.B.}} = 2 (20 /3) / 6 = 20 / 9 \text{ and } s^2_{\bar{y}_{.i.}} = 20 / 3 (6) = 10 / 9.$$

For the bib design in Example 10.4, the e_{ij} are already computed. Therefore,

$$s^2_e = \sum_{\text{all } ij} e^2_{ij} / (12 - 4 - 6 + 1) = 16 / 3 \text{ and}$$

$$s^2_{t_i - t_{i'}} = 2 (2 / 4 (1)) (16 / 3),$$

for all pairs i and i' of adjusted treatment means or effects.

10.6 COMPUTERS AND STATISTICS

Computers and their use for word-processing and computations are wide-spread. Within the last ten years, millions of people have become computer-literate even though they never visualized doing so 20 years before. The impact of computers on every-day operations and as a part of what is now considered normal in the last ten years is nothing short of phenomenal. The computer, whether a pocket calculator, a PC, an Apple, or a main-frame facility, is an integral part of scientific research, of statistical computations, and of every-day activities. When suitably programmed, it can be used to help construct experiment and treatment designs, to construct graphical displays, and to compute a variety of statistics. For large data sets, the drudgery associated with statistical computing can be mostly eliminated with appropriate use of the computer. Many software packages have been written for computers to perform various numerical, and even symbolic, computations. These packages are not fool-proof as the computer is a very intelligent idiot. It does only what it is told to do and then does *exactly* what it was told to do without any concern as to whether this is what the user intended to do. This means that the uniformed user of computer packages and programs may obtain results that are not what he thinks they are.

To do statistical computations on a PC or on a main-frame computer, it is necessary to:

a. find a terminal whether on a PC or in a terminal facility for a main-frame computer,
b. if a terminal facility, get connected to the computer you are to use,
c. identify yourself to the computer which is termed "logging on",
d. run a program software package that does the computations desired,
e. perform the numerical computations desired,
f. print or store the output of the software package computations and end the session, and
g. log off the computer (inform the computer that you have finished).

There are many packages that have been and are being developed to do various statistical computations. As might be suspected, the packages vary in quality, appropriateness, and correctness. For the novice, it is a good idea to seek the advice of an expert before purchasing and installing a statistical software package on a computer. Some statistical software packages and addresses where information about them can be obtained that come to mind are:

a. MINITAB	Minitab, Inc. 3081 Enterprise Drive State College, PA 16801
b. DATADESK	Datadesk Inc. 15 Lisa Lane Ithaca, NY 14850
c. GENSTAT	Numerical Algorithms Group 1400 Opus Place, Suite 200 Downers Grove, IL 60515
d. SAS	SAS Institute Inc. SAS Circle, Box 8000 Cary, NC 27512-8000.
e. SYSTAT	Systat Inc. 1800 Sherman Avenue Evanston, IL 60201

MINITAB was developed specifically for classes in Statistics using a computer. It is probably the most widely used package for statistics classes. It has been devised for the novice and made as "user-friendly" as possible. For example, once the MINITAB program is ready to operate, the following set of instructions are typed in the computer to show how to enter the data, how to obtain a box plot, a stem and leaf graph, and two-way plots of 11 responses for each of three variables, and how to print out the results:

note data are inflation and unemployment rates for the years 1959 to 1969
(The note command allows the user to insert comments, notes, and/or description of the data sets.)
set c1 1959 1960 1961 1962 1963 1964 1965 1966 1967 1968 1969
set c2 0.7 1.6 1.0 1.2 1.2 1.4 1.7 2.9 2.9 4.3 5.4
set c3 5.4 5.4 6.6 5.5 5.5 4.9 4.4 3.7 3.7 3.6 3.5
(The data may be entered leaving a space or two between each datum or a comma and space may be used to separate the numbers.)
name c1 'year'
name c2 'inflatn'   *(Name must be 8 characters or less.)*
name c3 'unemplmt'
info
print c1 c2 c3

```
plot 'unemplmt' by 'year'   (Or plot c2  c1.)
plot c3  c2
describe  c2
boxplot  c2
stem  c2
restart                  (Use only if you wish to go to another data set.)
stop                     (This gets you out of MINITAB and into console.)
console print xxx        (This prints a copy of your session at printer xxx.)
logoff                            (This gets you off the computer.)
```

The largest task is typing in the data sets but once they are in the computer, numerous calculations may be performed quickly on them.

DATADESK has been written specifically for the MacIntosh (Apple) computer. This is a very "user-friendly" package and quickly learned by students. In a very short time students will be spending their time on understanding statistical procedures rather than on understanding the software package. All producers of statistical software strive to do this but with varying degrees of success.

10.7 REFERENCES AND SUGGESTED READING

Bevan, J. M. (1968). *Introduction to Statistics.* Philosophical Library, Inc., New York, vii + 220 pp..
(Chapters 4 and 5.)

Campbell, R. C. (1967). *Statistics for Biologists.* Cambridge University Press, London, xii + 242 pp..
(Chapter 2.)

Fisher, R. A. (1950). *Statistical Methods for Research Workers, eleventh edition,* Oliver and Boyd Ltd., Edinburgh and Hafner Publishing Company, New York.
(Chapter 3.)

Huff, D. (1954). *How to Lie with Statistics.* W. W. Norton and Company, Inc., New York, 142 pp..
(Chapters 1, 2, and 4.)

Koopmans, L. (1987). *Introduction to Contemporary Statistical Methods.* Duxbury Press, North Scituate, Mass., 683 pp..

McCarthy, P. J. (1957). *Introduction to Statistical Reasoning.* McGraw-

Hill Book Company, Inc., New York, Toronto, and London, xiii + 402 pp..
(Chapters 3 and 4.)

Moroney, M. J. (1956). *Facts from Figures, 3rd edition.* Penguin Books, Baltimore and London, viii + 72 pp..
(Chapters 7, 8, and 9.)

Pearce, S. C. (1965). *Biological Statistics: An Introduction.* McGraw-Hill Book Company, New York, St. Louis, San Francisco, Toronto, London, Sydney, xiii + 212 pp..
(Chapters 1 and 2.)

Snedecor, G. W. and Cochran, W. G. (1980). *Statistical Methods, 7th edition.* The Iowa State University Press, Ames, Iowa, xiv + 507 pp..
(Chapters 1, 2, 3, and 8.)

Siegal, A. E. (1988). *Statistics and Data Analysis. An Introduction.* John Wiley & Sons, Inc., New York, Chichester, Brisbane, Toronto, Singapore, xxv + 518 pp..

Steel, R. G. D. and Torrie, J. H. (1980). *Principles and Procedures of Statistics, second edition.* McGraw-Hill Book Company, New York, Toronto, and London, xvi + 481 pp..

10.8 PROBLEMS

Problem 10.1. For the class survey data in Table 9.1, for the 12 girls who were in class, what are the median values for height (4th column), weight (10th column), and age (11th column)? What are the modal values? Compute the arithmetic means if a computing machine is available; otherwise, *show how* to compute the arithmetic mean.

Problem 10.2. A man travels by car from town X to town Y, and he gets only 20 miles to a gallon of gasoline. On the return journey, he gets 30 miles to a gallon. Then, by assuming that the distance from X to Y is 60 miles, verify that the harmonic mean is the correct average to calculate. Find the arithmetic average for comparison.

Problem 10.3. Groups of boys and girls are tested for reading ability. The forty boys make an average score of 40%. The sixty girls have an average score of 80%. Calculate the arithmetic average for boys and girls combined. Would you use a weighted or unweighted mean? Why, or why not?

Problem 10.4. On March 1st a baby weighed ten pounds. On June 1st, it weighed 22.5 pounds. Use the geometric mean to estimate its weight on April 15th. Also compute the arithmetic mean and compare it with the geometric mean.

Problem 10.5. The following data (about 1936-38) represent percentages of individuals enrolled in various types of schools. The base is the number of individuals in the age bracket, and the numerator is the number of individuals enrolled. Thus, more than 100% can be enrolled by this method since, for instance, children under six and over 13 are in elementary school.

Country	6-13 years % enrolled in elementary schools	14-17 years % enrolled in secondary schools	18-21 years % enrolled in college and universities
United States	119.5	63.6	14.6
Canada	113.0	55.8	6.2
Australia	119.6	21.1	8.5
New Zealand	111.0	27.9	5.6
EUROPE			
Belgium	92.5	55.7	3.6
Czechoslovakia	83.8	43.3	2.5
Denmark	102.0	44.3	3.0
France	114.8	33.5	2.6
Germany	83.4	43.5	3.0
Great Britain	104.3	50.8	3.6
Netherlands	88.8	40.9	2.0
Russia	99.0	32.9	6.0
Switzerland	92.3	42.6	3.8

Data from Appendix in *Education - America's Magic* by R. M. Hughes and W. H. Lancelot.

Find the median, range, and mode (if one exists) for each set (column) of the above data. Prepare bar and pie graphs of percentages with the countries arranged by increasing order of percentage. Where would you expect to find current data of this nature?

Problem 10.6. The following data were constructed for ease of computation to illustrate the computations for a completely randomized design. There are v = 4 treatments and r = 3 replicates on each treatment. Compute the arithmetic means ($\bar{y}_{i.}$) for the treatments, the sum of squares of the error deviations which is equal to $\sum_{i=1 \text{ to } 4} \sum_{j=1 \text{ to } 3} e^2_{ij}$ = 172, the error variance for all treatments, and the variance of difference between two treatment means.

Treatments

		1		2		3		4	
	Y_{11}	11	Y_{21}	13	Y_{31}	21	Y_{41}	10	
	Y_{12}	4	Y_{22}	9	Y_{32}	18	Y_{42}	4	
	Y_{13}	6	Y_{23}	14	Y_{33}	15	Y_{43}	19	
Total $Y_{i.}$	$Y_{1.}$		$Y_{2.}$		$Y_{3.}$		$Y_{4.}$		$Y_{..}$ =
Mean $\bar{y}_{i.}$	$\bar{y}_{1.}$		$\bar{y}_{2.}$		$\bar{y}_{3.}$		$\bar{y}_{4.}$		$\bar{y}_{..}$ =

Show all computations, as for example, $Y_{1.}$ = 11 + 4 + 6 = 21, $\bar{y}_{1.}$ = 21 / 3 = 7, etc., including computation of e_{ij} .

Problem 10.7. Given below is a set of data constructed to result in relatively simple computations for a randomized complete block design. Here v = 3 treatments were arranged in r = 3 complete blocks. The yields arranged in a systematic fashion are:

Treatment		Blocks						Treatment	
		1		2		3		Total	Mean
1	Y_{11}	6	Y_{12}	5	Y_{13}	4		$Y_{1.}$	$\bar{y}_{1.}$
2	Y_{21}	15	Y_{22}	10	Y_{23}	8		$Y_{2.}$	$\bar{y}_{2.}$
3	Y_{31}	15	Y_{32}	15	Y_{33}	12		$Y_{3.}$	$\bar{y}_{3.}$
Block total	$Y_{.1}$		$Y_{.2}$		$Y_{.3}$			$Y_{..}$	$\bar{y}_{..}$
Block mean	$\bar{y}_{.1}$		$\bar{y}_{.2}$		$\bar{y}_{.3}$				

Compute the various block, treatment, and grand totals and means. Verify
that $\Sigma_{i=1 \text{ to } 3} \Sigma_{j=1 \text{ to } 3} e^2_{ij} = 10$. Show how to compute (do not actually
compute) the variance of a treatment mean, the variance of a difference
between two treatment means, and the standard error of a difference
between two treatment means.

Problem 10.8. The following 3 × 3 latin square design was constructed for
ease of computation:

Row = week	Column = Store 1	2	3	Week Total	Mean
1	Y_{11B} 3	Y_{12A} 7	Y_{13C} 9	$Y_{1..}$	$\bar{y}_{1..}$
2	Y_{21C} 14	Y_{22B} 8	Y_{23A} 2	$Y_{2..}$	$\bar{y}_{2..}$
3	Y_{31A} 6	Y_{32C} 15	Y_{33B} 6	$Y_{3..}$	$\bar{y}_{3..}$
Store total	$Y_{.1.}$	$Y_{.2.}$	$Y_{.3.}$	$Y_{...}$	
Store mean	$\bar{y}_{.1.}$	$\bar{y}_{.2.}$	$\bar{y}_{.3.}$		$\bar{y}_{..}$

				Sum	
Treatment total	$Y_{..A}$	$Y_{..B}$	$Y_{..C}$	$Y_{...}$	90
Treatment mean	$\bar{y}_{..A}$	$\bar{y}_{..B}$	$\bar{y}_{..C}$	$3\bar{y}_{...}$	30

Show computations necessary to obtain s^2_e the error variance, the standard
error of a difference between any two treatment means, and complete the
computations in the above tables.

Problem 10.9. For the data of Problem 9.10, prepare box plots, fences,
and determine whether there are any minor and major outliers.

Problem 10.10. For the data in Problem 9.14, prepare a box plot, fences,
and determine if there are any minor or major outliers.

Problem 10.11. The following data set came from an experiment on
wheat, which was designed as a randomized complete block design. The

responses are ratios of dried weight of grain to green weight of grain. The four treatments were nitrogen fertilizer applied early, mid-season, and late in the growing season and no nitrogen. Four blocks were used. The data are:

Block	none	Nitrogen applied early	middle	late	Total
1	0.718	0.732	0.734	0.792	
2	0.725	0.781	0.725	0.716	
3	0.704	1.035	0.763	0.758	
4	0.726	0.765	0.738	0.781	
Total					

Obtain the block and treatment totals, means, and effects. Also, obtain the 16 residuals. Prepare a box plot of the data and of the residuals. Are there any outliers in this data set? Suppose the datum for block 3 and treatment early application of nitrogen had been replaced by the value 0.762; now compute the residuals for this situation. Compute the variances when the value of 1.035 is used and when the value 0.762 is used. What are the relative sizes of the variances? Do you believe the value of 1.035 is realistic? Why or why not?

Problem 10.12. The following set of Own weights were obtained from a 1979 class survey for 24 women and 22 men:

Y_{fj}	$e_{fj} = Y_{fj} - \bar{y}_{f.}$	Y_{mj}	$e_{mj} = Y_{mj} - \bar{y}_{m.}$
127	$0.71 = 127 - 126.29$	141	$-17.77 = 141 - 158.77$
112		150	
125		160	
116		168	
105		145	
130		185	
150		170	
140		150	
138		170	
135		150	
117		138	
138		162	

107	165
127	143
115	189
145	222
135	148
146	147
127	132
115	154
122	155
122	149
103	-------------------------------------
134	Sum 3,493

Sum 3,031

Verify that the sum of squares for weights of women is 4,046.96 and that
for weights of men is 8,507.86. Show how to compute
   each of the above sum of squares
   ranges for both groups
   arithmetic means for both groups
   medians for each group
   variances for each group
   standard deviations for each group
   standard error of a mean for each group
   pooled variance for both groups
   pooled standard deviation
   standard error of a difference of two means when variances are
   estimates of different parameters
   difference of means divided by the standard error of a difference of two
   means

Would you consider the two variances to be estimates of the same or
different parameters? Why or why not? Prepare box plots of each data
set. Are there any minor or major outliers?

Problem 10.13.   The following data set came from an experiment
involving eight different oils as the treatments. A simple random sample
of six sets of 24 doughnuts was allocated to each of the eight treatments.
The 48 sets of 24 doughnuts were cooked in random order. What is the
experiment design? The amount of oil absorbed by 24 doughnuts for each
of the 48 batches is given below (data were collected by Belle Lowe and
given in Iowa Agricultural Experiment Station Report of 1935):

| | Treatment (fat number) | | | | | | | |
	1	2	3	4	5	6	7	8
	164	173	177	178	163	175	178	155
	172	161	183	191	165	193	146	166
	168	190	197	197	144	178	141	149
	177	180	169	182	177	171	150	164
	156	197	179	185	165	163	169	170
	195	167	187	177	176	176	182	168
Total	432							
Mean								

The sum of squares of the 48 residuals is 5,666. Obtain the pooled variance, the standard error of difference between two oil means, and the standard deviation divided by the over-all mean (This ratio is called the **coefficient of variation.**) Prepare a stem and leaf diagram of the 48 residuals. Prepare a box plot of the 48 residuals. Are there any minor or major outliers? Prepare a two-way array of the differences between means. For the columns, order the oil means from highest to lowest and for the rows, order the means in reverse order, i.e., from lowest to highest. Then, compute all possible differences. Note that there 28 possible differences among the eight oil means and that only the upper part of the table above the diagonal needs to be computed. The part below the diagonal is simply the negative of the entries above the diagonal. How many of the differences are larger than the standard error of a difference between two means? How many are larger than twice the standard error of a difference?

CHAPTER 11. A LITTLE PROBABILITY

11.1 INTRODUCTION

Throughout the first ten chapters, we have alluded to several probablistic concepts. We have relied on the reader's intuitive definition of odds, chance, or probability concepts rather than on a formal presentation, since all that was needed was a general notion of what is meant by probability. Few readers would be able to develop a precise, definite, and consistent meaning of the term which would withstand scholarly criticism. For many, it falls in the same vague, relatively undefined category as do a large number of other terms such as "imperialist", "reactionary", "liberal", "conservative", "radical", "sub-standard housing", "poverty", and other terms current in present day news media and in printed handouts of certain campus groups. To some, the term probability brings to mind thoughts of betting on the horses, Las Vegas, Atlantic City, chances of survival from auto and airplane crashes, male-female ratio, flipping coins, chances of rain, throwing dice, something insurance companies and gambling institutions worry about, and so on; their interpretation may include such concepts as chance, random event, odds, a victim of Fate, personal belief, what happens to the other fellow, frequency or proportion, mechanistic, haphazard, and terms learned by rote from the statistician's language. We shall briefly discuss three bases for definitions of probability:

1. relative frequency and empirical probability,
2. *a priori*, classical, or analytic probability, and
3. personal belief or subjective probability.

The relative frequency concept is the one often utilized; the controversy among a segment of the scholarly does not affect the practical use of the frequency concept of probability.

11.2 SOME TERMINOLOGY

One of the first items to consider in establishing a definition of probability is the **scale** or **range** that probability values can take. So far, all persons dealing with probability have agreed to measure probability on a scale marked zero on one end and one at the other, with all possible values in between and including the end-points, that is, for a probability value p we may say $0 \leq p \leq 1$. A scale such as the one in Figure 11.1 indicates the possible range of values with various kinds of happenings. If something simply cannot happen, then its probability of occurrence is given a value of zero. If something is certain to happen, then it has a probability value of one. Happenings which have the endpoint values of zero and one usually cause little difficulty. The values in between, the uncertainties, cause many problems in everyday application, and this keeps the statistician busy.

The quantity to be measured on the above scale is the uncertainty of a future happening, such as the probability that it will rain next Sunday, or of a past occurrence, such as the probability that Sir Francis Bacon wrote the plays commonly attributed to William Shakespeare. In such situations, there will be certain *possible* outcomes which we shall denote as *events*. For example, in coin tossing, we might categorize the possible events as follows:

E_h denotes a head on the side facing up,

E_t denotes a tail on the side facing up,

E_s denotes the coin standing on edge,

E_a denotes the coin remaining suspended in the air, and

E_v denotes the outcome when the coin vanishes.

We would most probably decide that E_s, E_a, and E_v could happen only with probability zero and are figments of the imagination rather than possible outcomes. This would leave us with the two possible events E_h and E_t. We would be concerned about probability values of occurrence for these two possible outcomes.

Many situations in nature have two possible outcomes, dead or alive, student or non-student, married or not married, go/no-go, if and if not, and either / or are examples of such a dichotomous classification. Many actions also fall into two possible categories: whether or not to sleep in class, to skip class, to carry an umbrella, to go to the baseball game, to vote in an election, to shave in the morning, to go to the Ice Follies, and so on. Despite the abundance of examples involving only two possible events, there is an even larger set of examples with more than two outcomes. Suppose that we have a six-sided die with one, two,..., six dots respectively on its six sides. In terms of possible outcomes, resulting from a roll of the die and observing the side that is facing upward, the six possible events are shown in Figure 11.2. Each of these events is denoted as an **elementary event** in that there is no simpler or more basic manner of defining a

value of probability
↓

1.0	certain without doubt that one will die someday
0.9	that this University will have a "confrontation" during the year
0.8	that a randomly selected tens digit will not be 8 or 9
0.7	that student "imperialists" will not "take over" this university
0.6	that a randomly selected vowel will be a, e, or i
0.5	50:50 that an unbiased two-headed coin will show a tail when flipped
0.4	that it will rain today
0.3	
	one in four chance that the first card dealt from a well-shuffled 52-card deck will be a heart
0.2	
0.1	that the reader will win the top prize in the next "give-away" he enters
0	that the reader could swim non-stop from Oahu to Bali

Figure 11.1. The probability scale of measurement.

happening of rolling a die. We assume that there are no additional happenings of relevance such as the die standing on an edge or a corner, hence the six elementary events define the totality of all possible events for this activity.

Instead of considering the six possible events given in Figure 11.2, we might consider only the following two events as possible outcomes of the activity of casting a die:

E_6 denotes six dots showing on the top side of the die

$E_{/6}$ denotes other than six dots shown on the top side.

$E_{/6}$ (read E not 6) is composed of five elementary events, E_1, E_2, E_3, E_4, and E_5 and is denoted as a compound event. A **compound event** is an event containing more than one elementary event. The event $E_{/6}$ happens whenever there are one, two, three, four, or five dots showing on the side of the die facing upward.

As another example, let us consider the activity of drawing one of ten ping-pong balls numbered 0, 1, 2,..., 9 from a container, the ten possible elementary events are shown in Figure 11.3. If our events were considered to be the compound events of odd numbers and even numbers, say E_o and E_e, each compound event would consist of five elementary events.

Another concept that is useful in probability contexts is that of a **mutually exclusive event** which results when one event in the set of possible outcomes happens, and no other event can happen simultaneously. Thus, in the coin tossing example with the elementary events E_h and E_t, the events are mutually exclusive, because either one or the other happens. On the other hand, suppose that the possible outcomes in the ping-pong example are the even numbers $= E_e$, the odd numbers E_o, and the numbers less than 5 $= E_{<5}$; events E_e and E_o are mutually exclusive, because these compound events contain no elementary events in common. However, $E_{<5}$ is not mutually exclusive of either E_e or E_o, because $E_{<5}$ has three elementary events in common with E_e and two with E_o.

Another useful idea which has been utilized previously in drawing random samples is that of an **independent event**. In an activity, if the happening of one event does not influence the happening of a second event, the two events are said to be independent. To illustrate, suppose that the elementary events are represented by the selection of one of the six letters a, b, c, d, e, and f and suppose that three are to be selected to form a sample. Furthermore, suppose that our mechanistic process allows only the samples abc or def. The selection of a simultaneously selects b and c, and

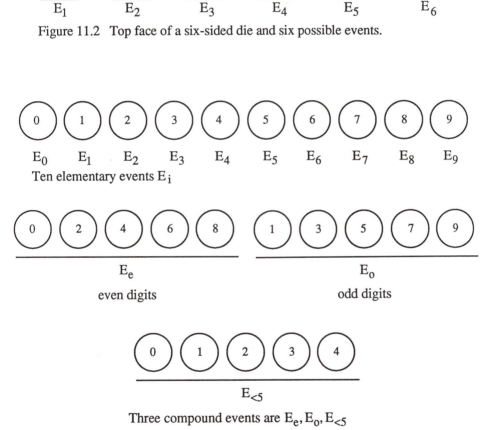

six-sided die

Figure 11.2 Top face of a six-sided die and six possible events.

Ten elementary events E_i

E_e

even digits

E_o

odd digits

$E_{<5}$

Three compound events are $E_e, E_o, E_{<5}$

Figure 11.3. Elementary and compound events.

hence these are not independent events. On the other hand, if our mechanistic process is one which allows no influence of selection of one on the other then the events are independent. One such process is to place the letters on six identical ping-pong balls, thoroughly mix the balls in a container, and blindly draw three balls from the hat with no replacement after each draw. Such a process results in one of the 20 possible combinations of three letters, as listed in Chapter 5.

Another example is to consider the activity of rolling a die with the events $E_{\div 2}$ meaning that the number on the top face is divisible by 2, $E_{/\div 2}$ meaning that the number on the top face is not divisible by 2, $E_{\div 3}$ meaning that the number on the top face is divisible by 3, and $E_{/\div 3}$ meaning that the number on the top face is not divisible by 3. Clearly, event $E_{\div 2}$ and $E_{\div 3}$ are not independent, because 6 is divisible by 2 and by 3, 2 and 4 are not divisible by 3, and 3 is not divisible by 2. No pair of the compound events given in this example is independent.

11. 3 EMPIRICAL PROBABILITY

Inherent in many practical applications of probability is the idea of a relative frequency of an outcome, a proportion or odds falls into this category. Associated with this is the idea of conceptually repeatable samplings or experiments. The samplings are conceived as being repeatable an infinite number of times. This would certainly fit a coin-tossing experiment or a die-casting experiment; conceptually, we could keep on throwing dice forever, not just 26,306 times as Mr. Weldon did (see Table 12.1). He threw 12 dice at one time, which resulted in 12(26,306) = 315,672 elementary events, of which 106,602 showed either a five or a six. The proportion of times that a five or a six occurred was 106,602 / 315,672 = 0.3376986 which is greater than the predicted proportion of 1 / 3, resulting in the conclusion that these 12 dice were biased. Then we say that the probability of obtaining a five or a six when these 12 dice are thrown is 0.3376986. More formally, we could write this as: $P(E_5$ or E_6 occurs) = 0.3376986. Also, $P(E_5$ or E_6 does not occur) = $P(E_1, E_2, E_3,$ or E_4 occurs) = 1 - 0.3376986, since the sum of the probabilities for the possible events must equal unity, that is, we must account for all the happenings. This is another property of probability which we may write more formally, as follows: let p_i = probability of the ith event for i = 1, 2,..., n possible events which exhaust all possible events, then

$$\sum_{i=1}^{n} p_i = 1.$$

To illustrate this, suppose that we had an unbiased die and each of the six possible events had a probability of occurrence of 1 / 6. Then we may write

$$P(E_1) + P(E_2) + P(E_3) + P(E_4) + P(E_5) + P(E_6)$$
$$= 1/6 + 1/6 + 1/6 + 1/6 + 1/6 + 1/6 = 1.$$

When all events have the same probability of occurrence, they are referred to as **equally likely events**.

All the above implies that probability can be defined as "the ratio of the number of times an event occurs to the total number of possible occurrences" or "as the ratio of the number of successes to the total number of trials." This relative frequency definition is quite useful in many walks of life and in statistical applications. The ratio $a / (a + b)$ may be used or the result may be stated in terms of odds as a:b (a to b).

To test a mechanism for drawing random samples, we would want to determine whether all possible samples of size n are equally likely events, that is, have equal chances of being selected. To do this, we would conduct an experiment to obtain a very large number of observations, and we would then compute the proportion of times each event occurred. If the proportions are considered to be equal, we could conclude that the process results in a random selection of samples.

To illustrate a familiar use of probabilities in the relative frequency sense, let us suppose that we have an alphabetical listing of the 15,000 students at University X, that the list contains 12,000 men's names and 3,000 women's names, and that we wish to draw a completely random sample of names from the list. The probability that a woman's name will be drawn is 3,000 / 15,000 = 1 / 5 = 0.2 and that a man's name will be drawn is 12,000 / 15,000 = 4 / 5 = 0.8, provided a truly random procedure is utilized in drawing the name from the list. Suppose that in addition to the above, we know that there are 600 freshman, 500 sophomore, 500 junior, 500 senior, and 900 graduate female students. We could then compute the various probabilities for randomly selecting a name for these categories as:

P(name belonging to a female freshman) = 600 / 15,000 = 1 / 25,
P(name belonging to a female sophomore) = 500 / 15,000 = 1 / 30,
P(name belonging to a female junior) = 500 / 15,000 = 1 / 30,
P(name belonging to a female senior) = 500 / 15,000 = 1 / 30,

A Little Probability

and
P(name belonging to a female graduate) = 900 / 15,000 = 3 / 50.

A random sample of 200 names, say, containing no names of women in any category would be considered a highly unlikely event, but we must remember that such an event does have a non-zero, albeit very low, probability of occurrence. Such an event should cause us to scrutinize the process utilized for drawing the sample. The probability that one of the above five categories would not be represented in the sample does not have such a low probability of occurrence. If we want all categories represented, we should use a form of stratified sampling as suggested in Chapter 5.

11.4 ANALYTICAL, A PRIORI, OR CLASSICAL PROBABILITY

In many situations we construct probabilities and a definition of probability from a theoretical consideration of an activity without any idea of conducting an experiment for collecting observations. From the very nature of the activity and of the associated possible events, we may be able to define the various probabilities of the various events. For example, we can observe that a coin has a head on one side and a tail on the other and that the coin is symmetrical, without any data from tossing coins, we state that $P(E_h) = 1 / 2 = P(E_t)$ where E_h = occurrence of a head and E_t = occurrence of a tail. Likewise, for the six-sided die we can observe by physical measurements that the die is symmetrical and unweighted, and hence we can state that the probability that any side will fall upward is 1 / 6, or that the six possible events are equally likely.

The use of theoretical knowledge of an activity to determine probability values such as those described above has been variously termed analytical, **a priori**, or **classical** probability. We actually used this basis previously when we considered an unbiased two-sided coin or six-sided die to state the expected probability values. This concept of probability is very useful for computing hypothetical probability values.

11.5 PERSONAL BELIEF, SUBJECTIVE, OR PSYCHOLOGICAL PROBABILITY

In the preceding two sections we have considered empirical and analytical bases for constructing probability values and concepts. Another basis is the "personal", "degree-of-belief", "psychological", "subjective", "egoistic", or "self-confidence" basis for probability. All relate to a person's belief. This approach has been formally studied by mathematicians, philosophers, and statisticians.

To illustrate the nature of the personal belief approach, consider such probability statements as:

P(my son's marriage will not end in divorce) = 0.75;

P(the next car I·buy will be a "lemon") = 0.61;

P(a "bloodbath" in South Vietnam in the event of a North Vietnamese takeover) = 0.999;

P(that the President's "honeymoon" will last 6 months longer) = 0.002;

P(that the cute girl sitting next to me in my genetics class has a date for next Saturday night) = 0.80;

P(I will spend more than $100 for washer and dryer repairs this year) = 0.50.

Aside from some "phony statistics" terms because of undefined terms, for instance "lemon", statements such as the above are based on the personal beliefs of *one* person; they would not be the same as those set by another person for the same statements. Hence the question of how such probabilities should be used in a scholarly sense is still in question.

In an article entitled "Life's Little Gambles" in the December 1989 issue of *Reader's Digest*, pages 61-64, several examples of using probabilities common to all of us are described and are given below. It should be pointed out that many of those given are of the unknowable type and the article should be considered for its entertainment value rather than as a source of valuable data. Did you know that two University of Louisville scientists reported that 13 percent of all coins and 42 percent of paper money carry infectious organisms? Who said money wasn't filthy? If the world is such a dangerous place to be because of all the toxic materials in our environment, drinking water, and foods, how come our life span keeps increasing? It is now 71.4 years for men and 78.3 for women. We all have fears of crashing in an airplane, being killed by a burglar, or dying on an operating table. But, these things are very unlikely events for each of us. The chance of dying in a commercial airplane is one in 800,000. We are more likely to choke to death on food than in being killed in an airplane. We are twice as likely of being killed in a sport than by a stranger. The chance of dying from a medical complication or a mistake is only one in 84,000. We have a far greater risk of dying in a car than any of the above, i.e., one in 5,000. The chance of dying in a car during one trip is only one in four million but since the average person takes about 50,000 trips during their lifetime, the possibility of coming to grief in a car accident is heightened. When risk is concerned most people are irrational in weighing probabilities. People believe that their chance of winning a jackpot is higher than being hit by lightning but they are 13 times more likely to be struck by lightning than hitting the jackpot!

The article goes on to discuss a variety of other items. For example, insurance companies make considerable use of probability theory in determining premiums for their various insurance policies. In 1989, they calculated that approximately 765,000 people in the United States will die of heart disease, 68,000 of pneumonia, 2,000 of tuberculosis, 200 in storms and resulting floods, 100 by lightning, 100 in tornadoes, and 50 from snakebites and bee stings. On the average, being 30 percent over-weight reduces our life span by three and one-half years, being poor reduces it by two years, being a single man reduces it by ten years, and being a single women reduces life expectancy by four years. Every cigarette one smokes cuts ten minutes off the life span. The *Reader's Digest* says that a "statistician" has calculated that if one wears both a belt and suspenders to hold up slacks, the chances of "coincident failure" (having both snap simultaneously) is once in 36,000 years. They say there is no way of escaping life's little gambles, and, hence, one should keep in mind that the laws of probability and risk are clearly on our side. They state that sometimes even the smallest probabilities are not enough and that we will be the "success" that makes odds what they are. For example, an American businessman tried to escape all the dangers of life by moving to a beautiful and balmy Pacific Island ringed with beautiful beaches in 1940. Unfortunately he picked the wrong island. The island was Iwo Jima!

11. 6 ADDITION LAW OF PROBABILITY

As we have illustrated in previous sections, the probability of a compound event is the sum of the probabilities of the elementary events making up the compound event. For example, in the last illustration in Section 11.3, the sum of the probabilities for the womens' names from the various classes is equal to

$$1 / 25 + 1 / 30 + 1 / 30 + 1 / 30 + 3 / 50 = 5 / 50 + 3 / 50$$
$$= 2 / 10 = 0.2,$$

which is equal to the probability of a woman's name being selected, that is 3000 / 15,000 = 0.2. In equation form, we could say

P(name belonging either to a freshman, sophomore, junior, senior, or graduate female student) = 1 / 25 + 1 / 30 + 1 / 30 + 1 / 30 +

$3 / 50 = 0.2 = $ P(name belonging to a woman).

This illustrates the **addition law of probability,** which states that the probability for an event is the sum of the probabilities for the elementary events making up that event. Note the "either" and "or" words in the probability statement. Also note that we could say that the probability of a specified outcome is the sum of the probabilities of the mutually exclusive events making up the outcome.

As a second illustration from Section 11.3, consider the example for E_6 = a six on the top face of a die and for $E_{/6}$ = not a six on the top face of the die. Then, for unbiased dice

$$P(E_6) = 1/6,$$
$$P(E_{/6}) = 5/6 = P(E_1) + P(E_2) + P(E_3) + P(E_4) + P(E_5),$$

and

$$P(E_6) + P(E_{/6}) = 1/6 + 5/6 = 1.$$

The last statement states that if a die is thrown, it is certain to show one of the six faces. We have ruled out the possibility of a die standing on an edge, that is, we have given this event zero probability.

11.7 MULTIPLICATION LAW OF PROBABILITY

Suppose that we have two unbiased coins each with a head on one side and a tail on the other, that the coins are to be tossed simultaneously, and that we wish to calculate the probabilities for the various events which are:

	Outcome			
coin 1	H	H	T	T
coin 2	H	T	H	T

where H represents the head on one side of a coin and T represents the tail on the other. Each of these four results is equally likely and each has a probability of occurrence of $1/4$. Note that the results for two consecutive tosses of the same coin produce the same set of four results. If we are given the probabilities of the events for coin 1 and for coin 2, we can compute the probabilities of the individual coins as follows:

$$P(\text{H on coin 1 } and \text{ H on coin 2}) = 1/2 \times 1/2 = 1/4,$$
$$P(\text{H on coin 1 } and \text{ T on coin 2}) = 1/2 \times 1/2 = 1/4,$$
$$P(\text{T on coin 1 } and \text{ H on coin 2}) = 1/2 \times 1/2 = 1/4,$$

and

$$P(\text{T on coin 1 } and \text{ T on coin 2}) = 1/2 \times 1/2 = 1/4.$$

Note the word "and" in the probability statement. Also note that the probability *either* of a head on coin 1 and a tail on coin 2 *or* a tail on coin 1 and a head on coin 2 is computed as follows:

P(*either* an H on coin 1 and T on the other coin *or* a T on coin 1 and an H on coin 2) $= 1/4 + 1/4 = 1/2$,

which illustrates the addition law of probability. Thus, the "either-or" statement is associated with addition of probabilities, and the "and" statement is associated with multiplication of probabilities.

More formally, we can say that the probability of an event which is the result of the simultaneous occurrence of two or more independent elementary events is the product of the probabilities of the elementary events. This is called the **multiplication law of probability.**

As a second illustration of the above, consider that we have a pair of six-sided unbiased dice one of which is red and the other white. What is the probability of a pair of sixes resulting from a throw of the dice? That is, we want to know what is the probability of the events E_{r6} = a six on the red die and E_{w6} = a six on the white die both occurring from one throw of the two dice. Such a combination is known as "boxcars" in the vernacular of "crap-shooters", that is, dice-throwers. We compute this probability as

$$P(\text{boxcars}) = P(E_{r6}) \times P(E_{w6}) = 1/6 \times 1/6 = 1/36.$$

Likewise, the probability of simultaneously obtaining a six on the red die, E_{r6}, and something other than 6 on the white die, $E_{w/6}$, may be computed as follows:

P(E_{r6} and $E_{w/6}$) $= P(E_{r6}) \times P(E_{w/6}) = 1/6 \times 5/6 = 5/36.$

The multiplication law may be utilized in many situations. For example, it can be used "to prove" that a 15-year old boy named Arthur with a registered cocker spaniel named Princess Taffy has a dog that is one in a billion. To do this, let us suppose that there are one-half as many boys as girls at age 15, that the name Arthur is given to one in 50 boys, that registered cocker spaniels constitute only 0.1% of the dog population, and that the name "Princess Taffy" is given to only 0.01% of the dogs. Then,

P(15 year old boy named Arthur owns a registered cocker spaniel named Princess Taffy) = P(boy) × P(name Arthur) × P(a registered cocker spaniel) × P(name Princess Taffy)

$$= 1 / 2 \times 1 / 50 \times 1 / 1,000 \times 1 / 10,000 = 1 / 1,000,000,000.$$

Another example that may have some financial consequences for the reader pertains to horse racing. We observe the odds posted on the board or given on racing forms, and if we believe these odds, we could make various kinds of bets. One method would be to place bets on two horses in one race. Our chances of picking a winner would be determined by the odds for the n horses running in the race and would involve the sum of probabilities. Likewise, we might select an unpredictable horse who wins occasionally and place a show, a place, and a win bet on the horse. These kinds of bets make use of the addition law. We make use of the multiplication law for the "daily-double" or accumulator type bet. For the latter, we pick a horse to win in the first race and bet all our winnings on a horse to win in the second race. For the "daily double", each of the two selected horses must win in their race. Our chance of winning is the product of the probabilities of each horse winning his own race.

11. 8 CONDITIONAL PROBABILITY

We need methods to compute a probability of an event happening given the condition that another event has already happened. These are available, as we shall illustrate with the red-white pair of dice example. Given that the red die has already been thrown and shows a six, E_{r6}, what is the probability that the white die will show a five, that is, event E_{w5} ? We know that the outcome on the throw of the white die is independent of what appears on the red die, and hence $P(E_{w5}) = 1 / 6$. Also, since we are given that E_{r6} has occurred, its probability of occurrence after the fact is unity. The probability that event E_{w5} will occur *given that* event E_{r6} has already occurred may be written as follows:

$$P(E_{w5} \text{ occurs given that } E_{r6} \text{ has already occurred}) = P(E_{w5} \mid E_{r6}) =$$

$$P(E_{w5}) \times P(E_{r6} \text{ has occurred}) = 1 / 6 \times 1 = 1 / 6.$$

As a second example to illustrate conditional probabilities, suppose that we have an urn containing five black and three white balls, thus:

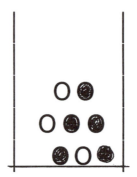

A random selection of a ball from the urn results in the following probabilities:

P(ball is black) = $1/8 + 1/8 + 1/8 + 1/8 + 1/8 = 5/8$

and

P(ball is white) = $1/8 + 1/8 + 1/8 = 3/8$.

Now suppose we wish to compute the probability of drawing a black ball given that the first ball is black and *not* replaced. This is sampling without replacement. The probability of drawing a black ball on the first draw is $5/8$, but since there are only four black balls and three white balls remaining in the urn, the probability that the second ball drawn will be black is now $4/7$. Hence the probability of the black ball on the first draw *and* a black ball on the second draw is

P(black on 1st draw) × P(black on 2nd draw | black on 1st) =

$5/8 × 4/7 = 20/56 = 5/14,$

whereas the probability of obtaining a black ball on the second draw *given that* a black ball was obtained on the first draw is $4/7$ as given above. Confusion about these two types of statements can lead to erroneous computations of the probabilities. At this stage we begin to see the reasons for making very precise statements and for understanding what is meant by the probability statements, one word can change the probability statement completely.

A practical application of conditional probability using known facts or previous data is the following. Suppose one student is flipping a coin and a second student names the side of the coin that will turn face up and does this while the coin is in the air. If the coin matches his call, he wins and if not, he loses. Suppose that the coin came up heads five times in a row. What should the caller do on the next call of the coin? In order to maximize his winnings he should call heads, that is, he should adopt a "play-the-winner" strategy. The reason for this is possibly to bias the odds in his direction. If the coin has a head on one side and a tail on the other and is unbiased, it does not matter whether he calls heads or tails, so why not call heads? If the coin is not two-sided or is biased in favor of heads, his best call is heads in order to bias the odds in his favor. As a precautionary measure, one had better continue to keep track of the number of heads, because the coin could actually be biased in favor of more tails even though five heads were obtained on the first five throws.

11.9 PERMUTATIONS AND COMBINATIONS

A teacher has five students and she wishes to set up a committee of three with a chairman = c, vice-chairman = v, and secretary = s. She wants all students to have an equal opportunity for all positions, but she does not want to be accused of bias. She rejects the idea of putting the five names into a hat,

D = Dick   Barb = B

E = Earl   Sara = S

Joni = J

and blindly drawing sets of three names, because she knows about the vagaries of random sampling and that she would not be able to conduct the drawing for a long enough time. She comes up with the following scheme, where pos. = position on committee:

Committee

SBJ	SBD	SBE	SJD	SJE	SDE	BJD	BJE	BDE	JDE
pos.	pos.	pos.	pos.	pos.	pos.	pos.	pos.	pos.	pos.
cvs	cvs	cvs	cvs	cvs	cvs	cvs	cvs	cvs	cvs

SBJ	SBD	SBE	SJD	SJE	SDE	BJD	BJE	BDE	JDE
SJB	SDB	SEB	SDJ	SEJ	SED	BDJ	BEJ	BED	JED
BSJ	BSD	BSE	JSD	JSE	DSE	JBD	JBE	DBE	DJE
BJS	BDS	BES	JDS	JES	DES	JDB	JEB	DEB	DEJ
JSB	DSB	ESB	DSJ	ESJ	ESD	DBJ	EBJ	EBD	EJD
JBS	DBS	EBS	DJS	EJS	EDS	DJB	EJB	EDB	EDJ

The top line gives the ten possible committees of size three; the six orderings under each committee give all possible orderings of the three

382   A Little Probability

individuals in the three positions on a committee. The teacher reasons that if she goes through all 60 orderings on 60 different days, then she should not be accused of prejudice.

We note that this scheme becomes somewhat unmanageable when there are more than n = 5 students. Therefore, we might wish to observe what happens in general for any n and to determine if some mathematical formula is available for schemes such as the above. In order to count or to compute the number of **possible orderings** or **permutations**, observe that any one of the n = 5 names may appear in the first position of chairman; then, given that a person is chairman, any one of the remaining n - 1 = 4 names may appear in the second position of vice-chairman. Thus all possible orderings of 5 = n names for the two positions yields n(n - 1) = 5(4) = 20 possible orderings. When we consider all three positions, we proceed for the first two positions as above and note that with a given chairman and vice-chairman there are only 3 = n - 2 names remaining from which to select a secretary. Therefore there are n(n - 1)(n - 2) = 5(4)(3) = 60 possible orderings of n = 5 individuals into committees of r = 3 names each as, indeed, there are 60 orderings as written down by the teacher.

The question now arises concerning a formula for the number of orderings or permutations. If we use n instead of five and r instead of three, we can write down the number of permutations as follows:

$$n(n - 1)(n - 2)(n - 3) \ldots (n - r + 1) =$$

$$\frac{n(n - 1)(n - 2) \cdots (2)(1)}{(n - r)(n - r - 1) \cdots (2)(1)} = \frac{n!}{(n - r)!}$$

where n! = n(n - 1)(n - 2) ... (2)(1) and (n - r)! = (n - r)(n - r - 1)(n - r - 2) ... (2)(1). Thus, if $_nP_r$ is the symbol used to express the number of permutations of n items taken r at a time, we can write the following simple mathematical formula:

$$_nP_r = \frac{n!}{(n - r)!},$$

where 0! is defined to be equal to one; for the example above for n = 5,

$$_5P_3 = \frac{(5)(4)(3)(2)(1)}{(5 - 3)(5 - 4)} = (5)(4)(3) = 60$$

possible permutations of five names into three committee positions.

To determine the total number of possible committees of n = 5 items into committees of size r = 3, we note that to form a committee or **combination**, the order of the r = 3 names is unimportant and that there are 3(2)(1) = 3! = r! permutations for any given combination of r = 3 items. If we divide the total number of permutations, 60, by the number of permutations of three names, we obtain the total number of combinations as follows:

$$_5P_3 / 3! = \frac{5!}{(5-3)!\ 3!} = \frac{5(4)(3)(2)(1)}{(2)(1)(3)(2)(1)} = 10 \text{ committees} = {_5C_3} = \binom{5}{3}$$

where $_5C_3$ or $\binom{5}{3}$ is the symbol used to denote the total number of combinations of five items taken three at a time. For any n, we may proceed to compute the number of combinations in a similar manner as for n = 5 and r = 3 and to write

$$_nP_r /r! = \frac{n!}{(n-r)!\ r!} = {_nC_r} = \binom{n}{r}$$

which is a simple and compact mathematical formulation for the total number of combinations of n items taken r at a time.

Some additional useful relations for permutations and combinations are the following. From

$$\binom{n}{r} = \frac{n!}{(n-r)!\ r!} = \binom{n}{n-r},$$

we see that we obtain the same number of combinations taking n items n - r at a time as we do taking n items r at a time. For the above example, we obtained ten committees of size three for the five students. We would also obtain ten committees of size 5 - 3 = 2. Another relation is

$$r\binom{n}{r} = \frac{r!\ n!}{(n-r)!\ r!} = {_nP_r}$$

for n different items taken r at a time. Again using the above example for n = 5 and r = 3, we note that there were ten combinations and that 3!(10) = 1(2)(3)(10) = 60 = the number of permutations that were obtained.

In another situation, suppose that there are n items such that p are of one kind and are identical, q are of a second kind and are identical, r are of a third kind and are identical, and the remaining items each occur once. For example, in the word "statistics", s and t each occur 3 times, i occurs twice, and the remaining two letters a and c each occur once, such that 3 + 3 +

2 + 1 + 1 = 10 letters. Suppose that we wish to know the number of different permutations of the ten letters taken ten at a time. The total number of permutations is

$$_{10}P_{10} = \frac{10!}{(10 - 10)!} = 10!$$

but not all permutations are different. Since there are p! = 3! permutations of s which are identical, q! = 3! permutations of t which are identical, and k! = 2! of the letter i which are identical, the number of permutations of these ten letters taken ten at a time is

$$\frac{10!}{3!3!2!} = \frac{1(2)(3)(4)(5)(6)(7)(8)(9)(10)}{1(2)(3)(1)(2)(3)(1)(2)} = 2(5)(7)(8)(9)(10) = 50,400.$$

By the same reasoning, we may write the general formula for p items of one kind, q of a second kind, k of a third kind, and n - p - q - k items which are of only one kind and for the total number of different permutations of n items taken n at a time as

$$\frac{n!}{p!q!k!} \, .$$

Furthermore, if we have n items such that p are of one kind, q of a second kind, k of a third kind, and n - p - q - k different single items, and if we wish to compute the number of permutations of r *different* items, we may use the following formula

$$\frac{[n - (p - 1) - (q - 1) - (k - 1)]!}{[n - (p - 1) - (q - 1) - (k - 1) - r]!}$$

Many more such relations need to be constructed to handle all types of combinations and permutations, but this is the subject of more advanced courses in statistics.

For the binomial distribution in Chapter 12, we note the term n! / (n - k)! k! which may be thought of in terms of combinations or in terms of permutations. For the latter, we want to know the number of permutations of n items taken n at a time with k of one kind and n - k of a second kind. For the example given, suppose that we have five unbiased two-sided coins which are tossed simultaneously. The total possible different permutations are:

event	coin 1	coin 2	coin 3	coin 4	coin 5	frequency
5 tails	T	T	T	T	T	1
4 tails, 1 head	H	T	T	T	T	
	T	H	T	T	T	
	T	T	H	T	T	5
	T	T	T	H	T	
	T	T	T	T	H	
3 tails, 2 heads	H	H	T	T	T	
	H	T	H	T	T	
	:	:				
						10
	T	T	T	H	H	
2 tails, 3 heads	H	H	H	T	T	
		:				
						10
	T	T	H	H	H	
1 tail, 4 heads	H	H	H	H	T	
		:				
						5
	T	H	H	H	H	
5 heads	H	H	H	H	H	1
Total						32

For all tails or all heads there is only one permutation of five of one kind: $5! / 5!$. For four tails and one head, or conversely four heads and one tail, we compute the number of permutations of five items with four of one kind and only one of the second kind as $5! / 4!\ 1! = 5$. Likewise, for n = 5 items with k = 3 of one kind and n - k = 2 of a second kind, the number of permutations is computed as $5! / 2!\ 3! = 10$ permutations. This illustrates the usefulness of the computational formulas for permutations and combinations in computing the coefficients for the binomial distribution and consequently the probabilities of the various events. The theoretical probability of occurrence of five heads or five tails from the toss of five unbiased coins is 1 / 32, the probability of either four heads and one tail, or conversely one head and four tails, is 5 / 32, and the probability of occurrence of three heads and two tails, or *vice versa*, is 10 / 32. Although it was easy to write out the 32 permutations for n = 5, it becomes more difficult as n increases. With the above formulas we need only compute the various probabilities as is done in Tables 12.1 and 12.2 to obtain the expected proportions or probabilities, say P_i. Then the various P_i were multiplied by the total number of observations to obtain the expected numbers for any given event.

This short introduction should be sufficient to acquaint the reader with some elementary notations of probability and to form a basis for some of the material in the next chapters. For further insight into more complex applications, the reader is directed to references listed below, especially Campbell (1974), Mosimann (1968), and Mosteller *et al.* (1961).

11.10 *REFERENCES AND SUGGESTED READING*

Campbell, S. K. (1974). *Flaws and Fallacies in Statistical Thinking.* Prentice Hall, Inc., Englewood Cliffs, New Jersey, viii + 200 pp.. (Chapters 11 and 12)

Chapman, D. G. and R. A. Schaufele (1971). *Elementary Probability Models and Statistical Inference.* Ginn and Blaisdell, Waltham, Mass., xix + 358 pp..

Goldberg, S. (1963). *Probability. An Introduction, 4th printing.* Prentice-Hall, Inc., Englewood Cliffs, New Jersey, ix + 322 pp..

McCarthy, P. J. (1957). *Introduction to Statistical Reasoning.* McGraw-Hill Book Company, Inc., New York, Toronto, London, xiii + 402 pp.. (Read chapter 7, giving particular attention to the examples.)

Moroney, M. J. (1956). *Facts from Figures.* Penguin Books, Baltimore and London, viii + 472 pp.. (Read chapter 2 and associated problems.)

Mosimann, J. E. (1968). *Elementary Probability for the Biological Sciences.* Appleton-Century Crofts, New York, xv + 255 pp..

Mosteller, F. (1965). *Fifty Challenging Problems in Probability with Solutions.* Addison-Wesley Publishing Company, Reading, Mass., viii + 82 pp..

Mosteller, F., R. E. K. Rourke, and G. B. Thomas (1961). *Probability with Statistical Applications.* Addison-Wesley Publishing Company, Reading, Mass., xv + 478 pp..

Subrahmaniam, K. (1990). *A Primer in Probability, second edition.*

Subrahmaniam, K. (1990). *A Primer in Probability, second edition*. Marcel Dekker, Inc., New York, 336 pp..

Wallis, W. A. and H. V. Roberts (1956). *Statistics: A New Approach*. The Free Press, Glencoe, Illinois, xxxviii + 646 pp.. (Read chapter 10.)

11.11 PROBLEMS

Problem 11.1. (Adapted from example 330B of Wallis and Roberts, 1956.) Suppose that license plates are numbered from 1 to 500 and suppose that the first digit is obtained for randomly drawn license numbers. What is your guess as to the frequency of the digits 0, 1, 2, 3, 4,..., 9? Compute the actual frequency for each digit for this set of numbers. What is the frequency of the last digit of numbers? Suppose that there had been 1000 license plates in your population. What would have been the frequency of the first and last digits in the license numbers?

Problem 11.2. (Adapted from example 331 of Wallis and Roberts, 1956.) The following parlor game may be used to increase your financial resources at the expense of your gullible friends. Take any book of statistical tables such as the *Statistical Abstract of the United States*. Open to a page in any haphazard manner. Then play the following game. Gullible (your friend) pays Sharpie (you) one dime every time a 1, 2, 3, or 4 appears and Sharpie pays Gullible a dime every time a 5, 6, 7, 8 or 9 appears. Why are the odds in favor of Sharpie?

Problem 11.3. A game with which the author has not had much success is a game called chuck-a-luck and is often played at carnivals and fairs. If you wish a carnival affair at your next party and perhaps want to make a little money, obtain a cage with see-through sides, and place three large six-sided dice into it. Then build a board, with numbers 1, 2, 3, 4, 5 and 6, divide it into six equal parts, and mark. You are now ready to play. The rules are as follows. Bets are placed on the numbers, and pay-offs are made if the dice numbers (of dots) correspond to the player's number. Otherwise, the operator keeps the amount wagered. If two of the dice show the same number, the pay-off is doubled, and if all three dice show the same number, the pay-off is tripled. The player often reasons that the odds are in his favor since his chances of winning are 50-50 when he plays three numbers and, in addition, he receives a double pay-off when two dice show the same number and a triple pay-off when all three dice show the same number. He erroneously believes that his chances of winning are greater than 50-50. Show that the odds of a player winning are 91:125.

Problem 11.4. (Adapted from example 330A, Wallis and Roberts, 1956.) A young Romeo wanted to give each of four girls, Miss East, Miss West, Miss North, and Miss South, equal time. He took a commuter train to see each of the girls. He reasoned that he would be fair and it would be more fun if he would randomly select the minute to arrive at the station and then take the train leaving next after his arrival. He kept track of the number of times that he saw each girl and to his chagrin he was not seeing them an equal proportion of the time. The train schedules were:
  Eastbound every 20 minutes starting on the hour,
  Westbound every 20 minutes starting one minute after the hour,
  Northbound every 20 minutes starting ten minutes after the hour,
  Southbound every 20 minutes starting twelve minutes after the hour.

What proportion of the time was he seeing each girl and why? Could you devise a selection of trains which would yield equal expected time with each girl?

Problem 11.5. Two coins are flipped and a "wheel of fortune" with 36 stopping places is spun. How many different results are there?

Problem 11.6. A pollen analyst counts grains of pollen found on a surface soil sample. He identifies the grains to be one of two types--tree pollen (T) or non-tree pollen (N). Suppose he records the order in which he counts the ten pollen grains. For example, he might obtain

TTTNNNTTNN, TNNNNTTTTT, NTTTTTTTTT, etc.

  a. If each of the above is a distinct sample, how many samples of ten grains are possible?
  b. How many different samples are possible if he counts 100 grains of pollen?
  c. How many different samples are possible if he counts n grains of pollen?

Problem 11.7. An octanucleotide is a chain of eight nucleotides. Each nucleotide contains either adenine (A), guanine (G), or uracil (U). For example, two different octanucleotides are

ACGUUUGC and CGUUUGCA.

a. How many different octanucleotides are theoretically possible?
b. A hexanucleotide consists of six nucleotides. How many hexanucleotides are theoretically possible?

Problem 11.8. A tripeptide chain consists of three amino acids which can be distinguished as first, second or third. Each is one of 20 naturally occurring amino acids. How many tripeptide chains are theoretically possible?

Problem 11.9. The fertilized cell of a certain organism has five pairs of chromosomes; no member of the pair is the same as its partner.
a. How many different germ cells can a single individual produce?
b. If a male mates with a female, and no chromosome has the same information content as any other for both the male and female together, how many different kinds of offspring could a single couple of organisms theoretically produce?

Problem 11.10. A log is across a stream. A wolf (W), a rabbit (R), and a man (M) use the log to cross the stream at night. Each crosses only once a night. When the wolf arrives, he waits at the log for his meal. If the rabbit arrives next, he is eaten by the wolf who then leaves. If the man arrives while the wolf is waiting, the wolf flees into the night. One order of arrival is WRM. (Here the wolf arrives first and the rabbit arrives next and is eaten.)
a. How many possible orders of arrival are there?
b. How many of these result in the wolf getting a meal?
c. How many result in the wolf fleeing without eating?
d. How many result in the wolf waiting until morning without eating?

Problem 11.11. A male deertick waits at the end of a blade of grass for a passing deer on which to attach himself. There are three deer in the glen-- one with a female tick (F), one with a male tick (M), and one with no ticks (N). There is also a ranger's truck (T) which goes through the glen once a day. If the truck drives through the glen, the waiting tick retreats to the bottom of the blade of grass and stays there, inactive, for the rest of the day. Suppose each deer and the truck pass exactly once by the tick's blade of grass during each day.

a. How many different orders of arrival are there? (A possible order is TMNF.)
b. In how many of these orders does the truck arrive before any of the deer? (In these cases the tick does not attach himself at all.)

with the female tick? (These are the cases when a mating of ticks is possible.)

Problem 11.12. A microscope slide preparation contains contains six pine (P) pollen grains and one each of oak (O), ash (A), chestnut (C). In how many possible orders can these nine pollen grains appear on the slide?

Problem 11.13. A path runs through the forest. A wolf (W). three rabbits (R), and a man (M) use the path once each night. If the wolf arrives first, he waits for a rabbit and eats all rabbits arriving after he does but before a man arrives in which case he flees and does not return.

   a. How many orders of arrival are possible? (One possible order of arrival is WRRMR in which case the wolf eats two rabbits and flees.)
   b. Enumerate all possible orders using a tree diagram.
   c. In how many cases are only one rabbit, two rabbits, or three rabbits eaten?
   d. In how many cases does the wolf end up neither eating nor fleeing from the man?

Problem 11.14. This situation is the same as the preceding one except that there are 100 rabbits (R) and one wolf with no man in the area. The wolf's capacity is ten rabbits and he leaves the path and returns to his den. All rabbits appear the same to the wolf.
   a. How many orders of arrival are possible?
   b. In how many cases are ten rabbits eaten in one night?
   c. In how many cases are exactly five rabbits eaten?
   d In how many cases are no rabbits eaten?

Problem 11.15. A sociologist was studying sex ratios in human families and was interested in whether there was a tendency for the first children born in a family to be in different sex ratios than later children born to the family. He studied families with five children. How many possible orders of birth are there for the following families?
            a. Families of two boys and three girls.
            b. Families of one boy and four girls.
            c. Families with five girls.
            d. Families with three boys and two girls.

Problem 11.16. A nucleic acid of 15 bases is known to contain four A's. four C's, four G's, and three U's.
   a. How many possible molecules (orders) are there with this

a. How many possible molecules (orders) are there with this composition?

b. If the molecule is known to begin with the bases AC and end with AA, how many possible molecules are there?

Problem 11.17. From ten different pine pollen grains and ten different locust pollen grains, how many different samples of six grains each can be chosen such that there are two pine and four locust grains in each sample? How many samples are there with three pine and three locust grains?

Problem 11.18. Mr. Adams, Mr. Brown, and Mr. Clay sit down to play a hand of poker. Accordingly each is dealt five cards from a 52 card deck. Suppose all possible deals are equally likely. Calculate the following probabilities:

a. Mr. Adams gets four aces.

b. Any one of the three gets four aces.

c. Mr. Adams gets four aces and Mr. Brown gets four kings.

Problem 11.19. A committee of six is to be drawn from a group consisting of 25 men and 15 women. Calculate the following probabilities:

a. The tallest man will be on the committee.

b. The tallest man and the shortest women will be on the committee.

c. The committee will have five female members.

d. The committee will have four male members given that the first two members selected are male.

e. The committee will have three female and three male members.

Problem 11.20. The Three Card Gambit - - The operator shows the pigeon three cards. One is white on both sides, one is red on both sides, and one is white on one side and red on the other side. He shuffles the cards thoroughly in a container and lets the pigeon select one at random and place it flat on the table without looking at, or showing to the operator, the under side. If the upper side turns out to be red, for example, the operator says, "It is obvious that this is not the card that is white on both sides. It must be one of the other two and the down side can be either red or white. Even so, I'll give you a break. I will bet you one dollar against your 75 cents that the other side is also red." Now this sounds like a good bet. After all, doesn't the pigeon have a 50-50 chance of winning? What are his odds of winning? What kind of probability is being considered here?

Problem 11.21. The License Plate Ploy - - Here the operator offers to bet

have one of the double digits 11, 22, 33, 44, 55, 66, 77, 88, 99, or 00 as
the last two numbers on its license plate. Although the bet might sound
reasonable at even odds, what are the chances of the operator winning?
What type of probability concepts are used to compute the odds?

Problem 11.22. The Eight Coin Con Game - - The operator takes eight
coins from his pocket and asks the pigeon how many heads are likely to
come up if he lets the pigeon flip all eight coins. The pigeon who is aware
that the odds of getting a head on any one coin is 50-50, will undoubtedly
say four. The operator then bets two to one, as a special inducement, that
four heads won't result. What are the odds of getting four heads when
eight coins are flipped? What are the odds of not obtaining four heads on
one flip of eight coins? What probability concepts are involved here?

Problem 11.23. Queens - - In this case, the operator takes two kings, two
queens, and two jacks from an ordinary card deck. He (or the pigeon)
shuffles the six cards thoroughly and places them face down on a table-top
so that neither knows the identity of any card. The operator then offers,
because of his inherent magnetism to attract fair ladies, odds of three to
two that if he selects two cards that at least one of them will be a queen. If
the pigeon objects because six cards are involved and this is not a fair bet,
the operator then offers the pigeon ten to one odds that he can't pick both
queens by drawing two cards. Compute the odds for both cases and state
what laws of probability are involved.

Problem 11.24 Student Cee oversleeps and has sock problems. Given that
he has two red socks and two yellow socks randomly disarrayed in his
dresser drawer and that being very sleepy in the morning, Cee randomly
selects a pair of socks from the drawer to wear each day.
   a. What proportion of the time will Cee wear socks of the same color?
   b. What are the possible outcomes and their frequencies?
   c. Given that his first selection is a red sock, what is the probability that
      Cee will be wearing red socks today?
   d. What laws of probability are involved in each of the preceding?

Problem 11.25. Student Dee also over-sleeps, is groggy in the morning,
and has problems in wearing the same color of socks each day. He has two
green socks, two red socks, and two yellow socks randomly distributed in
his dresser drawer. Each morning he randomly selects a pair of socks to
wear.

   a. What proportion of the time will he wear socks of the same color?

a. What proportion of the time will he wear socks of the same color?

b. What are the possible outcomes and their frequencies?

c. Given that a red sock was selected first, what is the probability that Dee will wear a red and a green sock today?

d. What laws of probability are involved in each case?

Problem 11.26 Mr. Eff is a student who never is really awake until noon. He never knows what color of socks he is wearing until after he has eaten lunch. His sock drawer contains two red socks, three green socks, one yellow sock, and one white one in random disarray. He randomly draws two socks from the drawer to wear today.

a. What is the probability that Eff will wear socks of the same color today?

b. What are the possible outcomes and their frequencies?

c. Given that the first sock drawn is a green one, what is the probability that Eff will be wearing two green socks on any given day?

d. What is the probability that he will be wearing white socks today?

e. What probability concepts are involved in each of the preceding?

Problem 11.27. The World Series of baseball is a seven game series unless one team wins four games earlier. Thus, a series can end in four, five six, or seven games. During the first 51 years, ten series ended in four games, ten ended in five games, nine ended in six games and 22 ended in seven games. Given that the probability of winning is equal for both teams, what are the proportions of the 51 games that should end in four, five, six, or seven games? Do you believe that the teams drag out the series to bring in more money? Given that the probability of an American League Team winning is 0.4, compute the expected proportions for each of the series outcomes with respect to number of games played. Obtain the outcomes for all World Series of baseball played to date and answer the above questions.

CHAPTER 12. ORGANIZED OR PATTERNED VARIABILITY

12.1 INTRODUCTION

So far, we have considered two types of statistics for summarizing information about a set of data arranged in a frequency distribution: an average and a measure of scatter. We have emphasized two particular statistics, the arithmetic mean and variance for single samples and for a number of experiment and survey designs. In certain situations, we can go further and define the variability completely when we know that the observations follow a prescribed pattern, that is, the variability is organized into a pattern rather than being completely unknown. When this pattern is known, we say that the distribution of all observations or data in the population follows a pattern or organization. Many types of organized variation have been described mathematically. Probability forms the basis for this description and for obtaining the mathematical form of the distribution. Some concepts of probability were discussed in Chapter 11.

In considering organized or patterned variation, i.e., distribution of random variables, we need to first know whether the variation is discrete or continuous. Counts are discrete events whereas measurements of height, weight, temperature, etc. are continuous variables even though our measuring instrument may give only discrete values like, e.g., centimeters. The possible values and not values obtained must be considered in determining whether a variable is continuous or not. Some of the discrete statistical distributions of random variables we shall consider in this chapter are the binomial, uniform, Poisson, and negative binomial distributions. The normal and Student's t are examples of statistical distributions with continuous variation and will be discussed in this chapter.

12.2 BINOMIAL DISTRIBUTION

If an event either occurs or does not occur in a series of n trials, if these two events are mutually exclusive, and if the probabilities of outcomes remain the same in repeated samplings, then the random variable Y is said to have a **binomial distribution**. The probability that there are k occurrences of an event in n trials (that is, the random variable Y takes the value k for k = 0, 1, 2,..., n occurrences of the event) is

$$P(Y = k) = n! \, \pi^{n-k} (1 - \pi)^k / (n - k)! \, k!;$$

this is the mathematical equation for the binomial distribution. To illustrate, suppose that we toss five unbiased coins and observe the number of heads which occurs on any one throw. This number could be $k = 0$, $k = 1$, $k = 2$, $k = 3$, $k = 4$, or $k = 5$ heads. The probabilities of the various outcomes for k equal to the number of heads and for $\pi = 1/2$ would be as follows;

$k = 0$: $[5!/(5 - 0)!0!] \, (1/2)^5 \, (1/2)^0 = 1/32$

$k = 1$: $[5!/(5 - 1)!1!\} \, (1/2)^4 \, (1/2)^1 = 5/32$

$k = 2$: $[5!/(5 - 2)!2!] \, (1/2)^3 \, (1/2)^2 = 10/32$

$k = 3$: $[5!/(5 - 3)!3!] \, (1/2)^2 \, (1/2)^3 = 10/32$

$k = 4$: $[5!/(5 - 4)!4!] \, (1/2)^1 \, (1/2)^4 = 5/32$

$k = 5$: $[5!/(5 - 5)!5!] \, (1/2)^0 \, (1/2)^5 = 1/32$

where $0! = 1$, $5! = (1)(2)(3)(4)(5)$, and $n! = (1)(2)(3)...(n - 1)(n)$. The total of the probabilities adds to unity, since $[\pi + (1 - \pi)]^5 = (1/2 + 1/2)^5 = 1^5 = 1$. In general terms,

$$1 = \{\pi + (1 - \pi)\}^n = \sum_{k=1 \text{ to } n} n! \, \pi^{n-k} (1 - \pi)^k / k! \, (n - k)!,$$

that is, the sum of the probabilities of all possible events equals one.

To obtain some idea of variability among parameter estimates for a sample of size n, let us consider a random number generator on a pocket calculator for drawing samples of ten random numbers between 0 and 9. In drawing ten sets of ten random digits, the following ten sets (rows) of numbers were obtained:

SET number 1 2 3 4 5 6 7 8 9 10		Set	Number is odd	Number is even	Number is 0-3	Number is 4-9
3 3 5 5 7 1 0 3 4 3		1	8	2	3	7
1 1 9 6 0 2 0 1 0 6		2	6	4	7	3
8 0 2 1 7 3 4 2 9 3		3	5	5	4	6
9 8 5 4 5 8 6 5 4 3		4	4	6	5	5

0 3 2 2 4 6 9 0 4 0			5	4	6	4	6
9 1 4 2 4 1 9 6 3 3			6	6	4	7	3
7 7 2 4 2 1 4 6 7 6			7	4	6	3	7
5 4 8 5 0 3 7 4 9 8			8	3	7	5	5
5 0 2 0 1 3 0 2 0 8			9	4	6	4	6
9 2 9 3 4 6 9 6 0 4			10	6	4	5	5
---------------------------		---------------------------------------				
		Totals	50	50	47	53

In the odd and even tabulations, the expected number for an unbiased random number generator would be 1 / 2. The proportion of odds varies all the way from 30% to 80% for the individual samples of size ten. For the entire set of 100 numbers the proportion of odds equals the expected proportion. For the second grouping of the numbers into 0 to 3 and 4 to 9, the expected proportion for the first group is 40%. The individual samples vary from 30% to 70%, and for the entire set of 100 numbers, the proportion of times 0, 1, 2, or 3 occurred is 47% whereas the expected percent is 40%. The larger the sample sizes, the less variation among individual sample estimates.

The data in Table 12.1 illustrate occurrences which follow the binomial distribution for p = 0.3376986. They do not follow the binomial for π = 1 / 3 as one would expect from unbiased dice, that is, 1 / 3 of the outcomes obtained from throwing a six-sided die should have a 5 or 6 face up on the average. Many other illustrations of the binomial distribution may be found in statistical texts. The data in Table 12.2 illustrate data on proportions which do *not* follow a binomial distribution. Note that there are deficiencies of families with three and with five boys based on the binomial distribution expectations.

12.3 UNIFORM DISTRIBUTION

The **uniform distribution** has n possible and independent outcomes each occurring with equal probabilities of 1 / n. The mathematical formulation for the uniform distribution is

$$f(Y) = 1 / n \text{ for } Y = 1, 2, 3, \ldots n.$$

Here again the sum of the probabilities for all possible events equals one, that is,

$$\sum_{k=1 \text{ to } n} 1 / n = 1.$$

Table 12.1. Weldon's data on 26,306 throws of 12 dice.

Number of dice with 5 or 6	Observed frequency	Expected true dice	Expected biased dice	Measure of divergence x^2/n True dice	Biased dice
0	185	202.75	187.38	1.554	0.030
1	1149	1216.50	1146.51	3.745	0.005
2	3265	3345.37	3215.24	1.931	0.770
3	5475	5575.61	5464.70	1.815	0.019
4	6114	6272.56	6269.35	4.008	3.849
5	5194	5018.05	5114.65	6.169	1.231
6	3067	2927.20	3042.54	6.677	0.197
7	1331	1254.51	1329.73	4.664	0.001
8	403	392.04	423.76	0.306	1.017
9	105	87.12	96.03	3.670	0.838
10	14	13.07	14.69 ⎤		
11	4	1.19	1.36	0.952	0.222
12	0	0.05	0.06 ⎦		
Total	26306	26306.02	26306.00	35.491 f = 10	8.179 f = 9

Taken from R. A. Fisher(1950) with permission of the author and publishers.

An example of a uniform distribution may be obtained by randomly selecting the ten integers 0 to 9. The theoretical expectation for any integer would be one-tenth. Any random number generator may be checked by drawing a very large number of random digits and observing how near the occurrences of each integer is to one-tenth.

A second example would be for n = 6, the number of outcomes on a six-sided die. In a large number of rolls of the die, each face should occur equally frequently. The following computer program written in BASIC may be used to simulate a die rolling experiment with N rolls of the simulated die. The text in parenthesis following a program step is for explanation of the step.

```
01          Rem Die Simulation: Print CHR$(147)
10          INPUT: "How many rolls";N
```

Table 12.2. Geissler's data on distribution of boys in 53,680 German
families of eight children each.

Number of boys	Number of families observed	Expected $p = 0.514677$	Excess x = observed minus expected	x^2 / m
0	215	165.22	+49.78	14.998
1	1485	1401.69	+83.31	4.952
2	5331	5202.65	+128.35	3.166
3	10649	11034.65	-385.65	13.478
4	14959	14627.60	+331.40	7.508
5	11929	12409.87	-480.87	18.633
6	6678	6580.24	+97.76	1.452
7	2092	1993.78	+98.22	4.839
8	342	264.30	+77.70	22.843
Totals	53680	53680.00		91.869

Taken from R. A. Fisher(1950) with permission of the author and
publishers.


```
20              FOR L = 1 to N
```
(This step establishes the loop to perform the next random roll and
increment the proper element of the array by one after each roll.)
```
30              R = INT(6*RND(1)) + 1
40              F(R) = F(R) + 1
```
(The array F, for face, is used to keep track of how many times a
particular face turns up. For example, every time a six shows up on the
top face F(6) is increased by one.)
```
50              NEXT L
60              Print "FACE", "Number of times"
```
(This prints the headings.)
```
70              FOR C = 1 to 6: Print C, F(C): NEXT
```

A sample run for N = 1,000 might look like this

  How many rolls: ? 1000
(The number 1000 is typed in after the question mark.)

FACE	Number of times
1	151
2	155
3	169
4	193
5	157
6	175

A second sample of 1,000 yielded the following:

How many rolls: ? 1000

FACE	Number of times
1	148
2	176
3	178
4	166
5	163
6	169

The expected number for each face is 1000 / 6, which is approximately equal to 166.7 or 167. For both runs the number 1 occurred 299 times whereas it was expected to occur 333 times. Whether or not this is a bias or sampling variation would need to be determined by extensive sampling. With an actual die, the experimenter would need to physically throw the die N times to check for possible bias.

For a simulated experiment which is not a uniform distribution, consider rolling a pair of dice N times and recording the sum of the two numbers appearing on the top faces of the two dice. A computer program for simulating N tosses or rolls of a pair of dice, say a red die (R) and a white (W) die, is

```
05   REM dice simulation for sum: Print CHR$(147)
10   DIM F(12)
```
(This statement is needed because some computers would only recognize the integers from 0 to 9. Any number equal to or greater than the number to be used could be inserted, say F(13). Since the largest sum we shall be considering is 6 + 6 = 12, we inserted 12 here.)
```
15   INPUT "How many rolls";N
20   FOR L = 1 to N
30   R = INT(6*RND(1)) + 1
40   W = INT(6*RND(1)) + 1
50   F(R+W) = F(R+W) + 1
```

```
60    NEXT L
70    PRINT "Number","Number of times"
80    FOR H = 2 to 12: PRINT H,F(H):NEXT
```

The statistical distribution for the sum of the integers on any number of dice, not just two, has been worked out (See Feller, 1968, e.g..) but the reader may make a two-way array of all possible outcomes from tossing a pair of dice and combine the outcomes for a particular sum. Then, the probability for each outcome may be obtained and the statistical distribution obtained.

12.4 POISSON DISTRIBUTION

In the binomial distribution we were concerned with success or failure of an event, whereas in the **Poisson distribution** we are concerned with *the number of times an event* occurs and not at all with the number of times it fails. Both of these distributions are concerned with discrete or discontinuous data ordinarily known as enumeration data. The probabilities of k = 0, 1, 2,... events are given by the following equation denoting the Poisson distribution:

$$P(Y = k) = \lambda^k e^{-\lambda} / k!$$

where λ is the arithmetic mean of the population, k is the number of times

the event occurs, and e is the base of natural logarithms. The variance is also λ. The probabilities for the various events are:

Occurrence of event	Probability of occurrence
0 time	$e^{-\lambda}$
1 time	$\lambda e^{-\lambda}$
2 times	$\lambda^2 e^{-\lambda} / 2$
3 times	$\lambda^3 e^{-\lambda} / 3!$
4 times	$\lambda^4 e^{-\lambda} / 4!$
etc.	

The sum of the probabilities of all occurrences of events in a Poisson distribution is unity, that is,

$$\Sigma_{k=1 \text{ to } \infty} \lambda^k e^{-\lambda} / k! = e^{-\lambda} \Sigma_{k=1 \text{ to } \infty} \lambda^k / k! = 1.$$

An example of data following a Poisson distribution is described by Fisher (1950). The data were obtained by Bortkewitch from records of ten army corps for 20 years; they refer to k = number of deaths per year per army corps resulting from horse-kicks. The data are:

Deaths per year per army corps	Observed number	P(Y = k)	nP(Y = k) = expected number
0	109	0.543	108.67
1	65	0.331	66.29
2	22	0.101	20.22
3	3	0.021	4.11
4	1	0.003	0.63
>4	0	0.003	0.08
Total	200 = N	1.000	200.00

The mean of the above observations is computed as

$$[0(109) + 1(65) + 2(22) + 3(3) + 4(1)] / 200 = 0.61 = m.$$

The parameter λ in the Poisson distribution is set equal to 0.61 and then the various P(Y = k) are computed for k = 0, 1, 2, 3, and 4. The last value of 0.003 is obtained by subtraction, likewise; the value 0.08 is obtained by subtraction. The various expected values are computed as:

$$N e^{-m} = 200 e^{-0.61} = 108.67$$
$$N m e^{-m} = 200(0.61) e^{-0.61} = 66.29$$
$$N m^2 e^{-m} = 200(0.61)^2 e^{-0.61} = 20.22$$
$$N m^3 e^{-m} = 200(0.61)^3 e^{-0.61} = 4.11$$
$$N m^4 e^{-m} = 200(0.61)^4 e^{-0.61} = 0.63$$
$$\text{Remainder} = 200 - \text{sum of above} = 0.08$$

There is close agreement between observed and expected values in the above table, illustrating a good fit to the Poisson distribution for the variable.

Feller (1968) gives several examples of Poisson distributed variables such as radioactive disintegrations, flying bomb hits on London during World War II, chromosome interchanges in cells treated with X-ray radiation, connections to the wrong telephone numbers dialed, bacterial and blood counts, and the number of misprints per page in a textbook. Many other examples could be cited.

An example of data with poor agreement between observed and expected values and not following the Poisson distribution, even when the zero class is omitted (See David and Johnson, 1952, and Rao and Chakravarti, 1956), is given below. The data represent number of decayed teeth found in 265 boys.

Number of decayed teeth	0	1	2	3	4	5	6	7	8	9	10	11	12
Frequency of boys	61	47	43	35	28	15	20	5	5	2	1	2	1

12.5 NEGATIVE BINOMIAL DISTRIBUTION

The **negative binomial distribution** (also called Pascal or waiting time distribution) has two parameters π and κ where π is the probability of occurrence of an event and the sampling continues until κ successes occur. For example, in a coin tossing experiment, one could continue tossing coins until *exactly κ heads* occur. The number of tosses, $n + \kappa$, in an experiment is unknown, and n could be a very large number. Now for n tails, say, and $\kappa - 1$ heads followed by a toss resulting in the κth head, the probability is

$$f(\kappa; n, \pi) = (n + \kappa - 1)! \, \pi^n (1 - \pi)^\kappa / \kappa! \, (r - 1)!.$$

This is the mathematical formula for the negative binomial distribution with parameters π and κ, where $0 < \pi < 1$ and $n = 0, 1, 2, \ldots$.

An example of a negative binomial distribution is for number of accidents, n_i, encountered by 647 women working on high-explosive shells over a period of five weeks in England during World War I (Greenwood and Yule, 1920). The data and some of the calculations for estimating the parameters are:

n_i	f_i	$f_i n_i$	$f_i n_i^2$	p_i	$647\ p_i$	$(f_i - 647\ p_i)^2 / 647\ p_i$
0	447	0	0	0.6846	442.9	0.038
1	132	132	132	0.2141	138.5	0.305
2	42	84	168	0.0686	44.4	0.130
3	21	63	189	0.0221	14.3	3.139
4	3	12	48	0.0072	4.7	0.615
5	2	10	50	0.0023 ⎤	2.2	0.018
>5	0	0	0	0.0011 ⎦		
Sum	647 301		587	1.0000	647	4.245

An estimate of π is obtained as:

$$p = \Sigma f_i\, n_i / \Sigma f_i = 301 / 647 = 0.465.$$

An estimate of the variance is equal to:

$$s_e^2 = [\Sigma f_i\, n_i^2 - (\Sigma f_i\, n_i)^2 / \Sigma f_i] / (\Sigma f_i - 1) =$$
$$[587 - 301^2 / 647] / 646 = 0.6919.$$

An estimate of κ is obtained as:

$$k = p^2 / s_e^2 = 0.4652 / 0.6919 = 0.09548.$$

As may be observed in the above table, the expected values, $647\ p_i$, are quite close to the observed frequencies, f_i, of women with n_i accidents, indicating a good fit to the negative binomial distribution. The last column above represents the computations for a chi-square statistic used for obtaining evidence as to the goodness of fit. A value greater than 4.245 for a chi-square with 6 - 2 = 4 degrees of freedom (six groups and two parameters estimated) has a rather large probability of occurrence and hence is not considered unusual in sampling from a chi-square distribution.

12.6 NORMAL DISTRIBUTION

The normal frequency probability function may be represented as a bell-shaped curve and is expressed mathematically by the following equation:

$$f(Y \text{ given } \mu \text{ and } \sigma^2) = [e^{-(Y - \mu)^2 / 2\sigma^2}] / (2\pi\sigma^2)^{1/2},$$

where Y is the particular value of the random variable from a normal distribution, μ is the center of gravity or arithmetic mean of the entire distribution, σ^2 is the population variance, and π is the constant 3.14159.... (The variance is strictly analogous to the square of the radius of gyration in the theory of moments of inertia in physics.) The mean and the variance are the only parameters associated with the normal frequency distribution. The development of this function is due independently to three persons: a French immigrant to England, De Moivre, a Frenchman, LaPlace, and a German, Gauss. The distribution is written as

$$\int_{-\infty \text{ to } \infty} [e^{-(Y - m)^2 / 2\sigma^2}] / (2\pi\sigma^2)^{1/2} \, dY = 1.$$

The limits of integration are from minus infinity to plus infinity. Note that for continuous variables, the integration performs the same function as the summation does for discrete variables.

A large proportion of statistical theory is based on the normal frequency distribution, and statistical techniques or procedures utilizing this theory are frequently used to summarize data from surveys and experiments. As has been amply demonstrated with examples for the binomial, uniform, Poisson, and negative binomial distributions, there are many instances of non-normal or anormal distributions. In fact, there are many types of distributions of various statistics such as, for example, Snedecor's F or variance-ratio, Fisher's z, Wishart's distribution of sum of squares, chi-square, beta, gamma, continuous uniform or rectangular, triangular, Cauchy, Weibull, contagious distributions, distribution of mixtures of binomials, distribution of mixtures of exponentials, exponential distribution, distribution of the correlation coefficient, distribution of confidence intervals, and hypergeometric, to name only a few. In many cases a simple transformation of yield data following one of the above distributions (say from Y to log Y or from Y to $\sqrt{Y}$) *may allow* normal theory to be utilized. This feature makes normal theory all the more popular among statisticians and users of statistics.

If a random variable Y is normally distributed with mean μ and variance σ^2, then the mean of a random sample of size n is normally distributed with the same mean μ and variance σ^2 / n. The usefulness of this result is increased by the fact that if the original distribution of the random variable is not normal and if the variance σ^2 is finite, the distribution of means from samples of size n tends to normality as the sample size n increases. (These results are embodied in the "Central Limit Theorem", a very important theorem in statistics from both the theoretical and applied points of view.) This result allows use of normal theory for practical applications in situations where the exact form of the sampled population is unknown. Even for relatively small values of n, say from five to ten, the approximation is usually close enough for practical applications. Since many summarizations of data involve the arithmetic mean, normal theory receives considerable attention is statistical textbooks.

The bell-shaped figure in Figure 12.1 illustrates the form of the normal frequency curve and the parameters as they relate to the curve. The point $Y = \mu$ centers and locates the curve on the Y axis. The parameter σ^2 determines the shape of the curve. For example, three different frequency distributions centered at μ with three different population variances, say $\sigma_1^2 \neq \sigma_2^2 \neq \sigma_3^2$, would look something like Figure 12.2, depending upon the actual values of the population variances.

The area under the normal curve has been tabulated for many values of $(Y - \mu) / \sigma$. The distribution of $(Y - \mu) / \sigma$ for Y having a normal distribution is again a normal distribution with a new mean of zero and a new variance of unity. This is called the **unit normal distribution** since it has unit variance.

Figure 12.3 indicates a use of the unit normal curve and its relation to an actual distribution of the variable Y = I.Q. (intelligence quotient). Here we note the average I.Q. is 100 and its variance is $13^2 = 169$. The proportion of the area under the curve between -1 and +1 is about 68% (the shaded area); the proportion of the area between 0 and +2 is approximately 47.5%; and the area between -2 and +2 is approximately 95%. Likewise, the area on the right of +2 plus that on the left of -2 is approximately 5%. The probability that a randomly selected observation Y falls in the interval between a = -2 and b = +2 is P(a $\leq$ Y $\leq$ b) = P(-2 $\leq$ Y $\leq$ +2) is approximately 95%.

It should be realized that when normal theory is used for most variables like height, weight, temperature, humidity, distance, intelligent quotient, and so forth, these variables are *not normally distributed.* The ranges of these variables are *not* from minus infinity to plus infinity. Weights, for example, usually start at zero and there is definitely an upper limit for the weight of an individual. It is possible that some individual could weigh 5,000 pounds (really gross) but it is impossible to ever find a person who weighed 100,000 pounds! Despite the fact that variables do not have the range of normally distributed random variables, it is often possible to use normal theory for variables or statistics having only a symmetrical distribution.

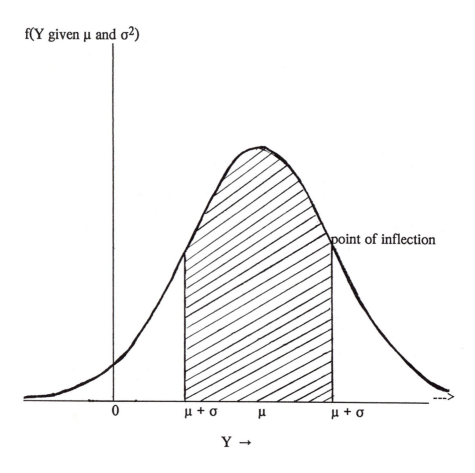

Figure 12.1. Normal frequency curve.

f(Y given μ and σ^2_i for i=1,2,3)

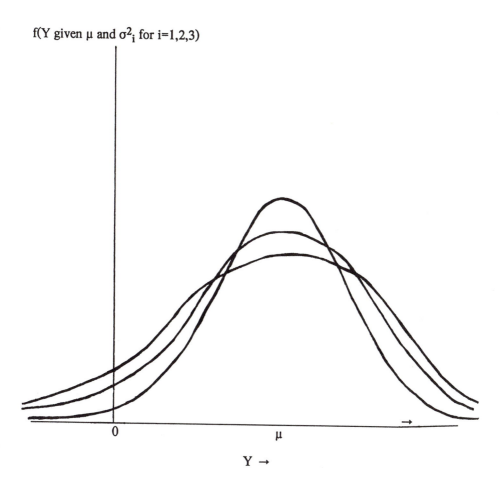

Figure 12.2. Normal frequency curves centered at a common μ and with three different variances.

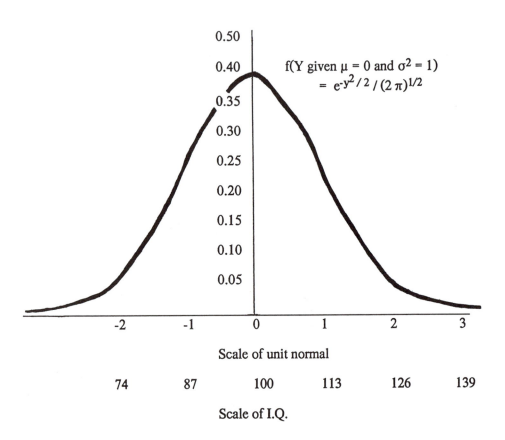

Figure 12.3. Relating unit normal deviates to I.Q. (intelligence quotient) scores.

12.7 INTERVAL ESTIMATE OF A PARAMETER ---- CONFIDENCE INTERVAL

Instead of obtaining a point estimate like the arithmetic mean as a measure of central tendency or of a variance as a measure of scatter or variability, it is often desirable to have an interval estimate. Instead of saying that the point described by the arithmetic mean is an estimate of the population mean, we say that the population mean is included in an interval and that the interval is an estimate of the population mean. We can even say what proportion of the time we would expect the estimated confidence interval to contain the population mean. Since we have "confidence" from theory that the population mean is contained in the estimated interval $1 - \alpha$ percent of the time and is not contained in the estimated intervals α percent of the time, then we speak of the interval as a $1 - \alpha$ **percent confidence interval** (with a specified error rate base).

To illustrate the computation of a confidence interval, let us consider that a random sample of $n = 5$ observations is available as follows:

$$Y_1 = 2, Y_2 = 3, Y_3 = 6, Y_4 = 7, \text{ and } Y_5 = 11.$$

The median value is $Y_3 = 6$. Consider the range from 2 to 11. What is the probability that all five sample values fall below the population median? Since any single Y_i has a probability of $1/2$ of falling below the population median, and since all five values were independently obtained, the multiplication law of probability holds, that is, P(all 5 sample values fall below the population median) $= (1/2)^5 = 1/32$. Likewise, the probability that all five sample values fall above the population median is $1/32$. Therefore the probability that the interval 2 to 11 contains the population median is $1 - 2(1/32) = 15/16$. We would have a $1 - \alpha = 15/16 = 93.75\%$ confidence that the sample interval contains the population median.

To further illustrate the construction of confidence intervals, suppose that the range from the second smallest to the second largest sample value above was used as the confidence interval, that is, from $Y_2 = 3$ to $Y_4 = 7$. From the binomial distribution with $p = 1/2$, the probability that all five values are below the population median is $1/32$, and the probability that four values are below and one is above the median is $5/32$. Likewise, the probability that all five values are above the population median is $1/32$, and the probability that four values are above and one is below the population median is $5/32$. Therefore the probability that the interval $Y_2 = 3$ to $Y_4 = 7$ contains the population median is equal to

$$1 - (1/32 + 5/32 + 1/32 + 5/32) =$$
$$1 - 12/32 = 5/8 = 62.5\%.$$

In order to compute a confidence interval for means of samples drawn from a normal distribution, we need to have Student's t-distribution tabulated in the same manner as was described for the unit normal case. **Student's t** is equal to

$$t = (\bar{y}. - \mu)/s_{\bar{y}.}$$

where $\bar{y}.$ = sample mean, $s_{\bar{y}.}$ = $(s_e^2/n)^{1/2}$ = standard error of a sample mean, the degrees of freedom for t are those associated with s_e^2 and μ is the population mean. Fortunately, this has been done. The **Student's t distribution** is a symmetric distribution with mean zero which depends only on the degrees of freedom. As degrees of freedom become large, the t-distribution approaches the unit normal distribution and is equal to it for infinite degrees of freedom. The mathematical formulation of the density function (the height above the abscissa for a value of t) for Student's t is

$$h(t \text{ given } f) = \{(f-1)/2\}! / (2\pi)^{1/2} \{(f-2)/2\}! \{1 + (t^2/f)\}^{(f+1)/2}$$

Thus the t distribution depends only on the parameter degrees of freedom f. The area under the curve (probability of observing a smaller value of t) from minus infinity to some value t may be obtained from

$$1 - \alpha = \int_{-\infty \text{ to } t} \{(f-1)/2\}! / (2\pi)^{1/2} \{(f-2)/2\}! \{1 + (x^2/f)\}^{(f+1)/2} \, dx.$$

Extensive tables have been tabulated for various values of α and f using the above equation. Tables have also been constructed for the limits of integration from -t to t. The former are called **one-tailed tables** and the latter are called **two-tailed tables**. For the same value of t, the α values for one-sided table are one-half of those for a two-sided table. This is so because the t distribution is symmetrical and the area to the left of -t is equal to that to the right of +t (See Figure 12.4). A one-tailed table is used when differences in one direction are desired. When the difference can be either positive or negative, a two-tailed table is used.

In considering differences between means, the following is another form of the t statistic:

$$t = \{(\bar{y}_{1.} - \bar{y}_{2.}) - (\mu_1 - \mu_2)\} / \{s_e^2(1/n_1 + 1/n_2)\}^{1/2},$$

where the degrees of freedom for t are those associated with s_e^2 and n_1 and n_2 are the sample sizes from which the two means were calculated. As stated before the mean of the t distribution is zero and the variance of the t statistic is $f/(f-2)$ which approaches one as the number of degrees of freedom f approaches infinity (See Wilks, 1962, for example.).

The tabulated values of t in Table 12.3 are taken from Table IV of Fisher (1950) (by permission of the author and publishers) or obtained elsewhere. A computer program may be utilized to obtain these values. As stated before, the t distribution has been tabulated for many values of the degrees of freedom and of the probabilities. The tabulated t value from the following table is denoted as $t_{\alpha,df}$ where α refers to the probability value in the table and df refers to degrees of freedom. Thus, $t_{0.05,30}$ is the value of t equal to 2.04, $t_{0.1,12} = 1.78$, etc..

Now, let us make use of the tabulated values of t to construct interval estimates of the difference between any two treatment means in Example 11.1. The e_{ij} must be normally and independently distributed with population mean zero and the same population variance. The sample variance $s_e^2 = 19/3$ with six degrees of freedom and the variance of a difference of two means, say 1 and 2, was computed as $s_{\bar{y}_{1.} - \bar{y}_{2.}}^2 = 2(19/3)/(r=4) = 19/6$. Now we compute a $1 - \alpha = 90\%$ interval estimate of the difference between any two means as follows:

Mean 1 minus mean 2

$$\bar{y}_{1.} - \bar{y}_{2.} \pm t_{0.1,6} \, s_{\bar{y}_{1.} - \bar{y}_{2.}} = 7 - 9 \pm 1.94 \,(19/6)^{1/2}$$

$$= 7 - 9 \pm 3.45 = -5.45 \text{ to } +1.45$$

Mean 1 minus mean 3

$$\bar{y}_{1.} - \bar{y}_{3.} \pm t_{0.1,6} \, s_{\bar{y}_{1.} - \bar{y}_{3.}} = 7 - 5 \pm 1.94 \,(19/6)^{1/2}$$

$$= 7 - 5 \pm 3.45 = -1.45 \text{ to } 5.45$$

Table 12.3. Tabulated values of t.

Degrees of freedom	α = probability that a larger value of | t |* is obtained:						
	two-sided	0.25	0.20	0.15	0.10	0.05	0.025 0.005
	one-sided	0.5	0.4	0.3	0.2	0.1	0.05 0.01
2		0.82	1.06	1.39	1.89	2.92	4.30 9.92
3		0.76	0.98	1.25	1.64	2.35	3.18 5.84
4		0.74	0.94	1.19	1.53	2.13	2.78 4.60
5		0.73	0.92	1.16	1.48	2.01	2.57 4.03
6		0.72	0.91	1.13	1.44	1.94	2.45 3.71
8		0.71	0.90	1.11	1.40	1.86	2.31 3.36
9		0.70	0.88	1.10	1.38	1.83	2.26 3.25
10		0.70	0.88	1.09	1.37	1.81	2.23 3.17
12		0.69	0.87	1.08	1.36	1.78	2.18 3.06
14		0.69	0.87	1.08	1.34	1.76	2.14 2.98
16		0.69	0.86	1.07	1.34	1.75	2.12 2.92
18		0.69	0.86	1.07	1.33	1.73	2.10 2.88
20		0.69	0.86	1.06	1.32	1.72	2.09 2.84
30		0.68	0.85	1.06	1.31	1.70	2.04 2.75
40		0.68	0.85	1.06	1.30	1.68	2.02 2.70
50		0.68	0.85	1.06	1.30	1.68	2.01 2.68
60		0.68	0.85	1.05	1.30	1.67	2.00 2.66
70		0.68	0.85	1.05	1.29	1.67	1.99 2.65
80		0.68	0.85	1.05	1.29	166	1.99 2.64
90		0.68	0.85	1.05	1.29	1.66	1.99 2.63
100		0.68	0.85	1.05	1.29	1.66	1.98 2.62
∞		0.67	0.84	1.04	1.28	1.64	1.96 2.58

---

* \| t \| denotes absolute value of t, i.e., the sign of t is ignored.

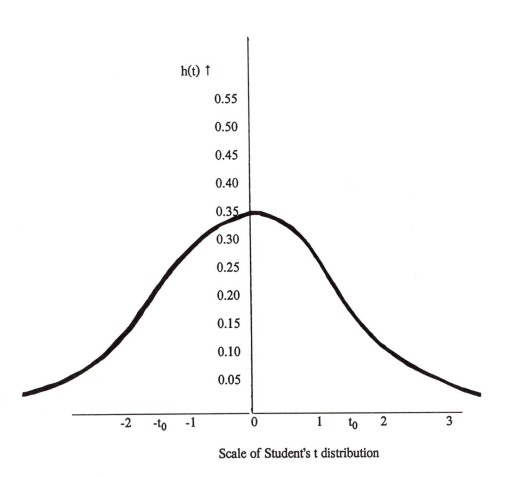

Figure 12.4. Student's t distribution for f = 10 degrees of freedom.

<u>Mean 2 minus mean 3</u>

$$\bar{y}_{2.} - \bar{y}_{3.} \pm t_{0.1,6} s_{\bar{y}_{2.} - \bar{y}_{3.}} = 9 - 5 \pm 1.94 \, (19 / 6)^{1/2}$$
$$= 9 - 5 \pm 3.45 = 1.55 \text{ to } 7.45$$

The previous two intervals contain the point zero, but the last one does not. Since the last interval does not contain the point zero, this is evidence that the hypothesis of no difference (the null hypothesis) between means for treatments 2 and 3 may not be true.

In the above, we have described a confidence interval for the difference between two treatment means. If it is desired to construct a $(1 - \alpha)\%$ confidence interval for a single mean with a sample size of r and with a variance s^2_e associated with f degrees of freedom, the following formula is used for treatment 1 in the above example:

$$\bar{y}_{1.} \pm t_{\alpha,f} \, (s^2_e / r)^{1/2}$$

which for $\alpha = 0.2$, $f = 6$, $r = 4$, and $s^2_e = 19 / 3$ is

$$7 \pm 1.44 \, [19 / 3 \, (4)]^{1/2} = 7 \pm 1.44 \, (1.26)$$

$$= 7 \pm 1.81, \text{ or from } 5.19 \text{ to } 8.81.$$

The interval 5.19 to 8.81 is an $80\% = (1 - 0.2)100$ percent confidence interval of the population mean for treatment 1.

The above procedure of constructing confidence intervals is for the normal distribution. The intervals may be constructed for any distribution for which appropriate tables or computer programs are available. Tables and charts for the binomial distribution are available in many places. Charts for constructing 95% and 99% confidence intervals have been developed and are known as Clopper and Pearson charts (See Figures 12.5 and 12.6.).

To illustrate the construction of confidence intervals with binomial variation, let us suppose that $\pi = 0.4$, that $n = 50$, and that we are sampling from a binomial distribution. The various $1 - \alpha$ confidence intervals for $\pi = 0.4$ as read from the charts are:

90%	29 to 52
95%	27 to 55
99%	23 to 59

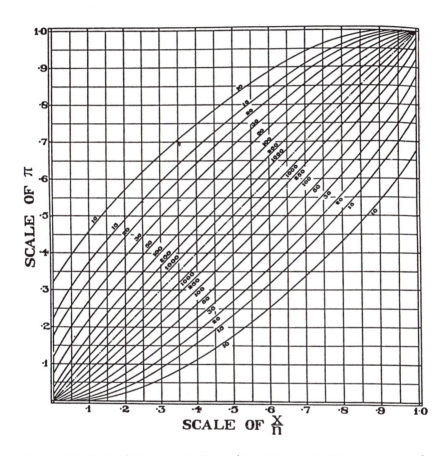

Figure 12.5. Confidence belts for π (Confidence Coefficient = 0.95).
Reprinted by permission of the authors, C. J. Clopper and E. S. Pearson, and the publishers, the Biometrika Office.

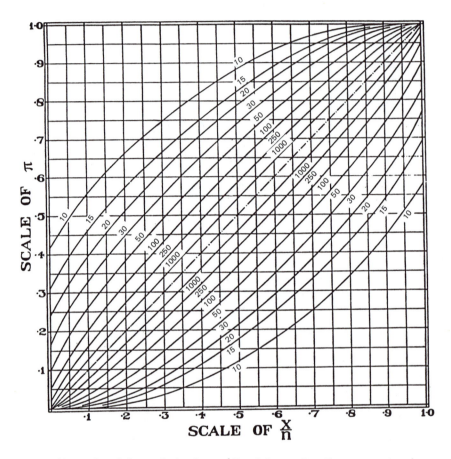

Figure 12.6. Confidence belts for π (Confidence Coefficient = 0.99)

Reprinted by permission of the authors, C. J. Clopper and E. S. Pearson, and the publishers, the Biometrika Office.

The 90% confidence values were obtained from page 414 of Dixon and Massey (1957). Here we note that the size of the confidence interval increases as $1 - \alpha$ increases. We may use these values to put $(1 - \alpha)\%$ confidence intervals on $1 - \pi = 0.6$ as follows:

90%	48 to 71
95%	45 to 73
99%	41 to 77

The values for $\pi = 0.4$ are subtracted from 100.

The confidence intervals on π from the binomial distribution may be approximated by using the t-distribution. The approximation becomes better and better as n, the sample size, increases. To illustrate, let us use the above example with a sample size of n = 50, $t_{0.10,\ 49\ df} = 1.68$, $(1 - \alpha)\% = 90\%$, and $p \pm 1.68\ [p(1 - p)\ /\ n]^{1/2} = 0.4 \pm 1.68\ [0.4(0.6)\ /\ 50]^{1/2} = 0.4 \pm 1.68\ (0.069) = 0.4 \pm 0.116$, or from 28.4% to 51.6%. This agrees within rounding errors and within errors in reading the chart. The number 2 used in the example in Chapter 5 corresponds to the tabulated value for $t_{0.05,\ 49\ df}$ as tabulated in t-tables.

Thus far we have been concerned with summarizing information from survey data or experimental data. In order to draw inferences about the population parameter, we need to characterize the interval estimate in terms of the parameter being estimated. We have noted the following characteristics about the interval estimate:

1. The average length of the confidence interval of a mean decreases as $n^{-1/2}$ where n = sample size.
2. The length of the confidence interval increases as $1 - \alpha$ increases.
3. The construction of a confidence interval depends upon the underlying distribution.
4. The point estimate of the parameter does not always fall at the midpoint of the confidence interval.
5. The length of the confidence interval is a statistic and has a distribution. Some confidence interval estimates are relatively short, and some are relatively long for the same sample size and the same $1 - \alpha$, depending upon the size of the sample variance.
6. In repeated sampling with the same sample size n and the same α, the true value of the parameter will be contained in the estimated intervals $1 - \alpha$ percent of the time and will not be α percent of the time.

The last item above is relevant in writing a definition of a confidence interval. A $(1 - \alpha)\%$ **confidence interval** is an interval which is said to bracket the value of the true unknown parameter and in repeated samplings the value of the parameter will be contained in the estimated intervals (1 - α)% of the time and will not be α% of the time. When we assert that the confidence interval contains the true value of the parameter, we are either right or wrong in making this inference.

We can increase the confidence proportion 1 - α, but this increases the width of the interval; all we have to do to be 100% certain that the true parameter lies within an interval is to make the interval large enough. For example, the interval from minus infinity to plus infinity contains 100% of the possible values of the population mean of normally distributed observations, the interval from zero to plus infinity contains all possible values of the population variances, the interval from zero to unity contains all possible values of a proportion in this interval, and so on. However, we would like to reduce the size of this interval without making 1 - α too small or the sample size n too large. The selection of α is determined by the investigator; odds of 4:1, 9:1, 19:1, and 99:1 are often used. The latter two odds are popular with statistical textbook writers and editors of scientific journals. The selection of α is, however, made by the investigator. Once he selects an α, then one of the ways to reduce the size of the confidence interval is to increase the size of the sample. (Other ways are to change the experimental procedure, to select an experiment design which controls all assignable variation, etc.).

12.8. REFERENCES AND SUGGESTED READING

David, F. N. and N. L. Johnson (1952). The truncated Poisson. *Biometrics* 8:275-285.

Dixon, W. J. and Massey, F. J., Jr. (1957). *Introduction to Statistical Analysis, 2nd edition,* McGraw-Hill Book Company, Inc., New York, xiii + 488 pp..

Feller, W. (1968). *An Introduction to Probability Theory and its Applications, Volume One, third edition.* John Wiley & Sons, Inc., New York, xviii + 509 pp..

Fisher, R. A. (1950). *Statistical Methods for Research Workers, eleventh edition,* Hafner Publishing Co,, New York, xv + 354 pp..

Greenwood, and Yule (1920). *Journal of the Royal Statistical Society, Series A* 83:255-

Moroney, M. J. (1956). *Facts from Figures, third edition.* Penguin

Books, Baltimore, Md., xii + 472 pp..

Rao, C. R. and I. M. Chakravarti (1956). Some small sample tests of significance for a Poisson distribution. *Biometrics* 12:264-282.

Wilks, S. S. (1962). *Mathematical Statistics.* John Wiley & Sons, Inc., xvi + 644 pp..

12.9 PROBLEMS

Problem 12.1. Show how to compute the 80%, 90%, and 95% confidence intervals for the differences between treatment means for the data given in Problems 10.6, 10.7, and 10.8.

Problem 12.2. In a genetic experiment, let us suppose that eye color in the common fruit fly segregates in a 3:1 ratio of red-eyed to white-eyed flies. Suppose that we observe 70% red-eyed flies and 30% white-eyed flies out of a sample size of n = 100 flies. What are the values of the 95% and 99% confidence intervals as read from the Clopper and Pearson charts (Figures 12.5 and 12.6)? Do these intervals contain the theoretical proportion of red-eyed flies? What would be the confidence intervals for the sample percentage of 70% if the sample sizes had been 50, 250, and 1000? Would these intervals contain the theoretical proportion of red-eyed flies? Would the same statements made about the theoretical proportion of red-eyed flies hold for the theoretical proportion of white-eyed flies? Illustrate.

Problem 12.3. From statistics or other types of textbooks, list three examples of data (not listed in this text) which are assumed to follow the binomial, uniform, Poisson, normal, and negative binomial distributions. Give complete reference citations, including page number.

Problem 12.4. Show how to compute the 90% confidence intervals for the data of Problem 12.2, using the t-statistic and tabulated values of t as follows:

$$t_{0.10, 49 \text{ df}} = 1.676, t_{0.10, 249 \text{ df}} = 1.65, \text{ and } t_{0.10, 999 \text{ df}} = 1.65$$

Problem 12.5. A graduate student performed the following experiment using a simple change-over design. He obtained ten sets of 15 two-digit numbers from a random number table, these are given below. He had two calculating machines, himself and a pocket calculator. He wished to observe the difference in mean length of time required to compute a sum of squares of 15 two-digit numbers on the two computers. Since there could be a learning process in computing a sum of squares a second time, he decided to compute five of the sets on himself first in random order and on the pocket calculator second, and the reverse for the other five sets.

Ten sets of size 15 of two-digits from random digit table

1	2	3	4	5	6	7	8	9	10
30	86	18	48	40	75	53	56	36	06
20	82	32	36	28	43	55	80	65	61
62	27	67	97	92	77	76	02	56	82
67	04	01	30	13	44	38	66	89	60
00	00	38	17	05	74	49	87	61	66
73	34	71	69	77	58	20	04	85	53
76	65	97	79	12	32	72	31	90	36
30	16	49	24	26	06	22	96	67	56
10	05	67	93	88	45	07	59	69	90
77	83	89	32	28	20	25	18	53	62
76	01	56	07	65	46	02	28	40	24
28	07	51	71	60	67	11	88	99	21
26	80	78	47	56	81	94	93	38	73
31	16	82	26	43	93	17	95	57	60
73	46	66	36	10	23	60	32	05	51

The design used and the time in seconds required to compute each sum of squares are given below where A = himself and B = pocket calculator:

1	2	3	4	5	6	7	8	9	10	Sum
A255	B115	A280	B107	B105	A240	A195	B110	A202	B 85	1694
B113	A200	B117	A238	A210	B104	B 90	A200	B105	A180	1557
Sum 368	315	397	345	315	344	285	310	307	265	3251

$$\Sigma\, e^2_{hij} = 2219.50 \qquad s^2_e = 277 \qquad \begin{array}{l} A\ total = 2200 \\ B\ total = 1051 \end{array}$$

Compute a 90%, a 95%, and a 99% confidence interval on the difference between the two means of the computers.

Problem 12.6. In addition to the 100 random numbers given in Section 12.2, suppose the following 60 were obtained:

```
5 2 6 4 6 6 8 0 9 3 6 0
2 0 5 0 9 6 0 1 8 9 9 1
0 3 2 8 0 9 9 5 7 7 7 5
8 8 7 4 8 6 0 7 4 8 8 6
2 9 9 1 4 3 1 9 1 8 6 7
```

Form 20 samples of five numbers from the numbers in Section 12.2 and use the above 12 sets of five numbers to form 32 sets of five numbers each. What are the five possible events for even numbers in a sample of five? Prepare a table of possible events, observed frequency for each event, and theoretical frequency. Discuss the results.

Problem 12.7. An experimenter obtained 50 samples of four plants each. The 200 plants came from a cross of two parents, one with a dominant gene and one with a recessive gene, for example, for a red flower and a white flower. The segregation of plants was expected to be simple Mendelian in that 3 / 4 of the plants were expected to have red flowers and 1 / 4 were expected to have white flowers. The following results were obtained:

Number of red flowered plants	Number of white flowered plants	Frequency
4	0	13
3	1	20
2	2	13
1	3	3
0	4	1

What are the expected numbers assuming a binomial distribution? Compute the differences between observed and expected frequencies.

Problem 12.8. The following data are for number of goals scored per team per match in English football (Moroney, 1956):

Number of times team scored that many goals	Number of goals							
	0	1	2	3	4	5	6	7
	95	158	108	63	40	9	5	2

```
----------------------- |----------------------------------------------------------
Expected no.        |
------------------ |----------------------------------------------------------
Observed            |
minus expected |
_____ |_____
```

Complete the computations in the above table assuming Poisson variation and discuss the results. Prepare histograms of frequencies for both observed and expected numbers.

Problem 12.9. Suppose the population consists of the five observations 1, 3, 5, 7, and 9. The population mean μ = (1 + 3 + 5 + 7 + 9) / 5 = 5 and the population variance σ^2 = {(1 - 5)^2 + (3 - 5)^2 + (5 - 5)^2 + (7 - 5)^2 + (9 - 5)^2} / (5 - 1) = 40 / 4 = 10. For the ten possible samples of size two (sampling without replacement), compute the means, variances, and confidence intervals using t = 1 and also for t = 2. In how many of the cases do the confidence intervals include the parameter μ = 5? Prepare a line histogram of the frequencies for value of the sample mean. Do likewise for the sample variances. Note that this gives the distribution of sample means and variances for this population and a sample of size two.

Problem 12.10. The coefficients in the binomial distribution function of n items taken k at a time may be computed from the following which is known as Pascal's triangle:

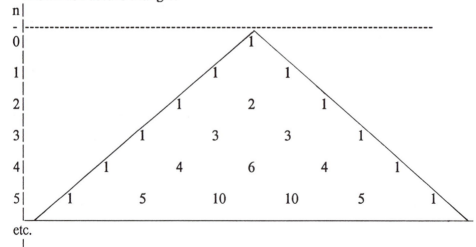

```
n|
- |-----------------------------------------------------------------
0|                              1
 |
1|                      1               1
 |
2|                  1           2           1
 |
3|              1           3           3           1
 |
4|          1           4           6           4           1
 |
5|      1           5          10          10          5           1
 L_____
etc.
 |
```

The coefficients are obtained from the two numbers above it, e.g., 4 = 3 + 1, 10 = 6 + 4, and so forth. Extend the triangle to include the numbers n = 6, 7, 8, 9, and 10.

Problem 12.11. The number of noxious weed seeds in 1 / 2 ounce samples of bluegrass seed obtained were:

Freq-ency	0	1	2	3	4	5	6	7	8	9	>9
						Number of noxious weed seeds					
obser-ved	3	17	26	16	18	9	3	5	0	1	0
exp-ected											

Assuming Poisson variation, compute the expected values and observed minus expected differences. Discuss the fit to the Poisson distribution.

Problem 12.12. The following data are for number of yeast cells in 400 haemocytometer counts with a mean of 536 / 400:

Frequency	0	1	2	3	4	5	6	7	>7
			Number of yeast cells						
observed	103	143	98	42	8	4	2	1	0
expected									

On the basis of a Poisson distribution, compute expected values, observed minus expected, and discuss the fit. Note that if clumping of yeast cells occurs, the observed data are not likely to be Poisson distributed.

Problem 12.13. Using the computer program given in Section 12.3, perform a simulation experiment simulating one die with random numbers for 100 runs, 1,000 runs, 10,000 runs, and 100,000 runs. Compare observed numbers with expected numbers. Repeat the experiment and compare the results of the two experiments.

Problem 12.14. Using the computer program in Section 12.3 for a pair of dice, perform the simulation for N = 100, 1,000, 10,000, and 100,000. Compare observed numbers with expected numbers. Repeat the experiment and discuss the results of the two experiments.

Problem 12.15. Draw ten sets of 100 random numbers and compute the number of odds and evens in each set. Do these frequencies fit the binomial distribution?

Problem 12.16. For the data in Problems 10.11 and 10.13, obtain 90% and 95% confidence intervals for the differences among treatment means.

CHAPTER 13. SAMPLE SIZE

13.1 INTRODUCTION

The problem of determining sample size or replicate number arises whenever an experiment or a survey is to be conducted. There are many criteria for determining sample size. The suitability of a given criterion depends upon the objectives and the nature of the investigation, and has to be determined by the investigator or by the administrator. Various criteria will be set forth and discussed individually, and the investigator will need to determine which criterion fits his situation.

The ensuing discussion relates equally well to experiments and to sample surveys. The results for one are applicable to the other type of investigation. In general, the number of samples utilized in an investigation is determined by the following interrelated factors:

1. the degree of precision desired,
2. the amount of variability present in the material under investigation,
3. available resources including personnel and equipment, and/or
4. size and shape of experimental and sampling units.

The nature of the material under investigation, the characters observed, and/or the expected magnitude of the treatment and category differences determine the degree of precision desired. The very nature of some treatments, strata, and categories may be such that large differences are expected or the observations may be such that low variability is expected. Some characters are more variable than others and therefore require more observations to attain the same degree of precision. In connection with determining the number of replicates for an experiment or the sample size for a survey, the characters of interest with their respective standard deviations should be listed. Replicate number or sample size should be determined for the most variable character to be measured. This number will be sufficiently large for all other characters. The relative importance of the characters will also affect the determination of sample size for the investigation in that sample size will be determined for the most important character.

The amount of variability present in experimental material is determined by the experimental conditions, the characters measured, and the treatments tested. The variance of a treatment or category mean relative to the mean or to the difference between means may be small for some characters and large for others. Larger experimental and sampling units tend to have smaller variation (on a unit basis) than smaller ones. Technique and other errors do not usually have as great an effect on large units as on small ones. An error in weight of 100 pounds on a small plot may have a relatively large effect, whereas an error in weight of 100 pounds on a large plot may be of little consequence. In fact, some of the scales used for weighing sugarcane from large experimental plots are only accurate to the nearest 200 pounds.

There are basically five ways to reduce the variation among experimental and sampling units. These are:

1. use of homogeneous material,
2. use of blocking or stratification to attain homogeneity within blocks or strata,
3. refinement of experimental or survey techniques,
4. measure related variables and take advantage of relationship to reduce variation through a procedure known as covariance, and
5. if none of the above suffices, then increase sample size or replicate number.

The shape of the experimental or the sampling unit usually has a relatively small effect on the variance of a treatment mean. In general, long narrow field plots tend to be less variable than square ones. In surveys, geographic nearness acts in the same manner; it has been found that for several characteristics of the population simple, geographic stratification into compact areas is as effective as complex stratification on known characteristics of the area. The preceding are generalizations and hence, the investigator should carefully study his own situation. In surveys on crops, wild life, forests, species occurrence, fisheries, and on characteristics of human populations, the size and shape of the unit can be of importance. In laboratory studies, the size of the sample may be very important for detection of an item, e.g., in taste testing. In sports tournaments, it is necessary to determine whether to have a three, five, or seven game series in order to have the best chance of determining which is the best team. In field experiments, the units need to be large enough and spaced correctly in order to simulate conditions in practice and to avoid competition between units with different treatments.

The following list of criteria for determining sample size is not all inclusive, but it does illustrate some of the diverse considerations used by investigators to determine sample size. For these criteria, it is assumed that the investigation will be conducted in a fixed period of time; this is called a one-stage investigation as opposed to multi-stage or sequential investigation where the results of the first stage determine whether additional stages are to be conducted.

13.2 THE "HEAD-IN-THE-SAND" APPROACH

As indicated by one researcher's statement, "Away with the duplicate plot! Give me one plot and I know where I am.", the use of only one observation could imply no variation since none is observed. Does the reader realize how often the results from only one sample or observation are used to determine a course of action for many individuals? Parents of a single (or a first) child often draw conclusions about all children from their observations on their child. A first experience in dining at a restaurant may color all future reactions to eating in this place. Our first impressions of a person, no matter how erroneous they may be, may completely dominate our future thoughts about that individual. Using a sample size of one to avoid variation in observations and to draw general conclusions is a "head-in-the-sand" approach to the problem. Variation remains, whether one or a dozen observations are made. Conclusions from one sample may be erroneous. The investigator is only deceiving himself with such an attitude and such an approach to investigation.

13.3 THE FAVORITE NUMBER APPROACH

A frequently used procedure for determining sample size for an investigation is to select a number without any particular goal concerning variation of sampling units. This is often done when experimenters use a number that others have used or that they have used in the past. The experimenter may use seven replicates because he likes the number seven, because a friend used seven replicates, or because seven is the number that he has been using in the past. At one large agricultural research station, a "grade A experiment" was defined to be one that utilized three replicates. As anyone knows, there are many items other than replicate number that determine a "grade A experiment". Using only three replicates can, at best, be considered to be a preliminary investigation searching for large effects.

Some surveyors in an Economics department used 100 samples for the specified reason that all others in their department use this number, that they like to use this sample size, or that this sample size was recommended by one of their earlier greats. This procedure can be denoted as the "favorite number" procedure. It is a subjective and unscientific criterion for determining replicate or sample number.

13.4 AVAILABLE RESOURCES

Some of the resources that might limit the number of replicates or samples for a given experiment or survey are:

1. amount of material,
2. available personnel or funds,
3. amount of experimental land, pots, equipment, etc. available for an experiment, or
4. number of enumerators available for a survey.

The amount of experimental seed or treatment material may be limited in the early stages of an investigation. This is quite often the situation with new varieties, new chemicals, new drugs, new products, and the like, and the experimenter must decide whether to run the experiment or to wait until more experimental material becomes available. In launching satellites, there will be only one at a time and very few will be launched in a short period of time.

Likewise, limited personnel or other experimental resources may result in insufficient replication or insufficient sample size. There may be only one piece of equipment in a hospital or laboratory, e.g., an electron microscope. When investigators wish to use this equipment, they may have to limit the number of observations they take in any given investigation. Greenhouse space, growth chambers, land, subjects, animals, etc. may be in short supply and this limits the number of samples for an investigation. The procedure usually followed is to conduct the experiment or survey with the available number of replicates or samples. Then, based on the results of the first experiment or survey, the experimenter determines whether further investigation is required. This "sequential" method of testing has long been used by experimenters.

If the experiment is to be conducted with the number of replicates dictated by **available resources** with no further testing, serious consideration should be given to the idea of not running the experiment; the limited resources may be allocated to other experiments. The same comment applies to sample surveys.

13.5 NUMBER OF DEGREES OF FREEDOM IN VARIANCE

A criterion that has some usefulness, especially for experiments with few treatments, is to use sufficient replication such that the variance is associated with at least 12 to 16, preferably 20, degrees of freedom. Denote this criterion as **number of degrees of freedom in error variance**. The reasoning here is that the variance of the estimated variance will not be so large, relatively, for 20 or more degrees of freedom. The estimated variance of a sample variance, s^2, is equal to

$$2 \text{ (variance)}^2 / \text{(degrees of freedom)} = 2 s^4 / \text{(degrees of freedom)}.$$

From the following table and Figure 13.1, we see that relatively small decreases in the variance of an estimated variance are made with greater than 20 degrees of freedom:

Degrees of freedom = df	2 / df
2	1.00
4	0.50
8	0.25
10	0.20
16	0.125
20	0.10
50	0.04
100	0.02

The error degrees of freedom may be increased by increasing the number of replicates or the number of treatments in an experiment and the sample size in a survey. Thus, with 21 or more treatments, two replicates would be sufficient to satisfy the criterion of at least 20 degrees of freedom for the error variance.

13.6 STANDARD ERROR EQUAL TO A SPECIFIED PERCENTAGE OF THE MEAN

Some experimenters use sufficient replication to obtain a standard error of a mean which is not greater than a specified percentage of the mean. For this criterion it is required that a relatively good estimate of the ratio of the standard deviation of a single observation to the mean, which is the coefficient of variation, be available. Given this estimate, a sample number is selected which yields the desired percentage. For example, suppose that the coefficient of variation is ten percent and that it is desired to have a standard error of a mean which is less than 2.5 percent of the mean. Since $10 / (16)^{1/2} = 10 / 4 = 2.5$, 16 observations would be required to obtain a standard error which is 2.5% of the treatment mean.

The requirements for this criterion are:

1. the estimated standard deviation, s, and
2. the estimated mean, $\bar{y}$.

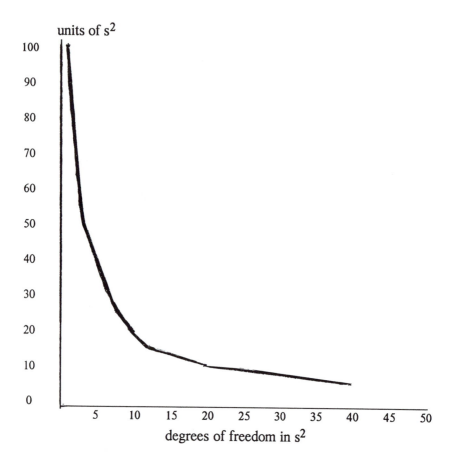

Figure 13.1. Estimated variance of a variance versus degrees of freedom.

With these two statistics, the coefficient of variation is computed as $s / \bar{y}$ = c. v..

Example 13.1. Let $s_e^2 = 25$, $s_e = 5$, sample size n = 9, and $\bar{y} = 20$. The coefficient of variation is equal to $5 / 20 = 25\%$ and the standard error of the mean in percent of the mean is equal to $(25 / 9)^{1/2} = 1 / 12 = 8.3\%$. Suppose that the investigator wishes to have a standard error of the mean which is 3% of the mean. The problem is to have a sample size which will accomplish this. We want $s_e (1/n)^{1/2} / \bar{y} = 5/n^{1/2}/20 = $ 3%. Squaring both sides of the equation and solving for n, we obtain 25 / $20^2 (0.03)^2 = 69$. The investigator needs to use a sample size of 69 or else needs to find some method for reducing the amount of variation in his material.

13.7 EFFECT OF SAMPLE SIZE ON THE LENGTH OF THE CONFIDENCE INTERVAL

Suppose that the samples are from a binomial distribution with an unknown π value and that our estimate p of π is 0.4. Then, from Figures 12.5 and 12.6 and from McCarthy (1957), pages 201 to 204, the 90%, 95%, and 99% confidence intervals on $\pi = 0.4$ are read from the charts as:

	(1 - α)% confidence intervals (%)					
sample	90%		95%		99%	
size = n	lower	upper	lower	upper	lower	upper
10	14	70	12	75	7	82
50	29	52	27	55	23	59
100	31	49	30	50	28	53
250	34	45	34	46	32	48
1000	38	42	37	43	36	44

From the above, we may note that the length of the interval decreases with sample size for any specified value of 1 - α. For a 90% confidence interval, the length of the interval for a sample size of ten to one of 1,000 decreases from 56% to 4%. For a 95% confidence interval, the decrease is from 63% to 6% and for the 99% confidence interval from 75% to 8%. In order to obtain relatively small lengths of intervals, say less than 10% for p in the range from 0.2 to 0.8, the sample size must be greater than n = 250 even for 90% confidence intervals.

Example 13.2. Let the estimated proportion p be equal to 30% and suppose that the (1 - α)% = 95% confidence interval is to be no larger than 10%. From Figure 12.5, we may note that for n = 250 the upper bound is 37% and the lower bound is 24%, or a confidence interval of length 13%. Interpolating between the bounds for n = 250 and n = 1,000, an n of about 400 would give a confidence interval of about 10%.

Likewise for the normal distribution, the sample size must be relatively large and/or the variance must be relatively small in order to have short lengths of confidence intervals on the average. The effect of sample size on length is the reciprocal of the square root of sample size in the estimated confidence interval. Thus, the estimated $(1 - \alpha)\%$ confidence interval of the population mean μ is:

$$\bar{y} = t_{\alpha,f}(s_e^2 / n)^{1/2},$$

where there are f degrees of freedom associated with s_e^2; the estimated confidence interval on the difference of two means with different sample sizes n_1 and n_2 is:

$$\bar{y}_1 - \bar{y}_2 \pm t_{\alpha,f}\{s_e^2(1 / n_1 + 1 / n_2)\}^{1/2}.$$

If $n_1 = n_2 = n$, then the estimated $(1 - \alpha)\%$ confidence interval is:

$$\bar{y}_1 - \bar{y}_2 \pm t_{\alpha,f}(2s_e^2 / n)^{1/2}.$$

In the above formulae, we note that the sample size is in the denominator of the estimated variance of a mean. Thus the variance of a mean, or of a mean difference, decreases with an increase in sample size; this in turn affects the length of the confidence interval.

Example 13.3. Suppose that we have an estimate of the error variance which is $s_e^2 = 25$ with 30 degrees of freedom, and suppose that we wish to have a 90% confidence interval which is less than or equal to 10. Now the estimated confidence interval between two treatment population means $\mu_1 - \mu_2$ is

$$\bar{y}_1. - \bar{y}_2 \pm t_{0.1,30}(2 s_e^2 / n)^{1/2}.$$

We want an n such that

$$t_{0.1,30} \{2 (25) / n\}^{1/2} \le 5. \text{ Or, } n \ge 1.70^2 \{2(25)\} / 5^2$$

where $1.70 = t_{0.1,30}$. Thus, if our estimate of σ_e^2 in the proposed experiment is less than or equal to 25, if the associated degrees of freedom are greater than equal to 30, and if n is equal to six, the desired degree of precision will be attained.

13.8   SPECIFIED PROBABILITY THAT THE ESTIMATED CONFIDENCE INTERVAL IS LESS THAN OR EQUAL TO A SPECIFIED LENGTH

Statisticians prefer a criterion for determining the sample size n that is based on more of a probability basis than any of those discussed in preceding sections. Two such methods are described below. Consider all possible estimated $(1 - \alpha)\%$ confidence intervals; these will have a known distribution for each specified sample size n and error distribution. Define the one-half length of a confidence interval for the difference of two means as $d = t_{\alpha,f}\, s_e\, (2\,/\,n)^{1/2}$ and then divide the lengths of all one-half confidence intervals into two parts, all those less than or equal to d and all those whose lengths are greater than d. We pick a d such that $(1 - \gamma)\%$, for a given γ, of the estimated $(1 - \alpha)\%$ confidence intervals is less than or equal to a specified value of d. Computationally, we proceed as follows for a confidence interval on the difference between two means. Let

$$d = t_{\alpha,f_2} (2\, s_1^2\,/\,n)^{1/2} = t_{\alpha,f_2}\, s_1\, (2\,/\,n)^{1/2},$$

where the variance s_1^2 has f_1 degrees of freedom. Squaring both sides of the equation, we obtain

$$d^2 = t_{\alpha,f_2}\, s_1^2\, (2\,/\,n) \text{ or, } n = 2\, (t_{\alpha,f_2})^2\, s_1^2\,/\,d^2.$$

If we multiply the right-hand side by an F-statistic with f_2 and f_1 degrees of freedom (F tables are available in many statistical texts, for instance, Snedecor and Cochran, 1980, Table A14, or from a number of computer programs; also see Table 13.1), we obtain:

$$n \geq 2\, (t_{\alpha,f_2})^2\, F_\gamma\, (f_2,f_1)\, (s_1\,/\,d)^2,$$

where s_1^2 is an available estimate of the variance and is associated with f_1 degrees of freedom, and f_2 equals the degrees of freedom to be used in the proposed experiment or survey. Thus, we specify s_1^2, f_1, α, γ, and d and then compute n. Since n specifies f_2, an iterative process must be used to

Table 13.1. F_γ (f_2, f_1) values for various probability values of γ, f_2 degrees of freedom for s_2^2, and f_1 degrees of freedom for s_1^2.

Degrees of freedom = f_1	γ	F statistic table of probabilities Degrees of freedom = f_2							
		5	10	15	20	30	40	120	∞
10	10%	2.52	2.32	2.24	2.20	2.16	2.13	2.08	2.06
	5%	3.33	2.98	2.84	2.77	2.70	2.66	2.58	2.54
	1%	5.64	4.85	4.56	4.41	4.25	4.17	4.00	3.91
20	10%	2.16	1.94	1.84	1.79	1.74	1.71	1.64	1.61
	5%	2.71	2.35	2.20	2.12	2.04	1.99	1.90	1.84
	1%	4.10	3.37	3.09	2.94	2.78	2.69	2.52	2.42
30	10%	2.05	1.82	1.72	1.67	1.61	1.57	1.50	1.46
	5%	2.53	2.16	2.01	1.93	1.84	1.79	1.68	1.62
	1%	3.70	2.98	2.70	2.55	2.39	2.30	2.11	2.01
40	10%	2.00	1.76	1.66	1.61	1.54	1.51	1.42	1.38
	5%	2.45	2.08	1.92	1.84	1.74	1.69	1.58	1.51
	1%	3.51	2.80	2.52	2.37	2.20	2.11	1.92	1.80
120	10%	1.90	1.65	1.54	1.48	1.41	1.37	1.26	1.19
	5%	2.29	1.91	1.75	1.66	1.55	1.50	1.35	1.25
	1%	3.17	2.47	2.19	2.03	1.86	1.76	1.53	1.38
∞	10%	1.85	1.60	1.49	1.42	1.34	1.30	1.17	1.00
	5%	2.21	1.83	1.67	1.57	1.46	1.39	1.22	1.00
	1%	3.02	2.32	2.04	1.88	1.70	1.59	1.32	1.00

obtain a solution for n.

Example 13.4. Let s_1 / d = 2, 1 - α = 0.90, γ = 0.10, f_1 = 40, and f_2 = g (n - 1) from a stratified survey design with g = 5 strata and n samples per stratum. As a first guess try n = 10. Then

$$n \geq 2 (t_{0.05,40})^2 F_{0.1} (45, 40) (2)^2 = 2 (2.02)^2 (1.48) (4) = 48.$$

Since our first try for n = ten was too small, try n = 41. Then f_2 = 200, and

$$n \geq 2 (t_{0.05,200})^2 F_{0.1} (200, 40) (2)^2 = 2 (1.97)^2 (1.40) (4) = 42.$$

In order to compute sample size more accurately, an extensive table of F values would be required. We conclude that around 40 to 45 samples per strata would satisfy the goal of the investigator.

Another procedure for determining the probability that the interval will be less than or equal to a specified value utilizes the t statistic (See Cochran and Cox, 1957, and Table 12.3.). Suppose that it is desired to have a one-half confidence interval which is less than or equal to some specified value δ, where δ is a difference expressed in percent of the means for treatments or groups from a survey or experiment. Let the standard deviation σ be expressed in percent of the mean, i.e., the coefficient of variation. In the proposed investigation, the proposed design will produce an error variance with f degrees of freedom. Let 1 - γ be the assurance that the one-half confidence interval will be less than or equal to δ. Then the sample size for a (1 - α)% confidence interval with an assurance probability of (1 - γ)% is determined as

$$n \geq 2 (\sigma / \delta)^2 (t_{\alpha,f} + t_{2\gamma,f})^2.$$

Example 13.5. Let σ = 5, δ = 10, α = 0.05, γ = 0.10, and let the statistical design be a randomized complete block with n replicates and v = 4 treatments and error degrees of freedom equal to (n - 1) (4 - 1) = 3 (n - 1) = f. In order to determine f, we try some value for n, say 6. Then for f = 15,

$$n \geq 2 (5 / 10)^2 (2.13 + 1.34)^2 = 3.47^2 / 2 = 6.02,$$

where $t_{0.05,15}$ = 2.13 and $t_{2(0.1),15}$ = 1.34 from Table 12.3. Since n = 6 was too small, try n = 7 and f = 18. Then,

$$n \geq (2 / 4)(2.10 + 1.33)^2 = 3.43^2 / 2 = 5.9.$$

The value of f for n = 6 gave values of t which were too large to satisfy the inequality but using n = 7 does satisfy the inequality. Therefore the investigator would use n = 7 in the forthcoming experiment designed as a randomized complete block.

These procedures for obtaining sample size are presented to illustrate that statistical procedures based on probabilities are available for computing sample sizes for a proposed experiment or survey. Other procedures are available, but many are mathematically more complex than the present ones. A statistician is frequently confronted with problems of sample size in the course of his consultation with research workers, particularly when grant or contract proposals are being submitted. Hence, when the investigator is faced with a sample size problem, it is suggested that a statistician who has had previous experience with the problem of sample size determination be consulted.

13.9 REFERENCES AND SUGGESTED READING

Cochran, W. G. and G. M. Cox (1957). *Experimental Designs, 2nd edition.* John Wiley & Sons, Inc., New York, xiv + 617 pp..

Huff, D. (1954). *How to Lie with Statistics.* W. W. Norton and Company, Inc., New York, 142.pp..

McCarthy, P. J. (1957). *Introduction to Statistical Reasoning.* McGraw-Hill Book Company, Inc., New York, Toronto, and London, xiii + 402 pp..

Snedecor, G. W. and Cochran, W. G. (1980). *Statistical Methods, 7th edition.* The Iowa State University Press, Ames, Iowa, xvii + 507 pp..

Wilson, E. B. [1952]. *An Introduction to Scientific Research.* McGraw-Hill Book Company, Inc., New York, Toronto, and London, x + 373 pp., Sections 3.10 and 4.7.

References in Chapters 7, 8, and 10.

Problems   437

13.10 PROBLEMS

Problem 13.1. Describe what your method of determining sample size was before reading this book. Did your method vary with the characteristic observed?

Problem 13.2. Obtain the 95% and the 99% binomial confidence intervals obtained for the proportion p = 0.60 for sample sizes of 10, 15, 20, 30, 50, 100, 250, and 1000. Plot the lengths of the confidence interval as the ordinate against the sample size as the abscissa and connect the points with straight lines. What conclusions may be drawn from the plots?

Problem 13.3. Given that s^2 = 25 with four degrees of freedom, that $\bar{y}$ = 10, and that n = 5 from a completely randomized design, compute $(1 - \alpha)$% confidence intervals on the population mean for $(1 - \alpha)$% equal to 50%, 60%, 70%, 80%, 90%, 95%, and 99%. Plot the results using $(1 - \alpha)$% as the ordinate and the end-points of the interval as the abscissa. What would be the shape of the curve through the end-points of the intervals if all $(1 - \alpha)$% intervals had been plotted?

Problem 13.4. Given that s^2 = 25 with f degrees of freedom, that $\bar{y}_1$ - $\bar{y}_2$ = 10, and that n = 10, compute the 80% and the 99% confidence intervals for f = 2, 3, 4, 5, 6, 8, 9, 12, and 30 degrees of freedom. Plot the results with the length of the interval as the ordinate and degrees of freedom as the abscissa. What conclusions do you draw from the plotted results?

Problem 13.5. Find examples in the media or other published literature of the "head-in-the-sand" and "favorite number" criteria for determining or using specified sample sizes.

Problem 13.6. In Problem 13.3, suppose that this was the first stage of an investigation and that the investigator wished to estimate how many more samples, n_2 would be needed in order for the one-half length of a 95% confidence interval to be less than or equal to 0.5 for s^2 = 25. Find n_2.

Problem 13.7. Repeat Problem 13.6 but for a standard error of a difference between two means.

Problem 13.8. Determine samples sizes for Examples 13.4 and 13.5 using two different values for γ and one different value for α.

CHAPTER 14. MORE ON STATISTICAL METHODOLOGY

14.1 INTRODUCTION

In Chapters 9, 10, and 12, some methods for summarizing data were presented. Graphs, frequency distributions, and two-way contingency tables were discussed in Chapter 9, measures of central tendency and measures of variation were discussed in Chapter 10, and interval estimates of a parameter were discussed in Chapter 12. A vast array of statistical procedures is available for summarizing and extracting the information from a set of data. The theoretical development and the application of statistical procedures form the basis for courses in Statistics. A statistics course following this text should be either an introductory statistical methods course or an introductory probability and mathematical theory course.

In the following sections, we present a few additional statistical procedures for summarizing the information contained in a set of data. The particular procedures discussed were selected because of their connection with procedures and concepts utilized in the preceding chapters.

14.2 ANALYSIS OF VARIANCE

In Chapter 10, a variance was defined to be the average of the squares of the deviations from a mean. The squares of the deviations were summed and divided by the degrees of freedom to obtain this average. The same concept will be used to obtain additional variances for a given survey or experiment.

Suppose that we have a sample survey design consisting of a random selection of v relatively large and equal-sized clusters with a random selection of r observational units (the sampling unit) in each cluster. If the model for a randomly selected observation is:

$$Y_{ij} = \text{true population mean} + \text{true cluster effect} + \text{random error}$$

$$= \text{estimated population mean} + \text{estimated cluster effect} + \text{estimated random error}$$

$$= \bar{y}_{..} + (\bar{y}_{i.} - \bar{y}_{..}) + (Y_{ij} - \bar{y}_{i.}) = \bar{y}_{..} + d_i + e_{ij}$$

where $\bar{y}_{..}$ is the arithmetic mean of the rv observations, $\bar{y}_{i.}$ is the arithmetic mean of the r observations in the ith cluster, $d_i = (\bar{y}_{i.} - \bar{y}_{..})$, and $e_{ij} = (Y_{ij} - \bar{y}_{i.}) = $ the estimated random error. Note that e_{ij} is the estimated deviation from the mean and that $s_e^2 = \Sigma_{j=1 \text{ to } r}(Y_{ij} - \bar{y}_{i.})^2 / v$ (r - 1) as described in Chapter 10.

The quantity $(\bar{y}_{i.} - \bar{y}_{..}) = d_i$ represents a deviation of the ith cluster sample mean $\bar{y}_{i.}$ from the overall sample mean $\bar{y}_{..}$. If these deviations are squared, summed, and divided by the degrees of freedom, (v - 1), a variance among cluster means is obtained as $\Sigma_{i=1 \text{ to } v}(\bar{y}_{i.} - \bar{y}_{..})^2 / (v - 1)$. The sum of the deviations of the v cluster means from the overall mean, $\Sigma_{i=1 \text{ to } v}(\bar{y}_{i.} - \bar{y}_{..})$, is forced to be zero, and thus, (v - 1) of the deviations $(\bar{y}_{i.} - \bar{y}_{..})$ may vary freely but the last deviation must be such that

$$\Sigma_{i=1 \text{ to } v}(\bar{y}_{i.} - \bar{y}_{..}) = 0.$$

Since the variance of a mean is s_e^2 / r the preceding variance must be multiplied by r in order to place the squares of the deviations on the same basis as those in s_e^2 as computed above. Thus the sum of squares comparable to

$$\Sigma_{i=1 \text{ to } v} \Sigma_{j=1 \text{ to } r}(Y_{ij} - \bar{y}_{i.})^2 = \Sigma_{i=1 \text{ to } v} \Sigma_{j=1 \text{ to } r} e_{ij}^2$$

is

$$r \Sigma_{i=1 \text{ to } v}(\bar{y}_{i.} - \bar{y}_{..})^2 = r \Sigma_{i=1 \text{ to } v} d_i^2 = \Sigma_{i=1 \text{ to } v} Y_{i.}^2 / r - Y_{..}^2 / r v.$$

Note that there are several algebraic formulas which may be used for computing the same sum of squares. Three equivalent forms have been presented in the last set of equations. The various forms all give the same numerical results within rounding errors. For example, whether one uses $rv\bar{y}_{..}^2$, $Y_{..}^2 / rv$, or $Y_{..} \bar{y}_{..}$ to compute the sum of squares due to the effect of the overall mean of the experiment, the numerical result will be the same within rounding error. The fact that there are several forms for computing a sum of squares is a source of confusion for beginning students. After they understand the equivalency of forms, the confusion disappears.

The above results may be summarized in an **analysis of variance** table as given below:

Source of variation	Degrees of freedom	Sum of squares	Variance
Due to mean	1	$Y_{..}^2 / rv = rv$	
Due to clusters	$v - 1$	$r \sum_{i=1 \text{ to } v} (\bar{y}_{i.} - \bar{y}_{..})^2$	$r\sum_{i=1 \text{ to } v} (\bar{y}_{i.} - \bar{y}_{..})^2/(r - 1)$ = s_d^2
Due to deviations from cluster means	$v(r - 1)$	$\sum_{i=1 \text{ to } v}\sum_{j=1 \text{ to } r} e_{ij}^2$	$\sum_{i=1 \text{ to } v}\sum_{j=1 \text{ to } r} e_{ij}^2/v(r - 1)$ = s_e^2
Total	vr	$\sum_{i=1 \text{ to } v} \sum_{j=1 \text{ to } r} Y_{ij}^2$	

As shown above, the total sum of squares may be partitioned into the same number of sums of squares as there are kinds of effects in the observational equation. The above is a simple algebraic partitioning of the total sum of squares into a number of parts since the equation

$$\sum_{i=1 \text{ to } v} \sum_{j=1 \text{ to } r} Y_{ij}^2 =$$
$$r \sum_{i=1 \text{ to } v} (\bar{y}_{i.} - \bar{y}_{..})^2 + \sum_{i=1 \text{ to } v} \sum_{j=1 \text{ to } r} (Y_{ij} - \bar{y}_{i.})^2$$

is an algebraic identity. For the survey design described, the two variances s_d^2 and s_e^2 are summary statistics for two sources of variation in the survey; s_d^2 summarizes the information about variation among cluster means, and s_e^2 summarizes the information about variation among observations within clusters. The analysis of variance represents a partitioning of the total sum of squares into a number of components and then computing the various variances. This is an analysis of the variation for the above survey design.

A comparison of the variance s_d^2 with s_e^2 may be of interest in certain situations. For example, the investigator may wish to determine whether or not he should stratify by clusters or take a completely random sample in his next survey on similar material. If the variance among clusters is relatively larger than the variance within clusters, say $s_d^2 / s_e^2 = 1.5$, he may decide to stratify; if $s_d^2 / s_e^2 < 1.5$, he may decide to use a simple random sample design. Given that the surveyor has used a cluster simple random sample design, he may compute the efficiency of the sample survey design used relative to a simple random sample design. The estimate of the variation when there is no stratification into clusters is

$$\{s_d^2(v - 1) + s_e^2 \, v(r - 1)\} / \{v - 1 + v(r - 1) = rv - 1\} = s_t^2.$$

The ratio s_t^2 / s_e^2 estimates the relative efficiency of a simple random sample design to a cluster simple random sample design. The ratio s_t^2 / s_e^2 is equal to

$$\{s_e^2 \, v(r - 1) + s_d^2 \, (v - 1)\} / (rv - 1) \, s_e^2 =$$
$$(vr - v) / (vr - 1) + (s_d^2 / s_e^2) \{(v - 1) / (rv - 1)\}.$$

In this form, we note that the ratio of the variances and the values of v and r as they affect the degrees of freedom, determine the relative efficiency. For $s_d^2 / s_e^2 = 1.5$, for example, the relative efficiency is equal to

$$v(r - 1) / (rv - 1) + \{(v - 1) / (rv - 1)\}(1.5) =$$
$$1 + (v - 1) / 2(rv - 1);$$

if r = 2, then

$$1 + (v - 1) / 2(rv - 1) = 1 + (v - 1) / 2(2v - 1)$$

is approximately equal to $1 + 1 / 4 = 1.25$ or 125% for $v \geq 20$. The selection of the value for the ratio s_d^2 / s_e^2, which we took to be 1.5, depends upon the cost of sampling under the two schemes, that is, it may cost 25% more to take a simple random sample than to take a random selection of clusters and of individuals within clusters or *vice versa*. The selection of the value 1.5 also depends upon the number of degrees of freedom associated with s_d^2 and with s_e^2.

The variance of a cluster mean is s_e^2 / r. The overall mean $\bar{y}_{..}$ contains variation among cluster effects as well as among observations within the cluster; hence, the variance of $\bar{y}_{..}$ is s_d^2 / rv. The variance of the total $Y_{..}$ is $rv \, s_d^2$. With these variances, the confidence limits may be computed for the overall mean $\bar{y}_{..}$ or for the total $Y_{..}$ (See Chapter 12.). Also, the population total may be estimated from the sample total, given that a specified proportion of the clusters and of the observations within a cluster have been sampled. This is accomplished by multiplying the sample total by the reciprocal of the proportion sampled. For example, if $1 / 100$ of the population has been sampled, then the sample total $Y_{..}$ multiplied by $1 / (1 / 100) = 100$ is the estimated population total. Likewise, estimated confidence limits on the population total are obtained by multiplying the confidence intervals for the sample total by the reciprocal of the proportion sampled.

The variances of an estimated treatment mean or differences between means from a completely randomized design are described in Chapter 10. The variance s_e^2 used in Chapter 10 may also be obtained from the above analysis of variance table. Likewise, the computation of confidence intervals follows the description given in Chapter 12 when s_e^2 is obtained from the above analysis of variance table.

For data obtained from a randomized complete block design, the sources of variation are the mean, the block effects, the treatment effects, and random error effects. Variances due to these sources may be computed as described in the following analysis of variance table:

Source of variation	Degrees of freedom	Sum of squares	Variance
Due to mean	1	$Y_{..}^2 / rv = rv \, \bar{y}_{..}^2$	
Due to block effects	$r - 1$	$v \sum_{j=1 \text{ to } r} (\bar{y}_{.j} - \bar{y}_{..})^2$	$v \sum_{j=1 \text{ to } r} (\bar{y}_{.j} - \bar{y}_{..})^2/(r - 1)$ $= s_b^2$
Due to treatment effects	$v - 1$	$r \sum_{i=1 \text{ to } v} (\bar{y}_{i.} - \bar{y}_{..})^2$	$r\sum_{i=1 \text{ to } v}(\bar{y}_{i.} - \bar{y}_{..})^2/(v - 1)$ $= s_t^2$

Due to deviations	$(r - 1)(v - 1)$	$\sum\sum(Y_{ij} - \bar{Y}_{i.} - \bar{Y}_{.j} + \bar{Y})^2$	s_e^2
Due to total	rv	$\sum_{i=1 \text{ to } v}\sum_{j=1 \text{ to } r} Y_{ij}^2$	

In the above table, the symbols are those described in Chapters 5, 7, 8, and 10. The total sum of squares is partitioned as indicated above, and the sum of the four sums of squares is equal to the total sum of squares. The variances s_b^2, s_t^2, and s_e^2 are summary statistics of variation in the experiment. In addition to constructing confidence intervals on treatment means and differences among means, one may compare this randomized complete block design to the completely randomized design using the ratio s_{cr}^2 / s_e^2, where s_{cr}^2 is computed as:

$$s_{cr}^2 = \{s_b^2 (r - 1) + s_e^2 (r - 1)(v - 1) + s_e^2 (v - 1)\} \div$$
$$\{(r - 1) + (r - 1)(v - 1) + (v - 1) = rv - 1\}.$$

If all treatments had been the same treatment, then the within blocks variance would contain only variation among units unaffected by treatment differences and would have $r(v - 1) = (v - 1) + (v - 1)(r - 1)$ degrees of freedom. The term $s_e^2 (v - 1)$ estimates what the treatment contribution to the within blocks sum of squares would have been had all treatments been the same treatment. Thus, s_{cr}^2 estimates what the residual error variance would have been for a completely randomized design. If, for example, s_e^2 were only $3 / 4$ as large as s_{cr}^2, then a completely randomized design with r = four blocks would give the same variance for a treatment mean as a randomized complete block with r = 3 replicates, that is, $s_e^2 / 3 = s_{cr}^2 / 4$. Thus if an experimenter used a randomized complete block design instead of the completely randomized design, he could use one replicate less to achieve the same value for the standard error of a mean or of a difference between two means. This also means that a randomized complete block design would be 33% more efficient than a completely randomized design for material like that in the experiment conducted.

   To illustrate the computations of the variances for a randomized complete block design, the data for v = 4 treatments and r = 3 blocks are:

Treat- ment	block 1	 2	 3	 Total	Mean $\bar{y}_{i.}$	 $\bar{y}_{i.} - \bar{y}_{..}$
A	$Y_{A1} = 1\ (0)$	$Y_{A2} = 3\ (0)$	$Y_{A3} = 5\ (0)$	$Y_{A.} = 9$	$\bar{y}_{A.} = 3$	-1
B	$Y_{B1} = 2(-2)$	$Y_{B2} = 4(-2)$	$Y_{B3} = 12(4)$	$Y_{B.} = 18$	$\bar{y}_{B.} = 6$	2
C	$Y_{C1} = 3(1)$	$Y_{C2} = 4(0)$	$Y_{C3} = 5(-1)$	$Y_{C.} = 12$	$\bar{y}_{C.} = 4$	0
D	$Y_{D1} = 2(1)$	$Y_{D2} = 5(2)$	$Y_{D3} = 2(-3)$	$Y_{D.} = 9$	$\bar{y}_{D.} = 3$	-1
Total	$Y_{.1} = 8(0)$	$Y_{.2} = 16(0)$	$Y_{.3} = 24(0)$	$Y_{..} = 48$		0
Mean	$\bar{y}_{.1} = 2$	$\bar{y}_{.2} = 4$	$\bar{y}_{.3} = 6$			$\bar{y}_{..} = 4$
$\bar{y}_{.} - \bar{y}$	-2	0	2	0		

The values in the parentheses are obtained as $(Y_{ij} - \bar{y}_{i.} - \bar{y}_{.j} + \bar{y}_{..}) = e_{ij}$ = estimated random error deviations. For example, $e_{A1} = (1 - 2 - 3 + 4) = 0$ and $e_{B1} = (2 - 2 - 6 - + 4) = -2$. The deviations of the block means are obtained as $(\bar{y}_{.1} - \bar{y}_{..}) = (2 - 4) = -2, (\bar{y}_{.2} - \bar{y}_{..}) = (4 - 4) = 0$, and $(\bar{y}_{.3} - \bar{y}_{..}) = (6 - 4) = 2$. The deviations of treatment means from the overall mean are similarly obtained. The various sums of squares are computed as:

$$r v \bar{y}_{..}^2 = 3 (4) (4^2) = 192 ,$$

$$v \sum_{j=1 \text{ to } r} (\bar{y}_{.j} - \bar{y}_{..})^2 = 4 \{(-2)^2 + (0)^2 + (2)^2\} = 32,$$

$$r\sum_{i=1 \text{ to } v} (\bar{y}_{i.} - \bar{y}_{..})^2 = 3\{(-1)^2 + (2)^2 + (0)^2 + (-1)^2\} = 18$$

$$\sum_{i=1 \text{ to } v}\sum_{j=1 \text{ to } r} (Y_{ij} - \bar{y}_{i.} - \bar{y}_{.j} + \bar{y}_{..})^2$$

$$= (0)^2 + (-2)^2 + (1)^2 + (1)^2 + (0)^2 + (2)^2 + (0)^2 + (2)^2 + (0)^2 + (4)^2 + (1)^2 + (-3)^2 = 40.$$

The above may be summarized in the following analysis of variance table:

Source of variation	Degrees of freedom	Sums of squares	Variance
Due to mean	1	192	
Due to blocks	$r - 1 = 2$	32	$16 = s_b^2$
Due to treatments	$v - 1 = 3$	18	$6 = s_t^2$
Due to deviations	$(r - 1)(v - 1)$ $= 2(3) = 6$	40	$40 / 6 = 20 / 3 = s_e^2$
Total	$r\,v = 12$	282	

The variance of a difference in treatment means is $2\,s_e^2 / r = 2\,(20 / 3) / 3 = 40 / 9$. The estimated variance for a completely randomized design is obtained as:

$$s_{cr}^2 = \{s_b^2(r - 1) + s_e^2(r - 1)(v - 1) + s_e^2(v - 1)\} /$$
$$\{(r - 1) + (r - 1)(v - 1) + (v - 1)\}$$

$$= \{16(2) + (20 / 3)(6) + (20 / 3)(3)\} / (2 + 6 + 3)$$

$$= (32 + 40 + 20) / 11 = 92 / 11.$$

$$s_{cr}^2 / s_e^2 = (92 / 11) / (20 / 3) = 3(92) / 11(20) =$$
$$69 / 55 = 1.25, \text{ or } 125\%.$$

It is estimated that this randomized complete block design is 25% more efficient than a completely randomized design would have been. This means that four replicates of a randomized complete block would result in the same variance of a difference between two treatment means as would have been obtained with five replicates of a completely randomized design.

Each of the above sums of squares, except that due to the mean, may be further partitioned if this is desired by the investigator. The total sum of squares for experiments designed as a latin square, a Youden design, a balanced incomplete block, or any other design may be partitioned into various components and the resulting variances computed. The particular partitioning utilized depends upon which summary statistics on variation are desired by the investigator. The partitioning of sums of squares is the same conceptually for all investigations regardless of the orthogonality or non-orthogonality of the effects. The computation of sums of squares is considerably more difficult arithmetically when the effects are non-orthogonal, but the concepts do not change.

The word "analysis" in the analysis of variance does not mean "interpretation". The term "reduction of variance" or "partitioning of variance" may be more meaningful to some than the term "analysis of variance". The last term is the one used in published literature utilizing this concept.

14.3 REGRESSION

A 19th century investigator, F. Galton, studied the heights of fathers and sons to determine the usefulness of a father's height in predicting his son's height. He noted that very tall fathers tended to have tall sons who tended to be somewhat shorter than their fathers. Likewise, short fathers tended to have short sons who were somewhat taller than their fathers. There was a tendency for heights to "regress" toward the mean height of all fathers, and hence the term **regression**. Of course, some sons were taller than their tall fathers but in general they tended to be shorter.

A father's height, say X, may be used to predict a son's height, say Y, by establishing a functional relation between these two variables X and Y. As described in Chapter 8, this functional relation may take many forms. One form which appears to describe the above situation is the simple linear regression equation

$$Y_i = \mu + \beta (X_i - x_.) + \varepsilon_i,$$

where X_i is the height of the ith father and Y_i is the height of his son, $x_.$ is the mean height of all fathers, μ is the population mean height of all sons, β is the slope of the line and is called the **regression coefficient**, and ε_i is the deviation of a son's height from the mean value of all sons' heights, $\mu + \beta (X_i - x_.)$, for a specified height of a father (for example, $X_i = 70$ inches). Thus, around each mean value $= \mu + \beta (X_i - x_.)$, for X_i specified, there is a distribution of heights of sons. If the population variance in the distribution for each X is equal to some constant σ^2, if the X_i are measured without error, and if a Y_i (or a set of Y_i) is randomly selected for each X_i, then we may estimate the parameters μ and β in the above linear regression equation by the equations:

estimate of μ: $\sum_{i=1 \text{ to } n} Y_i / n = \bar{y}.$

and

estimate of β: $b = \sum_{i=1 \text{ to } n} (Y_i - \bar{Y})(X_i - x) / \sum_{i=1 \text{ to } n} (X_i - x)^2$,

estimated value: $Y_i^* = \bar{Y} + b(X_i - x)$

deviation from regression: $e_i = Y_i - Y_i^*$

where there are n pairs of observations (X_i, Y_i) in the sample. That is, for each father's height, his son's height is obtained. In the equation for the estimated linear regression coefficient b , the numerator of b divided by n - 1 is called the estimated **covariance,** and the denominator divided by n - 1 is called the variance of the X_i. Therefore an estimated regression coefficient is the ratio of the estimated covariance of X_i and Y_i to the variance of the X_i. The variable Y is called the **dependent variable** since it depends upon the **independent variable** X. If the number of degrees of freedom in the covariance and the variance is the same, then the value for b is obtained as the ratio of the sum of the cross-products of the $(Y_i - \bar{Y})(X_i - x)$ to the sum of squares of the $(X_i - x.)$.

To illustrate the computations in linear regression, consider the following example:

Year $= X_i$	Y_i	$(X_i - x)$	$(Y_i - \bar{Y})$	$(X_i - x)^2$	$(X_i - x)(Y_i - \bar{Y})$	Y_i^*
		Speed records, Y_i, attained in the Indianapolis Memorial Day auto races				
1935	106	-3	-7	9	+21	108.5
1936	109	-2	-4	4	+8	110.0
1937	114	-1	1	1	-1	111.5
1938	117	0	4	0	0	113.0
1939	115	1	2	1	+2	114.5
1940	114	2	1	4	+2	115.0
1941	115	3	3	9	+9	117.5
13566	790	0	0	28	41	791
1938	113					113

The numbers in the next to the last line of the above table contain the totals of the columns and the last line contains the means. From the above,

$$b = \Sigma_{i=1 \text{ to } 7}(X_i - x_.)(Y_i - \bar{y}_.) / \Sigma_{i=1 \text{ to } 7}(X_i - x_.)^2 =$$
$$[41 / (7 - 1)] / [28 / (7 - 1)] = 41 / 28 = 1.5$$

and the estimated or predicted value of the ith Y is

$$Y_i^* = \bar{y}_. + b(X_i - x_.) = 113 + 1.5(X_i - 1938).$$

The mean $\bar{y}_.$ was calculated to the nearest whole integer. From the equation for Y_i^*, the regression line may be computed for the data in Figure 14.1. Computation of two values for Y_i^* is sufficient to draw the line of the equation in Figure 14.1, but three points are usually computed as an additional check on the computations. The line must pass through all three points. The Y_i values for $X_i = 1936$, 1938, and 1940 are:

$$Y_2^* = 113 + 1.5(1936 - 1938) = 113 - 3 = 110,$$

$$Y_4^* = 113 + 1.5(1938 - 1938) = 113 + 0 = 113, \text{ and}$$

$$Y_6^* = 113 + 1.5(1940 - 1938) = 113 + 3 = 116.$$

The estimated regression coefficient b = 1.5 indicates that speed records tend on the average to increase by 1.5 miles per hour each year. For each unit increase in X, Y is estimated to increase by 1.5 miles per hour. The estimated regression coefficient b = 1.5 is a summary statistic of the average slope or average increase in miles per hour (Y) per unit increase in X. If there were only two pairs of values (X_1, Y_1) and (X_2, Y_2), the slope would be $(Y_1 - Y_2) / (X_1 - X_2)$ which is a result used in high school geometry books and is sometimes called "rise-over-run". The formula for b given above is used when there are more than n = 2 pairs of values.

From the formula for b, two facts should be noted. The first is that the sum of the cross products, $\Sigma_{i=1 \text{ to } n}(X_i - x_.)(Y_i - \bar{y}_.)$ may be negative or positive and that its value is limited only by the values of $(X_i - x_.)$ and $(Y_i - \bar{y}_.)$. This means that the covariance may be negative or positive, whereas the variance is never negative. The second fact to note is that b may be positive, negative, or zero. Thus Y may tend to increase, decrease,

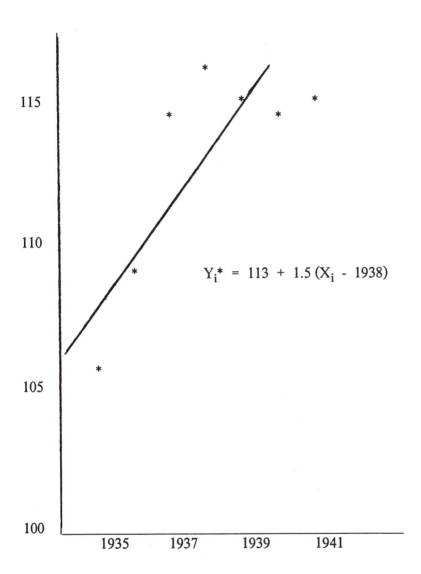

Figure 14.1. Speed records Y_i in miles per hour in the Indianapolis Memorial Day automobile races for 1935 to 1941.

or vary independently with X.

The deviations from the regression line may be obtained as $(Y_i - Y_i^*)$ = observation minus its computed value on the regression line. The total deviation $(Y_i - \bar{y}.) = (Y_i - Y_i^*) + (Y_i^* - \bar{y}.)$, that is, the total deviation is partitioned into two parts. The first is the deviation of the observation from the computed value on the regression line, and the second is the deviation of the value on the regression line from the overall mean. The values of the deviation from the regression line are -2.5, -1.0, 2.5, 4.0, 0.5, -2.0, and -2.5. (The deviations do not add to zero as they should because of rounding errors. At least one decimal in $\bar{y}.$ and two in b should have been retained.) The sum of squares is equal to 40.00 with n - 2 = 5 degrees of freedom. Therefore, $s_e^2 = 40.00 / 5 = 8.00 =$ the average squared deviation.

Additional independent variables, for example, the height of the mother and the height of a grandfather, may be useful in predicting the height of the son. A **multiple regression** equation of the form

$$Y_i = \mu + \beta_1 (X_{1i} - x_{1.}) + \beta_2 (X_{2i} - x_{2.}) + \beta_3 (X_{3i} - x_{3.}) + \varepsilon_i$$

may be useful to describe the height of a son, where $X_{1i} =$ height of father, $x_{1.} =$ mean height of fathers, $X_{2i} =$ height of mother, $x_{2.} =$ mean heights of mothers, $X_{3i} =$ height of grandfather, $x_{3.} =$ mean height of grandfathers, $Y_i =$ height of son of ith father, mother, and grandfather, $\beta_1, \beta_2,$ and β_3 are the respective population regression coefficients, and ε_i is the deviation of height of a son from the mean height for specified values of $X_{1i}, X_{2i},$ and $X_{3i}.$ Experience has shown that little is gained from using a grandfather's height to estimate a grandson's height if the heights of the father and of the mother are available, and that β_3 is small relative to β_1 and $\beta_2.$

Many other examples resulting in regression equations of the above form are available in published literature. In studying performance in college = Y , several variables such as $X_1 =$ high school grade average, $X_2 =$ amount of outside activities, $X_3 =$ Scholastic Aptitude Test score, and others are studied. Based on these studies, a student's performance in college is predicted, and based on this prediction, they may or may not be admitted to college.

We may summarize the information about sources of variation in linear regression in an analysis of variance table as follows:

Source of variation	Degrees of freedom	Sum of squares	Variance
Due to mean	1	$n \bar{y}_.^2 = Y_.^2 / n$	
Due to linear regression	1	$b^2 \sum_{i=1 \text{ to } n} (X_i - x_.)^2$	
Deviations from regression	n - 2	$\sum_{i=1 \text{ to } n} (Y_i - \bar{y}_. - b(X_i - x_.))^2$	s_e^2
Total	n	$\sum_{i=1 \text{ to } n} Y_i^2$	

In the above analysis of variance table, the total sum of squares is partitioned into three parts in the algebraic equation:

$$\sum_{i=1 \text{ to } n} Y_i^2 = n \bar{y}_.^2 + b^2 \sum_{i=1 \text{ to } n} (X_i - x_.)^2 + \sum_{i=1 \text{ to } n} \{Y_i - \bar{y}_. - b(X_i - x_.)\}^2$$

The variance $b^2 \sum_{i=1 \text{ to } n} (X_i - x_.)^2$ may be compared with the residual variance s_e^2 to determine relatively how much of the variation is attributable to regression. In the sum of squares due to the mean, $n \bar{y}_.^2$, n is the sample size. In the sum of squares due to regression, $b^2 \sum_{i=1 \text{ to } n} (X_i - x_.)^2$, the multiplier of b^2, i.e., $\sum_{i=1 \text{ to } n} (X_i - x_.)^2$, plays the same role as n in the preceding sum of squares. Thus, in selecting the X_i for regression, consideration should be given to making $\sum_{i=1 \text{ to } n} (X_i - x_.)^2$ as large as possible. In Chapter 8 it was suggested that n / 2 of the observations should be placed at the lowest value of X_i and n / 2 of the observations should be placed at the highest value of X_i among all the X_i considered. If we use this treatment design, the $\sum_{i=1 \text{ to } n} (X_i - x_.)^2$ attains its maximum value. This is also of importance for the estimated variance of the regression coefficient b which is equal to $s_e^2 / \sum_{i=1 \text{ to } n} (X_i - x_.)^2$. Here it may be noted that a maximum value of $\sum_{i=1 \text{ to } n} (X_i - x_.)^2$ minimizes the variance.

Other more complex forms of regression equations may be treated in a similar manner in an analysis of variance table. Since we showed that the variation in a regression situation may be treated in an analysis of variance table in the same manner as a set of data from an experiment designed as a completely randomized design, we shall now illustrate how the sums of squares in an analysis of variance table for the latter case may be treated from the standpoint of regression. The resulting sums of squares will be identical. Suppose that there are $v = 3$ treatments replicated r times in a completely randomized design. From simple geometric considerations, the slopes of the three non-horizontal lines in Figure 14.2 are computed as follows:

top line: slope $= (\bar{y}_1 - \bar{y}_.)/1 = (\bar{y}_1 - \bar{y}_.)$

middle line: slope $= (\bar{y}_2 - \bar{y}_.)/1 = (\bar{y}_2 - \bar{y}_.)$

bottom line: slope $= (\bar{y}_3 - \bar{y}_.)/1 = (\bar{y}_3 - \bar{y}_.)$

The slope is defined to be the increase in Y for each unit increase in X. Since all regression lines are required to go through the point $(0, \bar{y}_.)$, the sum of squares $\sum_{j=1 \text{ to } r}(X_j - 0)^2$ is equal to $\sum_{j=1 \text{ to } r} X_j^2 = \sum_{j=1 \text{ to } r} 1^2 = r$. Therefore the sum of squares due to regression for each of the $v = 3$ regression lines is

$$r(\bar{y}_1. - \bar{y}_..)^2 + r(\bar{y}_2. - \bar{y}_..)^2 + r(\bar{y}_3. - \bar{y}_..)^2 = r\sum_{i=1 \text{ to } 3}(\bar{y}_i. - \bar{y}_..)^2.$$

This is the same sum of squares obtained previously for this source of variation.

If the deviations e_{ij} follow a normal distribution, we may set confidence limits on various quantities in linear regression. The various variances are summarized below:

variance of $\bar{y}_.$: s_e^2/n

variance of b : $s_e^2/\sum_{i=1 \text{ to } n}(X_i - x_.)^2$

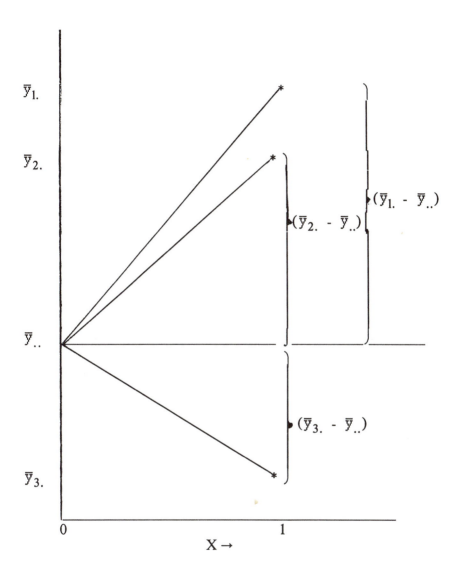

Figure 14.2. Graphical representation of means $\bar{y}_i$ of three treatments relative to the overall mean $\bar{y}_{..}$.

variance of $Y_i^* = \bar{y}_. + b(X_i - x_.)$:
$$s_e^2 \{1/n + (X_i - x_.)^2 / \Sigma_{i=1 \text{ to } n} (X_i - x_.)^2\}$$

variance of $Y_i = \bar{y}_. + b(X_i - x.) + e_i$:
$$s_e^2 \{1/n + (X_i - x.)^2 / \Sigma_{i=1 \text{ to } n} (X_i - x.)^2 + 1\}$$

The standard deviation is obtained as the square root of the corresponding variance. Then a standard error is multiplied by $t_{\alpha, n-2}$ from the t table in Chapter 12, Table 12.3, where n - 2 = degrees of freedom and α is the selected level for the (1 - α)% confidence interval.

14.4 CORRELATION

In the case of regression, an X_i was considered to be measured without error and to be the independent variate. Suppose that neither Y_i nor X_i are independent variates, that both are random variates, and that we wish to measure the *relationship* between the n pairs of measurements in a sample.

The interpretation of the measured relationship must be considered carefully and thoughtfully before conclusions are drawn. For example, the percentage of people speaking English in country i = Y_i and the number of cars owned per individual = X_i may be highly related variates but this does not mean that increasing the percentage of people speaking English in a country will increase the number of cars owned per individual, nor does it mean that an increase in the number of cars owned per individual will increase the percentage of people speaking English. The variates Y_i and X_i are related through some common factors and tend to increase together. The **correlation coefficient** is a measure of relationship and is defined as a covariance divided by a geometric mean of the variances, that is covariance / {(variance of Y)(variance of X)}$^{1/2}$. The estimated correlation coefficient is computed as

$$r_{xy} = \{\Sigma_{i=1 \text{ to } n} (X_i - x.)(Y_i - \bar{y}_.) / (n - 1)\} \div$$
$$\{[\Sigma_{i=1 \text{ to } n} (X_i - x.)^2 / (n - 1)] [\Sigma_{i=1 \text{ to } n}(Y_i - \bar{y}_.)^2 / (n - 1)]\}^{1/2}$$

The term (n - 1) cancels and is usually omitted. The correlation coefficient varies between plus and minus one. Correlations near zero indicate little or no relationship, and values near plus or minus one indicate a high relationship.

In order to set confidence limits on r and to use the t table in Chapter 12, the following function of r_{xy} is used:

$$z = \{\log_e (1 + r_{xy}) - \log_e (1 - r_{xy})\} / 2$$

Solving for r_{xy} in the above equation, we obtain

$$r_{xy} = (e^{2z} - 1) / (e^{2z} + 1).$$

Tables of r_{xy} in terms of z and z in terms of r_{xy} have been prepared as, e.g., Tables A.12 and A.13 in Snedecor and Cochran (1980).

The variable z is normally distributed with population variance $1 / (n - 3)$. Hence the $(1 - \alpha)\%$ confidence limits for the variable z are computed as $z \pm t_{\alpha,\infty} [1 / (n - 3)]^{1/2}$. Then z is transformed back to r_{xy} to obtain the confidence limits for r_{xy}.

To illustrate the arithmetic associated with computing the correlation coefficient, consider the first five values for own height and own weight in Table 9.1. These are (last line are the means and next to last line are the totals):

$X_i =$ $Y_i =$

own ht own ht |$(X_i - x)$ $(Y_i - \bar{y})$ $(X_i - x)^2$ |$(Y_i - \bar{y})^2$ $(X_i - x)(Y_i - \bar{y})$

X_i	Y_i	$(X_i - x)$	$(Y_i - \bar{y})$	$(X_i - x)^2$	$(Y_i - \bar{y})^2$	$(X_i - x)(Y_i - \bar{y})$
71	148	2	-5.6	4	31.36	-11.2
65	120	-4	-33.6	16	1128.96	134.4
72	175	3	21.4	9	457.96	64.2
70	173	1	19.4	1	376.36	19.4
67	152	-2	-1.6	4	2.56	3.2
345	768	0	0	34	1997.20	210.0
69	153.6	0	0			

$$r_{xy} = 210 / \{34 (1997.20)\}^{1/2} = 210 / 260.6 = 0.81.$$

In order to set a $(1 - \alpha)\% = 90\%$ confidence interval on the population correlation coefficient ρ for our sample value $r_{xy} = 0.81$, first transform r_{xy} to z as follows:

$$z = \{\log_e (1 + 0.81) - \log_e (1 - 0.81)\} / 2$$
$$= \{0.5933 - (-1.6607)\} / 2 = 1.127.$$

Then,

$$z \pm t_{0.1,\infty} / (5 - 3)^{1/2} = 1.127 \pm 1.160$$

Or,

$$z_l = 1.127 - 1.160 = -0.033$$

and

$$z_u = 1.127 + 1.160 = 2.287.$$

Then transforming z_l and z_u back to r values, we obtain:

$$(e^{2z_l} - 1) / (e^{2z_l} + 1) = (0.936 - 1) / (0.936 + 1) = -0.03$$

and

$$(e^{2z_u} - 1) / (e^{2z_u} + 1) = (96.93 - 1) / (96.93 + 1) = 0.98.$$

Note that with such a small sample size, i.e., n = 5, the 90% confidence interval with r_{xy} = 0.81 is very wide, i.e., from -0.03 to 0.98. This illustrates why it is necessary to have much larger sample sizes in order to have relatively narrow confidence intervals.

In all investigations involving small samples, interpretations should be made with caution. The example above and the one on regression contain only n = 5 and n = 7, paired observations, respectively. These are small sample sizes, and conclusions drawn from them are risky. The construction of confidence limits will aid in interpretation, but this does not overcome the vagaries of sampling associated with small sample sizes.

14.5 TESTS OF SIGNIFICANCE

In the manner used herein, a test of significance is a measure of the standardized distance or deviation of a sample value from a hypothetical or hypothesized value. To illustrate, let us consider a difference between the two sample means, $\bar{y}_1$ and $\bar{y}_2$ obtained from two randomly selected samples, and let us consider the test of significance known as a t-test which is

$$t = (\bar{y}_{1.} - \bar{y}_{2.}) / s_e (2 / r)^{1/2},$$

where $s_e / (2 / r)^{1/2}$ is the estimated standard error of the difference between the sample means $\bar{y}_{1.}$ and $\bar{y}_{2.}$. Suppose that the randomly selected sample values used to compute $\bar{y}_{1.}$ came from a population with mean μ_1 and that the randomly selected sample values used to compute $\bar{y}_{2.}$ came from a population with mean μ_2. Then, $\bar{y}_{1.} - \bar{y}_{2.}$ is an estimate of $\mu_1 - \mu_2$. If the two samples were from the same population, then $\mu_1 - \mu_2 = 0$. Suppose one hypothesizes that $\mu_1 - \mu_2 = 0$, i.e. the null hypothesis, then the larger the absolute value of t, i.e.$|t|$, the less likely it is that $\mu_1 - \mu_2 = 0$. In other words, the size of the value of the t-statistic measures the "strength of evidence" against the hypothesis that $\mu_1 - \mu_2 = 0$ (or some other specified value). The larger the value of $|t|$, the greater is the strength of evidence against the hypothesis.

In order to ascertain when a value of the t-statistic is large enough for the investigator to attach some significance to the result, one may compare the value of the t-statistic with the values computed from the distribution of t, such as those given in Chapter 12, Table 12.3. As an illustration, let $\bar{y}_{1.} = 10$, $\bar{y}_{2.} = 5$, $s_e^2 = 16$ with 28 degrees of freedom, and r = 8, then

$$t = (10 - 5) / 4 (2 / 8)^{1/2} = 5 / (4 / 2) = 5 / 2 = 2.5.$$

In Figure 14.3, we have constructed a "likelihood function", given that the error deviations are normally distributed. The $(1 - \alpha)\%$ confidence intervals are computed for the various values of α, and then a smooth curve (dotted line) is fitted through the points. The confidence intervals computed are

$$\bar{y}_{1.} - \bar{y}_{2.} \pm t_{\alpha,df} s_e (2 / r)^{1/2} =$$
$$5 \pm t_{\alpha,28} 4 (2 / 8)^{1/2} = 5 \pm t_{\alpha,28} (2),$$

where $s_e^2 = 16$ is associated with 28 degrees of freedom, the $t_{\alpha,28}$ values may be obtained from various sources (e.g., Fisher, 1950, Snedecor and Cochran, 1980, etc.). The $(1 - \alpha)\%$ confidence interval calculations for

Figure 14.3 are:

α	$\mid \bar{y}_{1.} - \bar{y}_{2.} \pm t_{\alpha,28} \, s_e / (2 / r)^{1/2}$	confidence limits lower - upper	
0.9	5 ± 2 (0.127)	4.746	5.254
0.8	5 ± 2 (0.256)	4.488	5.512
0.7	5 ± 2 (0.389)	4.222	5.778
0.6	5 ± 2 (0.530)	3.940	6.060
0.5	5 ± 2 (0.683)	3.634	6.366
0.4	5 ± 2 (0.855)	3.290	6.710
0.3	5 ± 2 (1.056)	2.888	7.112
0.2	5 ± 2 (1.313)	2.374	7.626
0.1	5 ± 2 (1.701)	1.598	8.402
0.05	5 ± 2 (2.048)	0.904	9.096
0.02	5 ± 2 (2.467)	0.066	9.934
0.01	5 ± 2 (2.763)	-0.526	10.526
0.005	5 ± 2 (3.047)	-1.094	11.094

In Figure 14.3, the proportion of the area under the curve between the values 3 to 7 indicates that these are likely values for $\mu_1 - \mu_2$. Values from -2 to 1 or 9 to 12 would represent unlikely values for $\mu_1 - \mu_2$ because the area under the curve to the left and to the right of these values is relatively small. The proportion of the area under the curve between two values gives the "likelihood" of $\mu_1 - \mu_2$ being included in the estimated interval in repeated samplings. The "strength of the evidence" that the hypothesized value for $\mu_1 - \mu_2$ may or may not be true, may be obtained by noting where it lies on the abscissa of a graph such as Figure 14.3.

A second test of significance used extensively is the **chi-square test** computed as

$$\chi^2(k - 1 \text{ degrees of freedom}) = \sum_{i=1 \text{ to } k} (Y_i - Y_i^*)^2 / Y_i^*,$$

where Y_i is the number observed in a particular category, and Y_i^* is the expected value computed for that category according to some expectation. In Table 9.1, the heights of recording team A were compared with the heights of recording team B, and the differences A - B were computed. Of these 85 differences, 51 were negative, 3 were zero, and 31 were

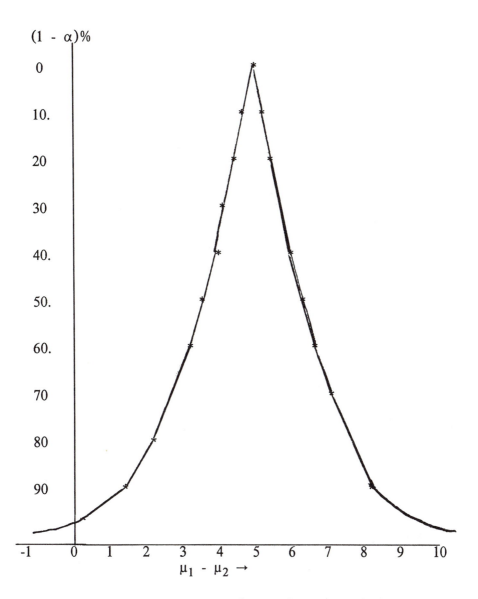

Figure 14.3. Likelihood function for $(\mu_1 - \mu_2)$ given that $(\bar{y}_{1.} - \bar{y}_{2.}) =$ 5, $s_e (2/r)^{1/2} = 2$, and degrees of freedom $= 28$.

positive. Based on a hypothesis of no difference, one would expect 85 / 2 = 42.5 = Y_1 negative and 42.5 = Y_2 positive differences. Place one and a half of the three zero differences into the negative group, making 52.5 = Y_1, and one and a half into the positive group, making 32.5 = Y_2. Since k = 2,

$$(52.5 - 42.5)^2 / 42.5 + (32.5 - 42.5)^2 / 42.5 = 200 / 42.5 = 4.71.$$

A comparison of 4.71 with tabulated values of χ^2 for one degree of freedom gives an indication of how much significance to attach to this value of χ^2. Table 14.1 has been reproduced for this purpose.

Table 14.1. Percentage values of the chi-square distribution.

Degrees of freedom	Probability of a larger value of χ^2						
	.90	.70	.50	.30	.10	.05	.01
1	0.02	0.15	0.46	1.07	2.71	3.84	6.64
2	0.21	0.71	1.39	2.41	4.60	5.99	9.21
3	0.58	1.42	2.37	3.67	6.25	7.82	11.34
4	1.06	2.20	3.36	4.88	7.78	9.49	13.28
5	1.61	3.00	4.35	6.06	9.24	11.07	15.09
6	2.20	3.83	5.35	7.23	10.64	12.59	16.81
7	2.83	4.67	6.35	8.38	12.02	14.07	18.48
8	3.49	5.53	7.34	9.52	13.36	15.51	20.09
9	4.17	6.39	8.34	10.66	14.68	16.92	21.67
10	4.86	7.27	9.34	11.78	15.99	18.31	23.21

Taken from table III of Fisher (1950) by permission of the author and publishers.

The relationship between the t-statistic and the chi-square statistic is that the square root of the values for chi-square with one degree of freedom is equal to the value of $t_{\alpha,\infty}$.

From the above table we note that the probability of exceeding a chi-square value equal to 4.71 with one degree of freedom is between 5% and

1%. Hence it would appear that a 50:50 split of the negative and positive difference is not a plausible hypothesis. The strength of the evidence indicates that team A records lower heights than does team B.

As another illustration, consider the number of students with brown hair in the first 20, the second 20, the third 20, and the fourth 20 reported under own measurements. The following results were obtained:

brown hair	Own hair color (no.)					total
	1st 20	2nd 20	3rd 20	4th 20		
observed Y_i	11	19	16	17		63
expected Y_i^*	15.75	15.75	15.75	15.75		63
$(Y_i - Y_i^*)$	-4.75	3.25	0.25	1.25		0
$(Y_i - Y_i^*)^2/Y_i^*$	1.43	0.67	0.00	0.10		2.20

Since the listing of names is alphabetical and since the number having brown hair should not be influenced by the first letter of one's name, one would expect one-fourth of the total, $63 / 4 = 15.75$, to be in each category. There are three degrees of freedom associated with this table, since three of the values $(Y_i - Y_i^*)$ may vary, but the fourth is required to be the negative sum of the previous ones. This is required because the sum of the observed values must equal the sum of the expected or computed values. Since

$$\chi^2 \text{ (3 d.f.)} = \Sigma_{i=1 \text{ to } 4} (Y_i - Y_i^*)^2 / Y_i^* = 2.20,$$

we would consider these data to be a good fit to a 1:1 hypothesis. From the tabulated values of χ^2 (3 d.f.), Table 14.1, we note that a value of 2.20 would be exceeded in more than 50% of random samplings from a chi-square distribution.

Let us consider the same set of data on hair color but from another viewpoint. Suppose that we divide all hair colors into brown and not brown hair as our first classification. Further suppose that we use the grouping given above as a second classification. Even though interest in the second classification would be hard to justify, we shall use to illustrate the use of the chi-square statistic to obtain evidence on the independence of the two classifications as given below:

Hair color = classification 1	Group = classification 2				total
	1st 20	2nd 20	3rd 20	4th 20	
brown	11	19	16	17	$63 = Y_{1.}$
not brown	9	1	4	3	$17 = Y_{2.}$
total	$20 = Y_{.1}$	$20 = Y_{.2}$	$20 = Y_{.3}$	$20 = Y_{.4}$	$80 = Y_{..}$

The above is a 2 × 4 contingency table with hair color as one classification and with groups of 20 names as the other classification. The observed number is Y_{ij}, and the computed or expected number is $Y_{ij}{}^* = Y_{i.} \ Y_{.j} \ / \ Y_{..}$. The border totals are regarded as fixed, and hence only $(r - 1)(c - 1)$ (where r = number of rows and c = number of columns) quantities in the table may vary, the remaining quantities are obtained by subtraction from the border totals. This results in $(r - 1)(c - 1)$ degrees of freedom for the resulting chi-square test computed as:

$$\chi^2 \{(r - 1)(c - 1)\} = \Sigma_{i=1 \text{ to } r} \Sigma_{j=1 \text{ to } c} (Y_{ij} - Y_{ij}{}^*)^2 / Y_{ij}{}^*$$

The computed values $Y_{ij}{}^*$ for the above table are:

Hair color	Group				total
	1st 20	2nd 20	3rd 20	4th 20	
brown	15.75	15.75	15.75	15.75	63
not brown	4.25	4.25	4.25	4.25	17
total	20	20	20	20	80

where 63 (20) / 80 = 15.75 and 17 (20) / 80 = 4.25. The deviations $Y_{ij} - Y_{ij}{}^*$ are:

Hair color	Group				total
	1st 20	2nd 20	3rd 20	4th 20	
brown	-4.75	3.25	0.25	1.25	0
not brown	4.75	-3.25	-0.25	-1.25	0
total	0	0	0	0	0

The quantities $(Y_{ij} - Y_{ij}^*)^2 / Y_{ij}^*$ are

Hair color	Group				total
	1st 20	2nd 20	3rd 20	4th 20	
brown	1.43	0.67	0.00	0.10	2.20
not brown	5.31	2.49	0.01	0.37	8.18
total	6.74	3.16	0.01	0.47	10.38

A chi-square value of 10.38 with three degrees of freedom has a probability between one and five percent of being exceeded in random sampling. Hence this is evidence that the grouping and hair color are not independent, even though we might believe that they should be. A chi-square test used in this manner results in a **test of independence**, whereas in the previous example the test was a **goodness of fit** test. In Chapter 11, observed frequencies were compared for goodness of fit with computed or expected frequencies utilizing the chi-square test for the binomial and Poisson distributions. Many forms of a chi-square test are used in statistical methodology.

Another frequently used test of significance is a **variance ratio test** which is a ratio of two variances and is denoted as **Snedecor's F-test** or simply as an **F-test**. The denominator in the ratio is usually an error variance and the numerator is the variance being compared with the error variance. The tabulated values of the F-statistic depend upon the degrees of freedom in the two variances. We denote this as $F_\alpha(f_n, f_d)$ where α is the probability of an F-value greater than the one observed, f_n is the number of degrees of freedom associated with the variance in the numerator, and f_d is the number of degrees of freedom associated with the variance in the denominator. Some values of the F-statistic are given in Table 13.1. To illustrate the computation of an F-statistic, use the variances in the analysis of variance table for the randomized complete block design example. The variance for treatments was $s_t^2 = 6$ and the variance due to deviations was $s_e^2 = 20/3$. The F-value is computed as

$$F(3, 6) = s_t^2 / s_e^2 = 6 / (20 / 3) = 18 / 20 = 0.9.$$

The probability of obtaining an F-value greater than 0.9 in random sampling from an F-distribution with 3 and 6 degrees of freedom and when the population variances are equal, $\sigma_t^2 = \sigma_e^2$, is greater than 50%. Hence the evidence suggests that the two variances are estimates of the same population parameter.

The relationships among the t-statistic, the chi-square statistic, and the F-statistic are:

$$t_{\alpha,f}^2 = F_\alpha (1, f), \text{ e.g.,}$$
$$(t_{0.1,10})^2 = 1.81^2 = 3.28 = F_{0.1} (1, 10).$$

$$t_{\alpha,\infty}^2 = \chi_\alpha^2 (1), \text{ e.g.,}$$
$$(t_{0.05,\infty})^2 = 1.96^2 = 3.84 = \chi_{0.05}^2 (1).$$

$$F_\alpha (f, \infty) = \chi_\alpha^2 (f) / f, \text{ e.g.,}$$
$$F_{0.1} (10, \infty) = 1.60 = 15.99 / 10 = \chi_{0.1}^2 (10) /10,$$

where f equals the degrees of freedom associated with the denominator variance of t and F and with the degrees of freedom associated with chi-square.

14.6 REFERENCES AND SUGGESTED READING

David, F. N. and N. L. Johnson (1952). The truncated Poisson. *Biometrics* 8:275-285.

Fisher, R. A. (1950). *Statistical Methods for Research Workers, eleventh edition.* Hafner Publishing Co., New York, xv + 354 pp..

Snedecor, G. W. and W. G. Cochran (1980). *Statistical Methods, 7th edition.* The Iowa State University Press, Ames, Iowa, xiv + 507 pp..

Also see references in Chapters 7, 8, and 10.

14.7 PROBLEMS

Problem 14.1. Show how to compute the analysis of variance for Example 10.1. (Remember that to show how to compute does not mean to compute.)

Problem 14.2. Show how to compute the analysis of variance for Example 10.2.

Problem 14.3. Show how to compute the analysis of variance for Example 10.3.

Problem 14.4. Show how to compute the analysis of variance for Problem 10.6.

Problem 14.5. Show how to compute the analysis of variance for Problem 10.7.

Problem 14.6. Show how to compute the analysis of variance for Problem 10.8.

Problem 14.7. Show how to compute the analysis of variance for Problem 12.5.

Problem 14.8. Show how to compute the analyses of variance for the first three examples of factorial treatment designs given in Section 8.6.

Problem 14.9. Show how to compute the linear regression coefficients for Y = grade-point average on X_1 = high school grade and on X_2 = aptitude test score for the data given in Section 8.6.

Problem 14.10. Show how to compute the correlation coefficient for X_1 and X_2 in the preceding problem.

Problem 14.11. The following data were obtained from Table 9.3 by grouping all ages less than or equal to 20.5 years and all those older than 20.5 years; the other classification was omitted. The regrouped data are:

Age class	fresh-man	sopho-more	junior	senior	graduate	total
< 20.5 years	13	32	8	4	0	57
> 20.5 years	0	1	10	19	4	34
total	13	33	18	23	4	91

From what we know about students, the two classifications should not be independent. Consequently, a large value of chi-square should be obtained. Show how to compute the chi-square test for independence for this two by five contingency table.

Problem 14.12. Show how to compute the goodness of fit chi-square test for the data in Tables 12.1 and 12.2.

Problem 14.13. Show how to compute a goodness of fit chi-square test for the data given by David and Johnson in Section 12.4 when the zero class is included and when it is excluded.

Problem 14.14. Show how to compute a goodness of fit chi-square test to the 1:1, the 3:1, and 7:1 ratios given that 80 red-eyed flies and 20 white-eyed flies were obtained in the experiment described in Problem 12.2.

Problem 14.15. The following data are weight gains of baby chicks in grams for four different feeding treatments A, B, C, and D where A was the standard method of feeding baby chicks and the others represent new feeding rations:

Treatment

Y_{Aj}	e_{Aj}	e_{Aj}^2	Y_{Bj}	e_{Bj}	e_{Bj}^2	Y_{Cj}	e_{Cj}	e_{Cj}^2	Y_{Dj}	e_{Dj}	e_{Dj}^2
55	11.2		61			42			169	26.2	686.44
49	5.2		112			97			137		
42	-1.8		30			81			169		
21	-22.8		89			95			85		
52	8.2		63			92			154		

$Y_{A.} = 219$

$\bar{y}_{A.} = 43.8$

$\sum_{j=1 \text{ to } 5} e_{Aj}^2 = 742.8$

$s_A^2 = 185.70$

range = 34

Complete the computations in the above table and prepare an analysis of variance table for the above data assuming that the design of the experiment was a completely randomized design. Compare the treatment means using the t-test as a test of significance for comparing treatments B,

C, and D with the standard treatment A. Use the tabulated t-values in Table 12.3. Compare the variance among the treatment means using the F-test and Table 13.1. Also, compare the variances for treatments B, C, and D with the variance for A using the F-statistic.

Problem 14.16. The following data are from a greenhouse experiment designed as a randomized complete block with five treatments (1 = LLL, 2 = LLH, 3 = HLL, 4 = LHH, 5 = HHL), which were nitrogen applications at three different dates of early, medium, and late and two levels L and H, and six replicates (blocks):

Treat-ments	|1	2	Blocks 3	4	5	6	|total	mean	$\sum_{j=1 \text{ to } 6} e_{ij}^2$
1	|8.8	12.9	11.7	31.2	22.0	9.9	|96.5	16.1	212.35
2	|23.5	26.3	21.6	15.6	24.4	23.3	|134.7		
3	|41.2	22.5	21.8	46.3	15.6	22.6	|170.0		
4	|28.4	48.4	16.4	44.5	38.8	43.6	|220.1		
5	|67.4	33.2	59.5	49.8	57.1	36.6	|303.6		
Total	|169.3	143.3	131.0	187.4	157.9	136.0	|924.9		

Compute the residuals e_{ij}, sum the squares of residuals for each treatment, prepare an analysis of variance table, compute 90% and 95% confidence intervals for the differences of treatment means, compute the t-test for differences of treatment means, compute the F-test for the variance among treatment means compared to the variance for residuals, and then plot the $\sum_{j=1 \text{ to } 6} e_{ij}^2$ against the treatment means. Do you believe that the error variances for each treatment are all estimates of the same parameter?

Problem 14.17. The following data are for pounds of Macintosh apples purchased per 100 customers in four stores on the four days Monday, Tuesday, Wednesday, and Thursday of a given week. The four treatments were A = grade-run apples, B = 2.25 inches in diameter apples, C = 2.5 inches in diameter carefully selected and handled apples, and D = 2.5 inches in diameter highly colored apples. The experiment design was a four by four latin square with stores as the columns and days of the week as rows. The data are:

Day of week	Store 1	2	3	4	total	mean
Monday	A 14	B 8	C 40	D 48	110	
Tuesday	B 20	A 22	D 48	C 25	115	
Wednesday	D 24	C 12	B 12	A 27	75	
Thursday	C 31	D 16	A 32	B 22	101	
Total						

Treatment totals:
Treatment means:

The letter next to the number in the above table indicates which treatment was in a given store on a given day. Compute the residuals for the above data, prepare an analysis of variance table, compute the F-statistic for comparing treatment variance against variance of residuals and compare the F-value with tabulated values in Table 13.1, compute 90% and 95% confidence intervals on differences between treatment means, and compute the t-test for differences between treatment means.

Problem 14.18. For the data used to construct Figure 14.1, compute the regression coefficient for the years 1935-1937 and the regression coefficient for the years 1937-1941. Do you think that the single regression equation used is the appropriate model? Why or why not? The use of two regression coefficients with a common intercept is know as **segmented regression**.

CHAPTER 15. STATISTICAL PUBLICATIONS

15.1 INTRODUCTION

Procedures for obtaining meaningful numbers have been described in the preceding chapters. The reader has been repeatedly cautioned to cast a critical and wary eye on all sets of numbers purported to contain information on particular characteristics of a population. With this attitude still uppermost in our minds, let us now turn our attention to some of the vast accumulations of statistical data available for the United States and for the World. The statistical publications for the United States are numerous, and this statement is documented by the fact that it took <u>1619 pages</u> of a *Guide to U. S. Government Publications*, Library of Congress number Z 1223 Z7 A5731, to list all of them. In the Reference room of Mann Library of Cornell University, there are eight sets of five shelves devoted to publications of the U. S. Bureau of the Census publications alone! Some of these special publications have been singled out for comment, and brief descriptions are given to illustrate some of the data available on the many and varied characteristics of the World's populations. The particular publications to be discussed are the

U.S.D.A. Agricultural Statistics,
Statistical Abstracts of the United States,
Statistical Yearbook of the United Nations,
FAO Production Yearbook,
FAO Trade Yearbook, and
Yearbook of International Trade Statistics.

15.2 STATISTICAL PUBLICATIONS OF THE UNITED STATES

Statistical publications of the United States are many and varied; each department of the United States Government issues many statistical publications on various items. For example, the Department of Labor issues unemployment statistics by month, by group, by section of the country, and so on; the Department of Commerce does likewise for items of interest to them, and this Department, which contains the Bureau of the Census, probably issues more statistical tables and summaries than any other United States Department.

The United States Department of Agriculture has its share of statistical publications with probably the most frequently used one being *U.S.D.A. Agricultural Statistics*, Library of Congress number HD 1751 A34. This is an annual publication of data on agricultural production, supplies, consumption, facilities, costs, and returns. Each annual publication contains the most recent ten years' data on a particular item. The 1982 issue contains historical tables dating back to 1866 for production figures on the principal crops and to 1867 for livestock numbers. Some international and world statistics are included where needed for comparison. The following major groupings of statistics were shown in the 1989 issue:

1. Grains (food and feed)
2. Cotton, tobacco, sugar crops, and honey
3. Oilseeds, fats, and oils
4. Vegetables and melons
5. Fruits, tree nuts, and horticultural specialities
6. Hay, seeds, and minor field crops
7. Cattle, hogs, and sheep
8. Dairy and poultry products
9. Farm resources, income, and expenses
10. Taxes, insurance, cooperatives, and credit
11. Stabilization and price-support programs
12. Agricultural conservation and forestry statistics
13. Consumption and family living
14. Miscellaneous statistics (weather, fishery, refrigeration, etc.)

Three types of data are included. These are
(1) actual counts and amounts of items covered,
(2) estimates from surveys of various types made by the U.S.D.A,
and
(3) census enumeration data.

The last was obtained with the cooperation of the Bureau of Census and published in their *United States Census of Agriculture*, Library of Congress number HD 1753. In addition to information for the entire United States, the same sort of information is given for each of the 50 States. Volume 2, Part 1, of this publication has the following major headings for statistical data covered:

Farms--number
Farms by value of sales
Farms by size
Market value of agricultural products sold
Farm-related income
Farm production expenses

Irrigation
Value of land and buildings
Machinery and equipment on place
Agricultural chemicals used
Tenure and characteristics of farm operator
Occupation of operator
Farms by type of organization
Livestock and poultry
Crops harvested
Vegetables
Fruits, nuts and berries
Nursery and greenhouse crops

In Volume 3, Parts 1 and 2, the major headings are:
Farm and ranch irrigation survey
Farms and land in farms
Market value of agricultural products sold, income and
production contracts
Assets of landlord and operator
Capital purchases and operating expenses
Real estate taxes
Farm operator characteristics and household assets
Landlord characteristics
Farms and landlords with debt
Owned land in farms

Many types of information are readily available from the tables. For example from the *1989 U. S. D. A. Agricultural Statistics*, the following six states are ranked in order of production of wheat in 1988, with their combined total being almost one-half the total production of wheat in the entire United States:

Total Production (1988)

Kansas	323,000,000 bushels
Oklahoma	172,800,000 bushels
Washington	124,620,000 bushels
North Dakota	103,390,000 bushels
Texas	89,600,000 bushels
Colorado	79,540,000 bushels
Total	892,950,000 bushels
U. S. Total	1,811,261,000 bushels

These results are further broken down into production of spring wheat, winter wheat, and durum wheat. Each crop in addition to wheat is broken down into similar detail.

The Bureau of the Census in the Department of Commerce has been publishing the *Statistical Abstracts of the United States* annually since 1878. It is a summary of statistics on the social, political, and economic organization of the United States. It is designed to serve as a reference and guide to other statistical publications and sources. Both government and private statistical publications are utilized in compiling each volume. Although the emphasis is on national statistics, many tables present data for regions, individual states, and outlying areas of the United States. The 1990 issue contains 31 sections with many tables in each section as follows:

  Population;
  Vital statistics, health, and nutrition;
  Health and nutrition;
  Education;
  Law enforcement, courts, and prisons;
  Geography and environment;
  Parks, recreation, and travel;
  Elections;
  State and local government finances and employment;
  Federal government finances and employment;
  National defense and veterans affairs;
  Social insurance and human services;
  Labor force, employment, and earnings;
  Income, expenditures, and wealth;
  Prices;
  Banking, finance, and insurance;
  Business enterprise;
  Communications;
  Energy;
  Science;
  Transportation - land;
  Transportation - air and water;
  Agriculture;
  Forests and fisheries;
  Mining and mineral products;
  Construction and housing;
  Manufacturers;
  Domestic trade and services;
  Foreign commerce and aid;
  Outlying areas; and
  Comparative international statistics.

In addition to these 31 sections, there are seven appendices as follows:

(1) Guide to sources of statistics, state statistical abstracts, and foreign statistical abstracts
(2) Metropolitan area concepts and components and population of metropolitan areas
(3) Statistical methodology and reliability
(4) Index to tables having historical statistics, Colonial times to 1970 series
(5) Listing of tables deleted from 1989 edition of *Statistical Abstracts*
(6) New tables
(7) Computers in the Office

A study of the first and second appendices will indicate the great diversity and the completeness on population characteristics in the United States.

To illustrate one type of data available from these tables, Table 15.1 and Figure 15.1 were prepared from Tables 14 and 17 and Table 15.2 and Figure 15.2 from Table 84 of the 121st annual edition of the *Statistical Abstracts of the United States, 1990.* The projections to future years represent extrapolations beyond the data. We were cautioned to be wary of extrapolation in Chapter 9, but sometimes this is necessary; demographers have devised and continue to devise extrapolation and projection procedures necessary for planning for the future welfare of a nation and of the World. Various series of projections are utilized, but we have selected the data for Series A projections simply because they appeared first in the table. Projections for Series B, C, and D may be obtained from this publication. The projections are smaller for these three series than they are for Series A.

Many other statistical publications of the federal government and of state governments are available in the various libraries in the country.

15.3 STATISTICAL PUBLICATIONS OF THE UNITED NATIONS

The first issue of the United Nations' *Statistical Yearbook* was published in 1948; the 23rd issue for the year 1971 was distributed in 1972. The Library of Congress number for this series is HC 57 A19. The series continues and expands the work of the *Statistical Yearbook of the League of Nations* in providing a convenient summary of international statistics on a World basis. The latter publication was discontinued in 1945; the 1948 issue of the *Statistical Yearbook* provides for continuity by

Table 15.1. Estimated and projected total, white, and black populations for the United States 1960 to 2010.

Year	Total	White	Black	Other
Estimated population in 1,000s				
1960	179,386			
1965	193,223			
1970	203,849	178,692	22,617	2,540
1975	214,931	186,955	24,602	3,375
1980	226,451	195,143	26,645	4,664
1984	235,961	200,989	28,391	6,580
1985	238,207	202,463	28,802	6,942
1986	240,532	203,990	29,223	7,319
1987	242,843	205,492	29,657	7,694
1988	245,231	207,054	30,104	8,073
Projected population in 1,000s				
1989	247,150	208,437	30,503	8,210
1990	249,331,	209,897	30,934	8,500
1995	259,238	216,267	33,000	9,971
2000	267,498	221,087	34,939	11,472
2005	274,884	225,048	36,816	13,020
2010	281,894	228,637	38,653	14,604

presenting data for several previous years, generally back to 1928. The official languages of this publication are English and French, with both being presented.

The *Statistical Yearbook* is an annual publication supplemented by several United Nations' publications. Current monthly data for many of the tables and series may be found in the *Monthly Bulletin of Statistics*. Population data in a comprehensive form and breakdown are published in a *Demographic Yearbook*. Only population data of general interest are included in the *Statistical Yearbook*. The order of presenting data for countries is first by continental groups, which are ordered as follows:

Africa;
America, North;

Table 15.2. Total fertility rate (TFR) and intrinsic rate of natural increase (IRNI) 1970 to 1987.*

Year	Total	TFR White	Other	Total	IRNI White	Other
1970	2,480	2,385	3,067	6.0	4.5	14.4
1971	2,267	2,161	2,920	2.6	0.8	12.6
1972	2,010	1,907	2,628	-2.0	-3.9	8.6
1973	1,879	1,783	2,443	-4.5	-6.5	5.7
1974	1,835	1,749	2,339	-5.4	-7.2	4.0
1975	1,774	1,686	2,276	-6.7	-8.6	3.0
1976	1,738	1,652	2,223	-7.4	-9.3	2.1
1977	1,790	1,703	2,279	-6.2	-8.1	3.2
1978	1,760	1,668	2,265	-6.8	-8.8	2.9
1979	1,808	1,716	2,310	-5.7	-7.7	3.8
1980	1,840	1,749	2,323	-5.1	-7.0	4.0
1981	1,815	1,726	2,275	-5.5	-7.4	3.3
1982	1,829	1,742	2,265	-5.2	-7.0	3.0
1983	1,803	1,718	2,225	-5.7	-7.5	2.5
1984	1,806	1,719	2,224	-5.6	-7.4	2.4
1985	1,843	1,754	2,263	-4.8	-6.6	3.1
1986	1,836	1,742	2,282	-4.9	-6.8	3.3
1987	1,871	1,767	2,349	NA	NA	NA

*The **total fertility rate** is the number of births that 1,000 women would have in their lifetime if, at year of age, they experienced the birth rates occurring in the specified year. A total fertility rate of 2,110 represents "replacement level" fertility for the total population under current mortality conditions, assuming no net immigration. The **intrinsic rate of natural increase** is the rate that eventually would prevail if a population were to experience, at each year of age, the birth rates and death rates occurring in the specified year and if these rates remained unchanged over a long period of time. A minus (-) sign indicates a decrease. NA means not available.

America, South;
Asia;
Europe;
Oceania (Australia, New Zealand, etc.);
United Soviet Socialist Republic (both Asian and European parts).

The order of countries within each continental group is in alphabetical order of the country's English name. Any changes in territory are described in the book. The reader must be aware of the many territorial changes, especially in the European and United Soviet Socialist Republic groups. The main groups of tables are the following:

Population;
Manpower;
Agriculture;
Forestry;
Fishing;
Industrial production;
Mining, quarrying;
Manufacturing;
Construction;
Electricity, gas;
Transport;
Communications;
Internal trade;
External trade;
Balance of payments;
Wages and prices;
National income;
Finance;
Social statistics;
Education, culture;
Appendices (conversion factors, index, etc.).

One may note that there is a vast amount of statistical information available on most countries of the World; however, many of these statistics are very crude, being little if any better than educated guesses, while others are relatively accurate. In the main, these statistics are extremely helpful to the various countries and allow such reports as the *World Economic Survey, 1965*, to be compiled. In this publication, considerable emphasis is placed on the developing countries of the world.

The Food and Agriculture Organization (FAO) of the United Nations

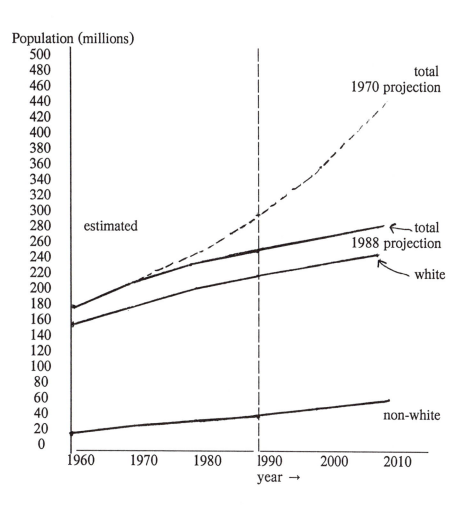

Figure 15.1. Estimated and projected size (in millions) of white, non-white, total population of the United States.

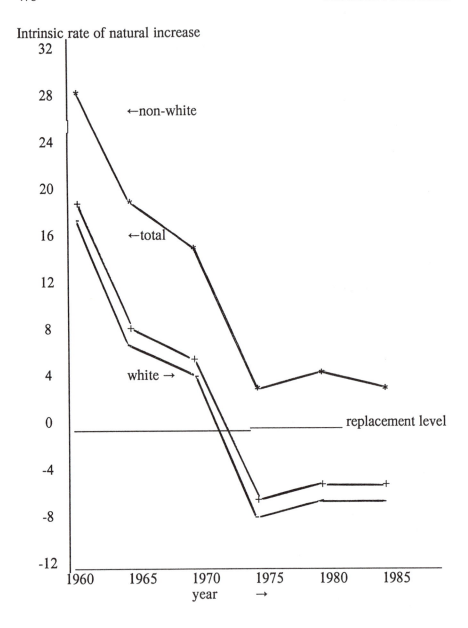

Figure 15.2. Estimated and projected intrinsic rates of natural increase for total, white, and non-white populations in the United States.

has been active in preparing statistical yearbooks on production and trade. The official languages for these are English, French, and Spanish; all three are presented in the *FAO Production Yearbook* and the *FAO Trade Yearbook.* The former was initiated in 1947 and the latter in 1948. The *Production Yearbook* continues the statistical series on crop acreages and yield and on livestock numbers formerly published by the International Institute of Agriculture at Rome. Current monthly statistics are available in a supplemental bulletin.

In connection with the above, a World census of agriculture is a relatively new concept, the first one having been taken in 1930, the second in 1950, and a third in 1960. FAO took a strong lead in the 1950 census of agriculture and developed the plans for the 1960 census. Sixty-eight countries cooperated in the 1930 census and 106 in the 1950 census; Communist country participation was almost lacking in the census for the 1950 program. The first two parts of a three-part publication of the 1960 census of agriculture were published in 1966 and 1967; the 3rd, 4th, and 5th parts were published in 1969, 1968, and 1971, respectively (Library of Congress number HD 1421).

In all these publications, the diversity of material is to be noted. If one wishes to know the number of camels in the United Soviet Socialist Republic or in Israel, for example, numbers are given in the tables. In the *FAO Trade Yearbook,* one can find items such as value of agricultural trade for tea and sesame seed or the commodity trade in groundnuts in cake, meal, and oil. Egg trade is broken down into eggs in the shell and eggs not in the shell in addition to classifications such as frozen, liquid, and dried egg.

The Statistical office of the United Nations compiles a *Yearbook of International Trade Statistics* among many other publications. In the tables for the year 1966 (published in 1968), there are tables of imports and exports for 142 countries by commodity groups in value and in amount. The total imports and exports are given as well as imports and exports to specific countries. The data for Figure 15.3 were obtained from the 1966 Yearbook. We may note from the graph that Cuba became an importing country in 1958 and that the total value of her exports has remained relatively constant since 1946. Russia makes up a sizeable amount of the difference between imports and exports.

It is hoped that the above illustrates the very many types of numbers available for many and sundry characteristics. More and more numerical information is being accumulated for more and more countries of the World. Some researchers spend all their time compiling statistics for a certain characteristic (as for instance, heavy industry and steel production in Russia and satellites in Europe) in order to obtain the most authentic information possible. These statistics are usable and quite valuable for

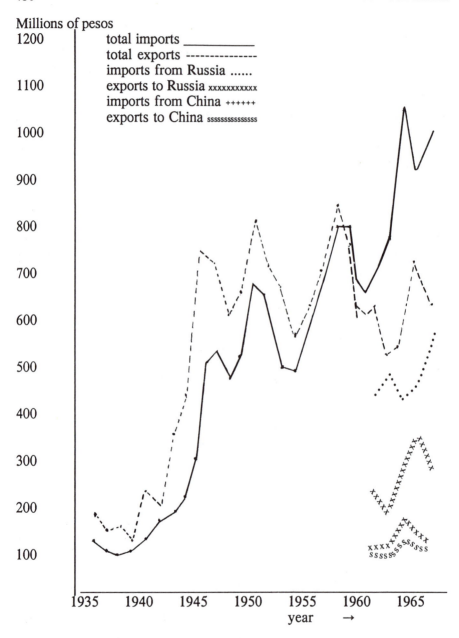

Figure 15.3. Total value of imports and exports for Cuba and for imports from and exports to Russia and China (mainland) 1937 to 1967.

for many studies, and are useless for others.

A complete list of publications under the auspices of the Statistical Office of the United Nations has not been compiled here. It should be noted that the publications are many and diverse. This active and energetic organization is serving a valuable need.

15.4 A COMMENT

Knowledge of available statistical data is important to politicians, social scientists, students, administrators, businessmen, and others for making decisions in everyday life of a Society. Ignorance of what information is available can lead to procurement of duplicate or triplicate sets of data by the survey method, with costly consequences. Also, in politics, social programs, economics, and other fields, ignoring available statistical data may lead to costly blunders for a Society.

The method of procuring and summarizing data for all of the nations of the World is a subject in itself. This brief introduction is intended to make the reader aware of this fact and to give an idea of the sources of statistical information on various population characteristics for a specified country.

15.5 SUGGESTED READING

Statistical Reporting Service [1969]. The story of United States agricultural estimates. United States Department of Agriculture, Miscellaneous Publication No. 1088, April. pp. v + 137.

15.6 PROBLEMS

Problem 15.1. Using a linegraph, plot the production of wheat, infant mortality rate, and number of passenger cars for the 20-year period 1969 to 1988 for the country listed on your card. The data are given in the *Statistical Yearbook* of the United Nations. (Note that some of these data may not be available for every country for all years; the countries were selected because they produced wheat. Each student is to be given a different country.)

Problem 15.2. From the United States Department of Agriculture *Agricultural Statistics* for 1988, prepare bargraphs of the characteristic listed on your card for the years given in the table. Note the diversity of information available. (Each student should be given a different characteristic.)

Problem 15.3. Prepare a graph, pictograph or a bargraph, depicting the number of camels in China, the United Soviet Socialist Republic, Egypt, India, Israel, Iran, and Pakistan for the years 1930, 1950, and 1960 as compiled in the *FAO Trade Yearbook*. How reliable do you think these statistics are? Give reasons for your answers.

Problem 15.4. Obtain current data for Figures 15.1, 15.2, and 15.3, and extend the graphs to include the current data.

Problem 15.5. Make a list of ten publications of the Bureau of Labor Statistics and briefly describe what each one covers.

APPENDIX I

EXAMINATIONS FOR ONE SEMESTER AND ADDITIONAL PROBLEMS

The sequence of chapters taught for the following set of examinations was Chapters 1, 2, 3, 4, 7, 8, 5, 9, 10, 12, and 15. Some of the material in Chapters 11 and 13 was included, but no formal discussion of probability or of sample size was given. None of the material in Chapter 14 was presented. The ordering of the chapters in the text is considered to be better pedagogically since sample survey design concepts are easier to grasp than are experiment design concepts. If the order given in the text is followed, the instructor may simply move the questions on sampling from the third examination to the front part of the second examination and then may include some of the material from the back part of the second examination on the third examination. For example, for the 1968 set of examinations, take questions 7 (15 points) and 10 (9 points) from the second examination and place them at the beginning of the third examination; then, take questions 17 to 28 (24 points) from the third examination and place them at the beginning of the second examination. Each of the revised examinations has 100 points. The examinations were given in the fourth, eighth, and twelfth weeks of the semester with the final examination covering the material in the last chapters as well as the material in the earlier chapters.

The instructor is urged to use questions which employ reasoning more than memory. In this way, the examination can be more of an educational tool. It is felt that the first examination relied too heavily on memory. Because of the nature of the material, it would have been difficult to write an examination relying more on the powers of reasoning and less on the powers of memory for the first examination. Both the preparation for an examination and the taking of an examination should provide an educational experience for the student.

It is highly recommended that an instructor using this text write the examination questions together with the answers prior to the time that the formal course work begins. With such a set of examinations available, the course will have a well-defined direction and pace. The presentation will be more comprehensible and less ambiguous to the student since they would be written at a time when the instructor was not under the strain of presenting the course and being forced to prepare an examination "the night before".

Also, it is suggested that the examination be given at a time when the student has sufficient time to complete the examination, such as evening examinations. This procedure was followed for the 1968 and later examinations, but the 1966 and 1967 examinations were given during a class period.

Additional problems for the individual chapters may be obtained from the examinations. If discussion periods are organized as was done for the course taught by the author, the questions on the examinations from previous years form an excellent medium for discussion. This has been found to be an effective means to aid the student in understanding the concepts of the course.

FIRST PRELIM (EXAMINATION)

Points
10   I. Define STATISTICS used as the name of a subject matter field. Distinguish between the subject of Statistics and "statistics" used as a noun.
10   II. Match the most appropriate term, by using the assigned letter, with the following definitions:
      Terms:
      (a) datum   (i) treatment
      (b) statistic   (j) absolute experiment
      (c) hypothesis   (k) comparative experiment
      (d) number   (l) single phenomenon experiment
      (e) research   (m) re-search
      (f) experiment design   (n) law
      (g) experiment   (o) parameter
      (h) theory   (p) postulation

Definitions:
      i) A _____ is one designed specifically to compare two or more treatments.
      ii) The quest or pursuit of new knowledge is defined to be

_____.

iii) An _____ is defined to be the planning and collection of measurements or observations according to a prearranged plan for the purpose of obtaining factual evidence on the plausibility of hypotheses.

iv) The arrangement of the treatments in an experiment is known as _____.

v) A number that is used to describe a characteristic of a population and that is derived from a consideration of all members in the population is called a _____.

vi) A number derived from a sample and used to estimate the value of a characteristic of the population is known as _____.

vii) A fact from which a conclusion is to be drawn is known as a _____.

viii) A tentative or postulated explanation of a phenomenon is known as a _____.

ix) A relatively well verified postulation which possesses some degree of generality is known as a _____.

x) A _____ is a single entity or phenomenon under study in an experiment.

30 III. Complete the following statements (2 points each).

1. _____ is the term applied to the procurement and systematization of scientific knowledge.

2. The method whereby scientific knowledge is acquired is known as _____.

3. An explanation of a phenomenon which has been verified beyond all reasonable doubt is known as a _____.

4. An experiment containing but one treatment or phenomenon is _____.

5. _____ inference is the process of determining the implications contained in a set of propositions.

6. A 100% sample from a population or universe of individuals is a _____.

7. An error in measurement is composed of _____ plus _____.

8. A _____ yields information on what is present in the universe.

9. The type of reasoning involved when we judge the temperature of the water of a lake by sticking our toe in the water at one spot in the lake is _____.

10. The process of reaching a conclusion from the data given in the following example is known as _____.
Example, "If all students who studied pass this test and if you studied, then you will pass this test."

11. Statistics is more concerned with _____ inference while mathematics is more concerned with _____ inference.

12. _____ inference is used to a considerable extent in formulating hypotheses.

13. A figure or number which pertains to some clear-cut characteristic and which is impossible to obtain is known as _____ (E.g., "dog-napping is big business, grossing $1,000,000 per year".)

14. A figure or number used in connection with an undefined term or an ambiguous thought in such a way that it is unclear what is meant is _____ . (E.g., "The National Safety Council estimates that traffic accidents caused 1.9 million disabling injuries and $11 billion in economic loss in 1967".)

15. In terms describing variability the variation among the true means of populations is defined as _____.

20 IV. Problem 3.2 was given as follows: (1) An experimenter used 1000 ripe peaches (from Georgia) in an experiment. (2) He wished to determine which quarter of a peach fruit was the most tender when squeezed. (3) He designated the peach fruit by quarters starting with the suture, or indentation, on the fruit as follows: (a) left front, (b) left back, (c) right back, and (d) right front. (4) (He did not cut the peach fruit.) (5) A device was utilized which measured the amount of pressure required to penetrate the peach skin; this was used to measure tenderness. (6) On each peach the experimenter measured the quarters in the order as numbered above. (7) He reached the conclusion that the least pressure was required to penetrate the right front quarter, the next lowest pressure was required to puncture the right back quarter, the next in amount of pressure required for penetration was the left back, and the left front required the greatest pressure for penetration. (8) He published in a scientific journal.

Show how each of the eight principles of scientific experimentation as listed in Chapter 4 of the text apply to this investigation, by listing principle and the sentence number which applies.

1. _____.
2. _____.
3 _____.
4 _____.
5 _____.
6. _____.
7. _____.
8. _____.

12 V. With arrows indicate the relationship between the steps discussed relative to data collection in Chapter 3 of the text with the steps in scientific experimentation as listed in Chapter 4:

Steps in Chapter 3	Steps in Chapter 4 (fill in blanks)
Why collect data? — →	1. Clear and precise statement of problem
What data are to be collected?	2. _____
How are data to be collected?	3. _____
Where and when are the data to be collected?	4. _____
Who collects the data?	5. _____
Complete description of data.	6. _____
Disposal of data.	7. _____
Conclusions.	8. _____

3   VI. What is the difference between precision and accuracy? Illustrate with an example.

5   VII. A characteristic is described for three populations (A, B, and C). The following numbers a, b, c, d, and e, as designated on the real line below, are defined as follows:

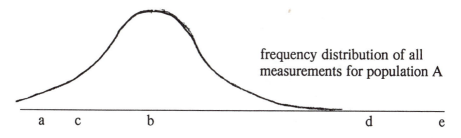

frequency distribution of all measurements for population A

a   c   b   d   e

a = true mean of population A
d = true mean of population B
e = true mean of population C
b = mean obtained from all the measurements for population A
c = a single measurement from population A

In the terms used for describing variability, the line segment ab is a measure of _____
the line segments ad, ae, and de are measures of variation due to

_____

the line segment ac is a measure of

_____

the line segment bc is a measure of
_____ and
the line segment ce is a measure of

_____.

SECOND PRELIM

Points
6 1. Miss I. C. Food was conducting an experiment on different kinds of pie thickeners to measure their effect on consistency as defined and measured by her. She was investigating v = 4 kinds of thickening agents (the treatments) and decided to use r = 3 repetitions of each treatment. She could prepare 12 pies at one time and then add the treatments. She could also bake 12 pies at one time in an oven which exhibited no gradients. Therefore, she decided to use a completely randomized design for the experiment. Describe the yield equation for this experiment when the effects are additive and enter the equation in linear form.

Are the assignable effects orthogonal to each other? Why or why not?

Construct an experiment design properly randomized using the attached random number table and starting at the top of column 1 and proceeding down the column. (Write down the sequence of random numbers used.)

8 2. Miss I. M. Fashion was conducting an experiment on length of time
   different colors of eye shadow stayed presentable which was a
   measurement made by her. She used $v = 5$ preparations of eye shadow
   which only differed in the color component. She was using five
   different time periods for which she knew that there were only random
   variations and no known gradients. Therefore, she decided to use $r =$
   4 different girls as the blocks and to use a randomized complete block
   design.

   Describe the yield equation for this experiment when the effects are
   additive and enter the equation in linear form.

   Which of the assignable effects are orthogonal to each other?

   Construct the plan for the experiment using the enclosed random
   number table and starting at the top of column 2 and proceeding down
   the column. (Write down the sequence of random numbers used.)

   3. Mr. Bunny Buck wishes to compare $v = 5$ feeding treatments on
   the growth of rabbits from three to six weeks of age. Since he has
   available to him only litters of three rabbits each he decides to use a
   balanced incomplete block design.

   Construct a balanced incomplete block design for $v = 5$ treatments in
   blocks of size $k = 3$ rabbits each by completing the following:

Rabbit No.	Litter number									
	1	2	3	4	5	6	7	8	9	10
1	E	D	A	E	E	C	B	B	E	E
2	B	C								D
3	A	A								A

The number of replicates is r = _____

The number of times any pair of treatments occur together in the b incomplete blocks is λ = _____

Which of the assignable effects are orthogonal to each other?

_____

Describe and illustrate the randomization procedure for this design using the attached random number table starting at the top of column 3 and proceeding down the column. (Write down the sequence of random numbers used.)

Random Digits (9,001 - 13,750)

(Published in Jour. Amer. Stat. Assoc., 48:931-934 (1953). Taken from *A Million Random Digits*, published by the Rand Corporation, Santa Monica, Calif.)

3780	28391	05940	55583	81256	38175	38422	64677	80358	52629
5325	05490	65974	11186	15357	21805	10371	95812	84665	74511
8240	92457	89200	94696	11370	75517	82119	09199	30322	33352
2789	69758	79701	29511	55968	19195	92261	44757	98628	57916
7523	17264	82840	59556	37119	77869	08582	63168	21043	17409
8853	59083	95137	76538	44155	79419	22359	65206	54941	95992
0274	79932	44236	10089	44373	59914	04146	01419	48575	77822
2805	21149	03425	17594	31427	43374	25473	60982	27119	16060
4971	49055	95091	08367	28381	22199	11865	26201	18570	72803
3606	46497	28626	87297	36568	13786	27475	31254	36050	73736
7286	28749	81905	15038	38338	45445	41059	55142	55585	39829
5670	72111	91884	66762	11428	21067	57238	35352	57741	98761
4262	09513	25728	52539	86806	30302	95327	12849	15795	97479
7375	85062	89178	08791	39342	70040	91385	96436	58982	91281
9483	62469	30935	79270	91986	13351	48321	28357	88526	74396
1206	65749	11885	49789	97081	15564	04716	14594	22363	85700
0908	21506	16269	54558	18395	30987	57657	33398	63053	46792
9944	65036	63213	56631	88862	79172	72764	66446	78864	96004
4963	22581	17882	83558	31960	57875	45228	49211	69755	27896
9286	45236	47427	74321	67351	57146	64665	31159	06980	64709
6075	20517	69980	18293	44047	42826	06974	61063	97640	13433
3375	62251	58871	70174	52372	93929	01836	36590	75052	89475
0487	38794	36079	23362	24902	83585	00414	62851	48787	28447

9473	45950	18225	09899	87377	27548	37516	24343	63046	02081
7703	83717	18913	66371	53629	32982	56455	53129	77693	25022
7612	72738	26995	50933	92936	30104	67126	76656	29347	28492
9042	37595	04931	73622	69902	35240	00818	09136	01952	48442
9609	35653	15970	37681	96326	94031	62209	43740	54102	76895
5354	65770	15365	41422	29451	99321	11331	06838	03818	77063
9452	71674	30260	97303	31002	78236	71732	04704	61384	57343
9867	89294	59232	31776	54919	43108	56592	42467	88801	91280
6719	06144	82041	38332	64452	91058	60958	20706	31929	57422
6970	45907	99238	74547	19704	98172	44346	60430	59627	26471
5747	78956	11478	41195	58135	12523	57345	41246	98416	08669
1838	07526	07985	60714	88627	66682	82517	33150	27368	53375
4361	34534	70169	24805	63215	01056	27534	23085	49602	74391
9278	17082	26997	32295	10894	18730	96197	64483	40364	90913
8124	84721	23544	88548	65626	07794	60475	49666	17578	12830
2025	16908	82841	24060	40285	48883	77154	74973	42096	34934
0326	86370	91949	19017	83846	70171	59431	76033	40076	20292
4855	27029	01542	72443	72302	48830	55029	10371	09963	85857
5434	12124	91087	87800	34870	73151	64463	50058	11468	93553
6800	16781	65977	65946	65728	06571	95934	09132	13746	82514
1233	81409	46773	69135	36170	76609	52553	47508	25775	91309
2933	77341	20839	36126	18311	32138	61197	95476	69442	54574

15 4. Mr. John Henry is studying $v = 3$ teaching methods (A, B, and C) with four different teachers in four different class periods. He wishes to include treatment A twice as frequently as treatments B and C. He knows that there are two sources of variation, i.e., teachers and class periods, so he selects a latin square design.

Construct an F-square square design for Mr. Henry.

Describe the yield equation for this experiment and demonstrate which effects are orthogonal to each other.

The number of replicates on treatment A equals _____

The number of replicates on treatments B and C equals_____

Construct an experimental plan, properly randomized, using the

attached random number table starting at the top of column 4 and
proceeding down the column. (Write down the sequence of random
numbers used.)

10 5. Suppose that Miss Fashion (problem 2) knew that there was a
gradient from the first period to the last period of the experiment. In
this case, she would have used a 5 row (the time periods) by 4 column
(the girls) Youden design.

Construct the 5 row by 4 column Youden design for v = 5 treatments.

Explain why each of various assignable sources of variation are or are
not orthogonal to each other.

Construct a properly randomized experimental plan for the experiment
starting with the top of column 5 and proceeding down the column.

Write down the sequence of random numbers used.

10 6. An experimenter used the following blocked design for v = 3
treatments, (A,B,C) replicated r = 3 times in b = 3 blocks of size k =
3 experimental units each.

Block

1	2	3
A	C	C
B	B	C
A	B	A

Are the occurrences of pairs of treatments within the blocks balanced?
Why or why not?

Illustrate which of the assignable causes of variation are orthogonal to
each other.

What design would have been more efficient than the one used above?

15 7. In Chapter 7 we studied five properties of experimental design after
introducing the subject by stating three characteristics that we desired
for our experimental arrangements. List the five properties in the
upper boxes below and list the three characteristics in the lowest three

boxes. Interrelate the properties and characteristics with each other by use of arrows.

```
 _____      _____      _____
|                |    |                |    |                |
|_____|    |_____|    |_____|

        _____           _____
       |                |         |                |
       |_____|         |_____|
```

Valid estimate of   Reduction of experi-
experimental error   mental error
variation   variation

characteristic (1)   characteristic (2)   characteristic (3)

```
 _____      _____      _____
|            |    |            |    |            |
|_____|    |_____|    |_____|
```

6   8. List the two purposes of randomization.
        i)
        ii)

3   9. One of the statements in each set is correct, circle the correct one.
       Given the following experimental arrangement without any additional
       information,   C   B   A, which of the following
                           A   C   B
                           B   A   C
    statements are correct:
    (a) The experiment design is a latin square design.
    (b) The experiment design is a randomized complete block design.
    (c) Nothing can be said about the experiment design until we know what
    randomization procedure was used.
3   An experimental unit is
    (a) The unit consisting of one comparative experiment.
    (b) The smallest unit to which one treatment is applied
    (c) The smallest unit on which an observation is made.
3   Sources of variation in an experiment refer to
    (a) The different assignable causes of variation and the random error

variation with each item being a source of variation.

(b) The measurement errors that arise in experimentation which are denoted as sources of variation.

(c) The variation caused by the experimenter and denoted as sources of variation.

10. In the problems below, 3 of the 6 statements in each problem are correct. Circle the letter of the correct statement.

3 A placebo is:

(a) A remedy for any kind of headache.

(b) A control in an experiment.

(c) May be, a pharmacologically inactive treatment.

(d) Sometimes used to divide participants in a study into a group that responds and a group that does not respond to the stimulus.

(e) A necessity in all medical experiments.

(f) Not of any value in experimentation.

3 With regard to the conditions under which an experiment is conducted, which of the following statements are correct?

(a) The conditions of the experiment must be such as to allow an expression of differences of treatment responses if in fact they exist.

(b) The conditions of the experiment must conform to ease with which the experiment can be conducted.

(c) For practical inferences, the experiment must be conducted under the same conditions as found in practice.

(d) For theoretical (not practical) inferences only, it is necessary to

(e) All experiments must be conducted in the same manner.

(f) All experiments must be conducted in the laboratory.

3 Relative to the inclusion of a control in the treatment design

(a) A control is included to have a point of reference.

(b) Often more than one control is required.

(c) A control is included to enlarge the experiment.

(d) A control is included to obtain estimates of differences in effects between non-control treatments.

(e) A control is necessary to complete a factorial experiment.

(f) A placebo is a form of a control treatment.

THIRD PRELIM

Points

Miss I. M. Fashion wishes to use three colors (red, yellow, and blue) with two shades (light and dark) of each color; she also wishes to observe the effect of two oils (lanolin and olive oil) in two different amounts (12% and 14% by weight). She wishes to use all possible combinations of these ingredients as listed above in eye shadow preparations. The response to be measured is number of hours the eye shadow stays presentable (a measurement she makes reliably) on Cornell coeds. The experiment is to be conducted during the academic year using $r = 20$ girls such that all v treatments are used on each girl. The differences between treatment periods or experimental units is known to represent only random sampling variation. Therefore, the v treatments are randomly allotted to the v time periods for each girl. There is a real difference between girls with respect to the length of time an eye shadow preparation stays presentable.

3 1. What is the treatment design?

3 2. What is the experiment design?

6 3. Write out the treatment combinations that Miss Fashion will use.

3 4. Suppose that the two treatments involving 14% olive oil in combination with the light shades of blue and yellow were omitted from the experiment because it was known that they gave unsatisfactory results. What would the resulting treatment design be?

3 5. Suppose that eight experimental periods are available for each of 12 girls and that there is a real (assignable) effect of period as well as for girls. What experiment design should Miss Fashion use?

5 6. Given the following treatment design

$a_0\,b_1\,c_1\,d_0$ $a_0\,b_2\,c_1\,d_0$ $a_0\,b_3\,c_1\,d_0$
$a_0\,b_1\,c_2\,d_0$ $a_0\,b_2\,c_2\,d_0$ $a_0\,b_3\,c_2\,d_0$
$a_0\,b_1\,c_3\,d_0$ $a_0\,b_2\,c_3\,d_0$ $a_0\,b_3\,c_3\,d_0$

can the experimenter estimate effects for a and d? Why or why not?

What is an alternative way to write the above treatment design?

10 7. Given the following data and solutions for effects

level of factor a	level of factor b 0	1	effect of factor a
	estimated interaction effects		
1	$ab_{10} = 2$	$ab_{11} = -2$	$a_1 = 3$
2	$ab_{20} = -2$	$ab_{21} = 2$	$a_2 = -3$
effect of factor b	$b_0 = 4$	$b_1 = -4$	$\bar{y}_{..} = 10$

Show how to compute the responses Y_{ij} given the above effects

$Y_{10} =$ _____ $Y_{11} =$ _____

$Y_{20} =$ _____ $Y_{21} =$ _____

Note that the sum of the estimates of main effect levels in the above 2 ×
2 factorial add to zero and that the estimated interaction effects add to
zero across rows or across columns. Is this true for any p × q factorial?
        yes   no (circle one).

10 8. Show how to compute the remaining main effects and remaining
interactions for the following data for a 2 × 3 factorial:

Level of factor a	Level of factor b b_0	b_1	b_2		sum	mean
a_0	3	4	5		12	4
a_1	1	8	15		24	8
sum	4	12	20		36	
mean	2	6	10			6

$a_0 = -2$ $a_1 =$ _____

$b_0 = -4$ $b_1 =$ _____ $b_2 =$ _____

$ab_{00} = -3,$ $ab_{01} = 0,$ $ab_{02} = -3,$ $ab_{10} = -3$

$ab_{11} =$ _____

$ab_{12} =$ _____

Given the same means in the above table what numbers would need to replace numbers 3, 1, 5, and 15 in the above table in order to have all interaction effects equal to zero?

6   9. The following plan involves 8 treatments with r = 6 blocks in which the treatments have been randomly allotted to the 8 experimental units within each block. Treatments 1, 2, 3, and 4 represent one level of a factor, say color, and treatments 5, 6, 7, and 8 represent the second level of this factor. For the second factor treatments 1 and 8 represent one level, treatments 2 and 5 represent a second level, treatments 3 and 7 represent a third level, and treatments 4 and 6 represent a fourth level.

What is the treatment design?

What is the experiment design?

In the problems below one-half of the statements are correct and one-half are incorrect. Check the correct statements in problems 10-12.
3   10. Which of the following statements pertain to a factorial treatment design?
a) Involves two or more factors.
b) Involves one or more factors.
c) Involves a single factor with equally spaced levels.
d) Involves two or more levels of each factor being studied.
e) Is useful only for discussion purposes in statistics classes.
f) May have the levels of one factor equally spaced and the levels of the remaining factors unequally spaced.

3   11. Relative to a biological assay or bioassay treatment design check the correct statements:
3   a) It relates to all laboratory experiments.
b) It relates to experiments involving reactions on living material.
c) It relates to estimation of potency of materials and relates to identification of constitution of materials.
d) It does not include analytical assays which estimate the potency of a test preparation relative to a standard.
e) It includes slope ratio assays and parallel line assays as two special types.
f) It is used only in human drug experiments.

5   12. In selecting a treatment design involving one or more factors which
    of the following conditions are pertinent?
    a) The nature of the experimenter.
    b) The level of factors not varied in the experiment.
    c) The specification of factors to be varied.
    d) The range over which the factors are to be varied.
    e) The experiment design.
    f) The form of the material to be used in varying the levels of the
    factors.
    g) The sample survey design.
    h) The number and spacing of the levels of each factor studied.
    i) The ease with which the results can be explained.
    j) Whether or not the levels of all factors can be ordered.

4   13. Define bioassay.

5   14. Define relative potency and illustrate how it is estimated in slope
    ratio and a parallel-line treatment design.

5   15. Define a simple random sampling procedure and illustrate your
    definition with a population of $N = 5$ objects A,B,C,D,E and with a
    sample size of $n = 2$ objects.

A student researcher, Mr. I. C. E. K. Facts, wishes information on several
characteristics, say Y_1, Y_2, Y_3, Y_4, Y_5, on the population of coeds at
Cornell University during the Spring term 1968. The population of
undergraduate coeds at Cornell University live in one of the following
residence units (no girl will be classified as living in two of the following
residence units) within each of two groups:

Non-sorority (group 1)	Sorority (group 2)
1. Comstock - A	13. Alpha Epsilon Phi
2. Comstock - B	14. Alpha Phi
3. Clara Dickson - V	15. Delta Delta Delta
4. Clara Dickson - VI	16. Delta Gamma
5. Balch - 1	17. Delta Phi Epsilon
6. Balch - 2	18. Kappa Delta
7. Balch - 3	19. Kappa Kappa Gamma
8. Balch - 4	20. Phi Sigma Sigma

9. Risley
10. Mary Donlon - 1
11. Mary Donlon - 2
12. Other

21. Pi Beta Phi
22. Sigma Delta Tau

Let character Y_1 be the proportion of outside area of stone (brick, mortar, and stone) to other materials in the residence unit.

Let character Y_2 be the number of girls living in the residence unit. Let Y_3 be the hair color of a coed. Let Y_4 be the cumulative grade point average of a coed.

Let Y_5 be the attitude of girls toward dating boys who have waxed, handle-barred, red mustaches like the one Mr. Facts was wearing.

Mr. Facts does not have the time nor the resources to perform a complete census. Consequently, he decides to sample the population. He randomly selects six non-sorority and five sorority residence units. He then randomly selects ten girls from each of the 11 selected residence units.

For character Y_1, Mr. Facts photographs the structures and then measures the area of stone to non-stone materials making up the outside area. For character Y_2, he counts the number of girls living in each room of the residence unit. For characters Y_3 and Y_4, he determines hair color and cumulative grade point average only on his sample of ten girls from each of 11 residence units. He does likewise for character Y_5.

2 17. What is the sampling design for character Y_1?

2 18. How should he have obtained information on character Y_2?

2 19. What is the sampling design for character Y_3?

2 20. What is the sampling design for character Y_4?

2 21. What type of allocation of sample size was utilized within each of the 11 selected residence units?

2 22. What is the sampling design for characters Y_3 and Y_4 within each stratum?

2 23. What type of allocation of sample sizes was utilized for characters Y_1 and Y_2?

2 24. How should the sampling design as described above be changed in order to obtain proportional allocation of sample size for characters Y_3, Y_4, and Y_5?

2 25. Suppose that the ten girls selected in each unit were not randomly selected, but were selected because Mr. Facts had dated them. He believed that they were representative of the population sampled. What type of sampling design would this be for characters Y_3, Y_4, and Y_5?

2 26. Suppose that Mr. Facts selected the girls to be interviewed for characters Y_3, Y_4, and Y_5 as in question 25 but he did this because it simplified his procedure. What type of sampling design would this be?

2 27. Suppose that Mr. Facts did not select the ten girls within each residence unit as described above, but selected them as random except with the proviso that there be two freshmen, two sophomores, two juniors, two seniors, and two others to make up the sample within each residence unit. What type of sampling procedure would this be?

2 28. Suppose that an alphabetical list of the girls names was available in each of the 11 selected residence units. Mr. Facts decided to number the names from 1 to 1500 (there were 1500 girls in the 11 selected residence units). He selected a random number between 1 and 12, which was 3. He then took the 3rd, 15th, 27th, 39th, 51st, 63rd, etc., names on the list.
   What type of sampling design would this be?

2 29. State one question that you think should have been on this prelim.

FINAL EXAMINATION (250 points)

An experiment was conducted to compare the relative speeds of three calculating "machines" for computing the sum of squares of a set of numbers. To do this 21 sets of five numbers each were obtained. The 21 sets were divided into seven groups of three sets each. The three sets in each group possessed a common characteristic. For example, one group of numbers involved only numbers between 1 and 14, a second group involved numbers between 15 and 28, a third group involved numbers between 29 and 42, a fourth group involved numbers between 43 and 57, a fifth group involved numbers between 58 and 71, a sixth group involved numbers between 72 and 85, and the seventh group involved numbers between 86 and 99. Three calculating machines, the three treatments, were to be compared for relative time in seconds for computing sums of squares. The three calculating machines were C = the Curta (hand operated), M = Monroe (electrically operated), and H = human (self operated). The observation or response was the number of seconds required to obtain the sum of the squares of five numbers. The three sets of numbers in each group were randomly allocated to the three treatments. The experiment as conducted and the observations (time in seconds) are given below:

	Set 1	Set 2	Set 3	Set 4	Set 5	Set 6	Set 7
	C - 23	M - 16	C - 20	H - 52	H - 56	M - 33	H - 38
	H - 4	H - 16	M - 15	C - 29	C - 38	H - 62	C - 31
	M - 15	C - 28	H - 38	M - 21	M - 26	C - 40	M- 21
Mean	14	20	27	34	40	45	30

Treatment means: H = 38 seconds = 266 / 7
                    C = 31 seconds = 217 / 7
                    M = 21 seconds = 147 / 7
Mean of all 21 observations: 30 seconds = 630 / 21
Sum of squares of all error deviations = 1100 = $\Sigma_{i=1 \text{ to } 3} \Sigma_{j=1 \text{ to } 7} e_{ij}^2$

Points For the experiment described above answer the following questions (1-24)

3 1. The experiment design is _____

3 2.   The number of degrees of freedom for error is _____

3 3. Show how to obtain the first error deviation in terms of the numbers above _____

3 4. Show how to compute s_e^2 = estimated variance of a single observation in terms of the numbers above _____

3 5. For the above data the estimated variance of a treatment mean equals
_____

3 6. For the above data the estimated variance of a difference between any two treatment means is _____

3 7. Show how to compute the 90% confidence interval on the difference between any two treatment means above. The tabulated t value required is in the following table:

Degrees of freedom	α = proportion of time a larger value of $\mid t \mid$ is observed				
	0.3	0.2	0.1	0.05	0.01
2	1.39	1.89	2.92	4.30	9.92
8	1.11	1.40	1.86	2.31	3.36
9	1.10	1.38	1.83	2.26	3.25
10	1.09	1.37	1.81	2.23	3.17
12	1.08	1.36	1.78	2.18	3.06
20	1.06	1.32	1.72	2.09	2.84

The 90% confidence interval or interval estimate for the above data is from _____ to _____

3 8. How do the computations in (7) change in computing the 70% confidence interval? _____

3 9. The sources of variation in the above experiment are
_____.

3 10. What effects are orthogonal to each other in the above design?
_____

3 11. The experimental unit is _____

3 12. The type of experiment described above is one of the following (circle appropriate one):
   factorial   comparative   basic   absolute   technical

3 13. In order for the variance of a difference between any two means as computed above to be appropriate the error deviations must be
_____

3 14. The sources of assignable error in the above experiment are
_____

4 15. Blocking or "local control" is one of the properties of an experiment design discussed in this course. Describe how blocking was u s e d   i n   t h e   a b o v e   e x p e r i m e n t .
_____

4 16. Randomization is a second principle of experiment design. Describe how it was used in a design of the above type, and why.
_____

4 17. Replication is a third principle of experiment design which was discussed. Describe how it was used in the above design, and why.
_____

4 18. Orthogonality is a fourth principle of experiment design. Describe how this property applies for the above experiment and tell why it is useful. _____

The seven principles of scientific investigation may be given as:
1. A clear statement of the problem requiring solution.
2. Formulation of a trial hypothesis.
3. A careful, logical, and critical evaluation of the hypothesis.
4. The planning or design of the experimental investigation.
5. Selection of measuring instruments and control of bias and other errors.
6. Complete and critical summarization and interpretation of results in terms of the hypotheses.
7. Preparation of a complete, accurate, and readable report of the investigation.

Describe briefly how each principle was utilized in the above experiment.
4 19. (1st principle)
4 20. (2nd principle)
4 21. (3rd principle)
4 22. (4th principle)
4 23. (5th principle)
4 24. (6th principle)

An experimenter was considering the following two experiment designs for conducting an experiment (questions 25-26):

Design I (schematic arrangement)

Block 1	Block 2	Block 3	Block 4	Block 5	Block 6
A	A	A	A	A	A
B	B	B	B	B	B
C	C	C	C	C	C
D	D	D	D	D	D

Design II (schematic arrangement)

Block 1	Block 2	Block 3	Block 4	Block 5	Block 6
A	A	A	B	B	C
A	A	A	B	B	C
B	C	D	C	D	D
B	C	C	C	D	D

5 25. What properties are possessed by the above two designs?
5 26. What design would you recommend to the experimenter, and why?

An experimenter used eight experimental units (for instance, 8 rats) in his experiment. He grouped the eight experimental units into two groups of four each. He wished to compare three treatments A, B, and C (for example, three nutritional levels) to measure their effectiveness on yield (for example, increase in weight). He randomly allotted two of the experimental units to treatment A and one each to the other two treatments B and C in each group. Thus treatment A was applied to four experimental units and treatment B and C were applied to two experimental units each. (Questions 27-29):

5 27. What is the experiment design _____

10 28. Show whether or not the treatments are balanced with respect to occurrence with each other in the groups.

10 29. Show whether or not treatment and group effects are orthogonal. Given the following set of numbers for questions numbers 30 to 34:

     1, 2, 8, and 9

3 30. The arithmetic mean is computed as _____

3 31. The median is computed as _____

3 32. The mode(s) is _____

3 33. The harmonic mean is computed as _____

3 34. The geometric mean is computed as _____

3 35. Given the following frequency distribution locate the relative positions of the mean, median and mode.

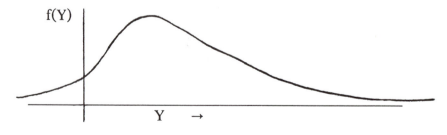

Complete the following statements:

3 36. The _____ is an appropriate average when dealing with rates or prices.

3 37. The _____ is an appropriate average when dealing with observations which increase in a geometric or exponential manner.

3 38. Suppose that we have the population of a city in 1950 and 1960, that we wish to estimate the population size in 1955, and that birth and death rates stay constant (not equal) and that immigration equals emigration. The appropriate average for estimating the 1955 population would be _____

3 39. The difference between the largest member in a sample and the smallest member is the _____

3 40. Another measure of variation which averages the squared deviations between the observations and the arithmetic mean is

_____

3 41. If an event either occurs or does not occur in n trials and if the two events are independent and neither includes the other in any way, then the random variable is said to have a _____ distribution.

3 42. In the above distribution we are concerned with a success or failure of the event whereas in the _____ distribution we are concerned with the number of times an event occurs and not at all with the number of times it fails.

10 43. Using the 99% confidence interval for the binomial distribution as given on the following figure, construct the 99% confidence intervals for the following pairs of values.

(Copy of Figure 12.6 p. 512)

Estimated proportion equals x / n	Sample size	lower limit	upper limit
p = 0.2	n = 50		
p = 0.8	n = 50		
p = 0.4	n = 10		
p = 0.4	n = 100		
p = 0.4	n = 1000		

3 44. Circle the correct statement below. One is correct and the other is incorrect.

The probability is 99% that the true proportion π falls in the above 99% confidence intervals.

On the average, 99% of confidence intervals such as those above will include the true proportion π.

5 45. Suppose that the following results were obtained from a 3 × 2 factorial experiment:

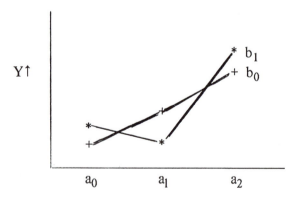

Show with a dashed line what responses Y would be necessary in order to have zero interaction effects.

5 46. Compute the main effects and interaction effects for the following 2 × 2 factorial treatment design.

| Level of factor a | Level of factor b | | Mean |
	b_0	b_1	
a_0	2	16	14
a_1	16	8	12
mean	14	12	13

Give the names for the following graphs (questions 47-51):

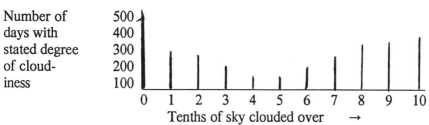

Probability of the stated number of defectives

Number of defectives in sample of 4 items →

Number of defectives in random samples of four items drawn from a population containing 10% defectives.

2 47. _____

Number of days with stated degree of cloud-iness

Tenths of sky clouded over   →

Distribution of cloudiness at Greenwich. Based on data given by Gertrude Pearse (1928) for month of July 1890-1904 excluding 1901) and quoted in M. G. Kendall, *Advanced Statistics Vol. 1.* Note tendency for sky to be either very clear or very cloudy.

2 48_____

Author
clinic   Resurrection mortality →
_____0_____8

Finsterer |_____5% of 202 cases (1929)
Vienna |
Balfour |_____5% of 200 cases (1934)
Rochester |
Truesdale |_____ 6% of 17 cases
Fall River | (1922)
Hanssen |_____ 7% of
Christiana | 51 cases (1923)

Lowest reported resection mortality percentages taken from a statistical
report by Liningston and Pack: End Results in the Treatment of Gastric
Cancer (Paul B. Hoeber, Inc., New York).

2 49. _____

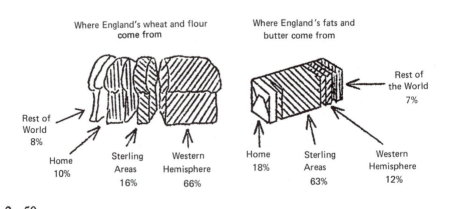

2 50. _____

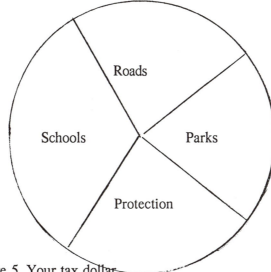

Figure 5. Your tax dollar

2 51. _____

2 52. The following graph appeared in a paper. What comments are relevant to the graph? _____

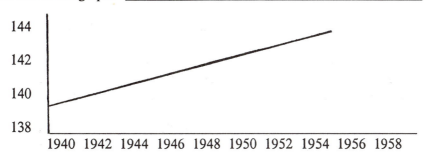

2 53.   What comment is relevant for the following graph?

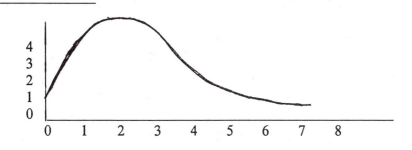

2 54. Circle the appropriate graph for presentation given that Y and X
are defined:

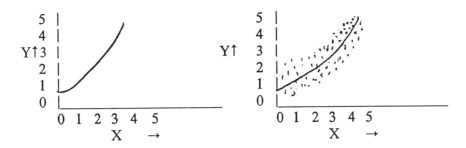

A list of N households in an area is available to a marketing researcher,
Mr. John Doe, for the purpose of conducting a survey. The measurement
on a household, say Y_1, for the first character has no known assignable
error other than a common effect and a random error. Mr. Doe wishes to
draw a random sample of n households and to obtain the measurement Y_1
on each household. (Complete the following statements including type of
allocation for questions 55-59):

6 55. What sample design is appropriate for the above situation?

6 56. If Mr. Doe wishes to obtain measurements Y_2 for a second
character on all persons living in the n selected households, what is the
name of the sample survey design for the second character?

6 57. If Mr. Doe randomly selects one individual from among those
living in a household to obtain the Y_2 measurements, what is the name
of the sample survey design?

6 58. For a third character, the measurement Y_3 on a household is to be
obtained on each of the n households; it is known that the area
containing the N households can be sub-divided into four sub-regions
which are relatively homogeneous. Mr. Doe selects n_1 observations
from sub-region one, n_2 observations from sub-region two, n_3
observations from sub-region three, and n_4 observations from sub-
region four such that $n_1 + n_2 + n_3 + n_4 = n$. Then, the households
are randomly selected within each sub-region. What is the name of this

design for the third character?

6 59. For the fourth character, measurements Y_4 are to be made on a random sample of individuals residing in the selected households in 58 above. What is the name of this sample survey design?

5 60. Define a simple random sample.

Miss I. M. Fashion wishes to use four materials (nylon, rayon, cotton, silk) with two gauges (light and heavy) of each material; she also wishes to observe the effect of two colorings (flesh and white) in two different amounts (12% and 14% by weight). She wishes to use all possible combinations of these items as listed above in stocking samples. The response to be measured is number of hours the stocking stays "wearable" (a measurement she makes reliably). The experiment is to be conducted during the academic year using r = 20 girls such that all v treatments are used on each girl. The difference between treatment periods or experimental units is known to represent only random sampling variation. Therefore, the v treatments are randomly allotted to the v time periods for each girl. There is a real difference between girls with respect to the length of time a sample stocking stays wearable.

2 61. What is the treatment design?

2 62. What is the experiment design?

4 63. Write out the treatment combinations that Miss Fashion will use.

2 64. Suppose that the four treatments involving light gauge silk were omitted from the experiment because it was known that they gave unsatisfactory results. What would the resulting treatment design be?

2 65. Suppose that only 8 experimental periods are available for each of 32 girls and that there is a real (assignable) effect of periods as well as for girls. What experiment design should Miss Fashion use?

8 66. Miss I. M. Fashion was conducting an experiment on length of time different colors of eye shadow stayed presentable which was a measurement made by her. She used v = 5 preparations of eye shadow

which only differed in the color component. She was using five different time periods for which she knew that there were only random variations and no known gradients. Therefore, she decided to use r = 4 different girls as the blocks and to use a randomized complete block design.

Describe the yield equation for this experiment when the effects are additive and enter the equation in linear form.

Which of the assignable effects are orthogonal to each other?

Construct the experimental plan using the following set of random numbers and describe how you obtained the plan.

3780	28391	05940	55583	81256	38175	38422	64677	80358	52629
5325	05490	65974	11186	15357	21805	10371	95812	84665	74511
8240	92457	89200	94696	11370	75517	82119	09199	30322	33352
2789	69758	79701	29511	55968	19195	92261	44757	98628	57916
7523	17264	82840	59556	37119	77869	08582	63168	21043	17409

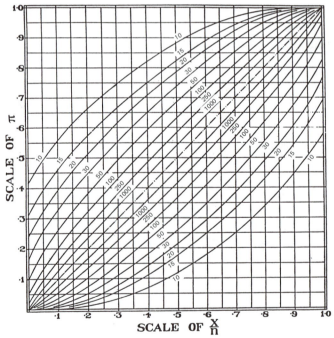

Figure 12.6. Confidence belts for π (Confidence Coefficient = 0.99)

APPENDIX II

EXAMINATIONS FOR ONE SEMESTER AND ADDITIONAL PROBLEMS

The three examinations given below were constructed so that a large proportion of the examination could be computer graded. The use of multiple choice questions lends itself to computer grading. A "standard answer sheet", form D51 7183B (which allows up to five choices), put out by the Optica Scanning Corporation together with a Number 2 pencil has been used successfully for this purpose. In addition to saving a considerable amount of grading time for the instructor, many diagnostics are available such as, e.g., the number of students selecting each of the five choices, a ranking of the grades from highest to lowest, etc..

With large numbers of students in a class, a large amount of the instructor's time can be devoted to grading examinations. In order not to let this happen, the computer, our robot, should be utilized. The author realized something had to be done after spending 60 hours grading one Prelim and this lead him to using the above form for a large proportion of each of the Prelims. It was deemed desirable to have a part of the examination which was not multiple choice in order that the author could ascertain how his students were comprehending the material presented and to become acquainted with a student's progress. This procedure allowed the author to grade 100 students' papers in only six to eight hours. One additional advantage accruing from using multiple choice questions is that much more material can be covered on an examination than when using discussion type questions.

FIRST PRELIM

Questions 1 to 85 are multiple choice. Write your answers on the mark sense sheet. Be certain to mark your name and your I.D. number on the sheet as well as putting your name on this prelim. Questions 86 to 90 need to be answered in the spaces indicated. Each question from 1 to 85 has a value of one point.

1. An error of measurement is made up of the following components
   a) bias plus random error
   b) assignable causes plus non-assignable causes
   c) mistakes plus blunders plus foolish errors plus prejudices
   d) treatment design plus experiment design
   e) physical measurements plus chemical measurements

2. Variability in a set of measurements consists of the following sources of variation (select answer from choices in question 1).

3. A fact or number derived from all members of the population and used to describe a characteristic about a universe or population is a) statistic, b) hypothesis, c) parameter, d) universal fact, e) inferential reasoning

4. A number or fact derived from a sample and used to estimate a population characteristic is (select answer from choices in question 3).

5. A precise statement of fact about a tentative explanation of a population characteristic is a
a) law,   b) proposition,   c) theory,   d) statistic,   e) hypothesis

6. A statement of fact about a population characteristic that has been verified beyond all reasonable doubt is (select answer from choices given in question 5).

7. The type of scientific investigation that introduces conditions which may or may not be present in the general population is
   a) a survey   b) an experiment   c) a scientific method
   d) observational study   e) re-search

8. The orderly quest for new scientific knowledge is known as
   a) Statistics   b) inference   c) creativity
   d) steps in an investigation   e) research

9. A single entity or phenomenon under study in an experiment is known
as   a) a point of interest   b) a treatment   c) observational unit
          d) an experimental unit   e) a comparative experiment

10. An investigation involving the comparison of two or more treatments

is known as   a) a comparative experiment   b) a treatment design
c) an absolute experiment   d) an experimental design
e) a re-search design

11. The statement, "On the average, there are 20 immigrating fruit flies in each room of Warren Hall each week", is known as
a) a valuable statistic   b) an unknowable statistic
c) a meaningless statistic   d) an eccentric theory statistic
e) an absurd statistic

12. "Of the 1986 Spring Term Graduating Class of Cornell Seniors, 31.9% have health deficiencies even though 18.1% had some form of medical treatment and medication" is an example of (select answer from choices in question 11).

13. The selection of treatments for a proposed experiment is
a) an experiment design   b) a comparative experiment
c) a treatment design   d) an absolute experiment
e) an arrangement scheme

14. The arrangement of the treatments in an experiment is (select answer from choices in question 13).

15. "In Sorority S, two mice were observed in the pantry room between 7 p.m. and 11 p.m.; it is estimated that there are 6(2) = 12 mice in the pantry on a given day since there are six four-hour periods in a 24-hour day; also, since there are 50 rooms in Sorority S, it is estimated that there are 50(12) = 600 mice in Sorority S." This statement is an example of
a) meaningless statistics   b) eccentric statistics
c) unknowable statistics   d) preposterous statistics
e) far-fetched statistics built up from a dubious cluster

16. The survey design that allows selection of the most readily accessible or more easily enumerated part of the population is known as
a) probability   b) judgement   c) accessible
d) convenience   e) representative

17. Based on the surveyor's knowledge and expertise, a sample is selected as representative of the population. This type of design is known as (select answer from choices in question 16).

18. The following are three possible definitions for a simple random sample (srs)
  i) A srs is one in which all sampling units in a population have an equal chance of being selected.
  ii) A srs is one in which all possible samples of size n have an equal chance of being selected.
  iii) A srs is one in which all sampling units have an equal and independent chance of selection.
Select the correct answer:
  a) i   b) i, iii   c) ii, iii   d) i, ii, iii   e) i, ii

19. When the probability of selecting each sampling unit in a sample of size n from the N units of a population is known, we say this is
  a) a probability sample   b) an area sample   c) a population sample
  d) a simple random sample   e) an unknown sample design

20. If one first divides the population into sub-populations or groups and if one then selects a simple random sample from each group, the sample survey design is
  a) stratified-simple random sample   b) sub-population sample,
  c) group sampling   d) cluster-simple random sample   e) divided

21. If the population consists of sub-populations, or groups, if one selects a simple random sample of the groups, and if one obtains a simple random sample within each of the selected groups, the sample survey design is known as (select answer from the choices in question 20).

22. If the sampling units of the population are in clusters or groups which are areas of land, and if a simple random sample of areas is obtained, the sample survey design is known as
  a) cluster   b) group   c) land   d) stratified   e) area

23. Given that the sampling units of a population have a natural ordering and given that every kth element of the population is selected after a random starting point between 1 and k, this sample survey design is
  a) natural sample design   b) ordered sample
  c) convenience sample   d) quota sample,
  e) every kth element with a random start sample design

24. Given that the fraction of items selected from each sub-group is a constant, this type of allocation is
a) equal   b) proportional   c) fractional   d) constant   e) unknown

25. Given that the number of items selected from each sub-group is constant, this type of allocation is known as (select answer from choices in question 24).

26. If one obtains a simple random sample within all clusters or groups of a population, the sample survey design is
   a) an area-simple random sample   b) a simple random sample,
   c) a probability sample   d) a cluster-simple random sample,
   e) a stratified-simple random sample

27. A list or a description of every sampling unit in a universe is known as a
   a) sampling unit   b) list   c) probability sample
   d) descriptive statistic   e) sampling frame

28. The smallest unit used in obtaining a sample survey design is
   a) an observational unit   b) smallest unit   c) sampling unit,
   d) experimental unit   e) probability element

29. The smallest unit on which a record or measurement is made is known as (select answer from choices in question 28).

30. The term used to define repeatability of measurements is
   a) on target   b) accuracy   c) precision   d) reliability,
   e) error of measurement.

31. The term used to define repeatability plus bias is (select answer from choices in question 30).

32. The process of drawing conclusions about population characteristics from a part to the whole of the universe is known as one of the answers a to e. To illustrate, suppose that in a representative sample (of size 2,000) of U. S. women ages 16-24, 30% expressed interest in playing competitive basketball. We conclude from these data that 30% of the entire U. S. population of women ages 16-24 have an interest in playing competitive basketball.

a) conclusion making   b) deductive inference   c) generalizing
d) inductive inference   e) the scientific method

33. The process of reaching a conclusion based upon a major and a minor premise is illustrated below and is known as (select answer from choices in question 32).

Example: Major premise: All living things require water.
             Minor premise: A dog is a living thing.
             Conclusion: Therefore, a dog requires water.

34. The U. S. Bureau of Labor Statistics gave the following definition of an average family in the U. S. An average family is composed of a husband 38 years old, a girl age 12, a boy age 8, and a wife (no age specified) with an income of $9,131. Since the surveyors believe this definition, they decide to select at random 100 families who fit this definition. Since this is an average family, the results from this small sample should portray the characteristics of the U.S. population. The type of sample survey design is
a) a simple random sample   b) a probability sample
c) a convenience sample   d) a quota sample   e) a judgement

Suppose that our universe or population consists of all individuals living in dormitories owned and operated by Cornell University during a specified time period. A simple random sample of students is obtained in each dormitory. The size of the sample in each dormitory is a constant percentage of the number of students in a dormitory. N is the total number of individuals sampled. (Questions 35-39)

35. The sampling unit is
a) dormitory   b) a room   c) student   d) time period
e) constant percentage

36. The name of the sample survey design is
a) simple random sample   b) cluster-simple random sample
c) judgement-simple random sample
d) stratified-simple random sample   e) dormitory sample design

37. The type of allocation is known as
a) equal   b) unequal c) constant   d) proportional   e) unknown

38. The chance that any individual is selected is

a) unequal   b) equal   c) constant d) random e) proportional

39. For every possible sample of size N, the chances of selecting any particular sample of size N is
    a) equal   b) unequal   c) proportional
    d) depends upon value of N   e) unknown

Suppose that there are 7 classes of students at Cornell University (freshman, sophomore, junior, senior, graduate, extramural, and part-time). In addition, sex is considered to give 14 classes of students. The sub-populations are the various age groups (16, 17, 18, etc.) within each class. A simple random sample of 10 students is selected from each and every age group within each of the 14 classes.(Questions 41-43)
40. The sampling unit is
a) a class b) sex c) an age group d) a student e) a student's opinion

41. The name of the sample survey design is
    a) stratified-simple random sample b) cluster-simple random sample
    c) simple random sample d) stratified-stratified-simple random sample,
    e) quota sample of ten students per group

42. The procedure described results in a simple random sample
    a) for the universe   b) of age groups   c) of classes,
    d) of students   e) of students within each age group

43. The type of sampling for age groups is
    a) unknown   b) a simple random sample   c) a census
    d) quota   e) judgement

The individuals on a list of patients from Clinic Z are designated as males and females. Using a random starting point, every 20th name of males and a 5% random sample of females are selected from the list.
44. The sample survey design is   a) a combination of designs
    b) stratified - 5 % simple random sample for females and every 20th name for males   c) judgement   d) simple random sample   e) quota

45. The type of allocation is
a) variable b) equal c) proportional d) unknown e) disproportionate

46. The information desired is the number of visits for each patient. The

sampling unit is
a) number of visits   b) a visit of a patient   c) an illness
   d) a patient   e) sex

47. The observational unit is (select answer from choices in question 46).

48. The list of patients is known as the
   a) sample   b) probability sample   c) sampling procedure
   d) descriptive statistics   e) sampling frame

49. If, instead of designing the survey as above, suppose our surveyor is instructed to obtain the number of visits for 50 males and 50 females. The sample survey design is
   a) quota   b) convenience   c) judgement
   d) simple random sample   e) proportional

50. Suppose that the surveyor decides to obtain number of visits per patient for those patients who enter Clinic Z on February 27 because that is her day off. This type of sample survey design is (Select answer from choices in Question 49).

51. If a 2% sample is taken from the entire list in such a manner that the random selection of one name is unrelated to the selection of another name and if each name has an equal chance of being selected, the sample survey design is
   a) stratified   b) proportional   c) convenience
   d) simple random sample   e) non-probability

52. Personal interviewing is a form of
   a) indirect questioning   b) direct questioning
   c) obtaining an anonymous response   d) used for job applications
   e) all of the preceding.

53. Four of five methods listed below are for obtaining anonymous responses to questions. Which one of the following is not?
   a) secret completion of questions on unmarked paper
   b) randomized response   c) balanced incomplete block
   d) supplemented block   e) telephone interviews.

54. The form of obtaining anonymous responses when a specific question is always included with one or more other questions and a total for all questions in the set is given is known as
   a) randomized response   b) randomized complete block
   c) balanced incomplete block d) supplemented block
   e) a randomized form of one of the preceding.

55. Two types of "control charts" used most frequently in quality control are
   a) quality and product   b) quality and mean   c) quality and ex-bar
   d) averages and range   e) quality and range.

56. The type of sampling used to determine whether or not a lot of material should be rejected or purchased at a lower price is known as
   a) price sampling   b) rejection sampling   c) acceptance sampling
   d) quality assurance sampling   e) product improvement sampling.

The following is a report of an investigation. Sentences are numbered from 1-26. Start with question 57 and proceed to question 62.

1. The following material was reported in the Journal of Irreproducible Results in an article entitled "The Data Enrichment Method' by Henry R. Lewis; it also appeared in Operations Research, 1957, volume 5, pages 551-4. 2. Related literature on the subject is scattered and confused. 3. Mr. Lewis' objective was "to show" how a small amount of data can be used to replace a large amount of data in certain types of investigation. 4. It has been known to those interested in psycho-physical phenomena that a man's tendency to flip a coin in such a way that when it lands he will be faced by an Indian's head rather than by a buffalo's tail increases with the altitude at which the experiment is performed. 5. The effect is small but a vast number of trials conducted on Mount Everest, from base to summit, have shown that the effect indeed exists. 6. With due respect to the hardy band of men who invested so many years and Sherpas in this effort, it is of interest to show how the same result can be obtained by one man with no more athletic ability than that required to climb a flight of stairs and no more equipment than an unbiased nickel. 7. Our advantage over the pioneers in this field lies, of course, in our knowledge of the "enriched-data method". 8. We shall consider a set of stairs with ten levels in the Pentagon, say, and number them in the order of their increasing altitude.

9. The experimenter has been trained in the art of coin flipping, especially with buffalo nickels. 10. The experimenter climbs the stairs, slowly, and at each level flips a coin ten times and records a head as a success and a tail as a failure. 11. The results of an actual test are summarized in Table 3.

TABLE 3
RAW DATA: COIN EXPERIMENT

Step number	Number of success	Number of failures
1	4	6
2	5	5
3	7	3
4	4	6
5	6	4
6	5	5
7	6	4
8	6	4
9	3	7
10	4	6

12. Now the results of Table 3 are not conclusive, and the altitude effect may be present but it is not evident, at least to a naive observer. 13. The method of data enrichment to be used makes use of our prior knowledge; we know how the experiment would have come out if it had been performed at a lower level; for example, there were 6 failures to detect a head on step 10, and a failure to detect a head would also have resulted in a failure to detect it at steps 9, 8, 7, etc. 14. Therefore, we compute "virtual failures" at step 9 to be 6 + 7 = 13, at step 8 to be 6 + 7 + 4 = 17, etc.; we do likewise for successes, that is a head on step 1 would have been a head on step 2, etc. 15. The "virtual successes" are computed in the same way using the altitude principle; for example, the virtual successes on step 2 would be 4 + 5 = 9, on step 3 it would be 4 + 5 + 7 = 16, etc.. 16. These results are presented in Table 4, where it is shown how the data enrichment method can be used to increase the size of the experiment. 17. A glance at Table 4 shows that the altitude principle, which was skulking almost unnoticed in the raw data of Table 3, has been fully brought forth by the data enrichment method. The probabilities in Table 4 are shown in Fig. 1 to further emphasize the point. 18. It might be mentioned in passing that the altitude effect in the Pentagon appears to be 105 times as large as that found in the Himalayas. 19. Whether this is a temperature effect, a geographical effect, or the result of psychical factors as yet unknown

should be the object of further study. 20. A final remark on the strength

TABLE 4

ENRICHED DATA: COIN EXPERIMENT

Step number	Number of virtual successes	Number of virtual failures	Probability of throwing a "head"
1	4	50	4 / 54
2	9	44	9 / 53
3	16	39	16 / 55
4	20	36	20 / 56
5	26	30	26 / 56
6	31	26	31 / 57
7	37	21	37 / 58
8	43	17	43 / 60
9	46	13	46 / 59
10	50	6	50 / 56

Probability of coin falling "heads"

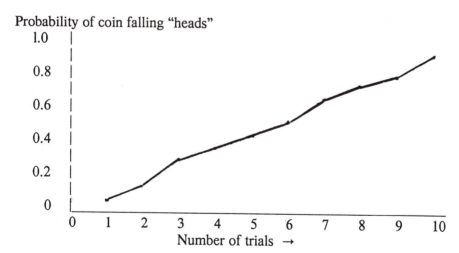

FIGURE 1. ALTITUDE EFFECT IN THE PENTAGON

and weakness of the method is in order. 21. As mentioned earlier, its strength lies in its breadth of applicability, and the method is as pertinent to experiments in classical physics as it is to experiments in psychical phenomena. 22. In short, the method will give new meaning to data quite without regard to the status of the hypothesis used to increase the sample size. 23. Despite its evident power, however, the method requires further study. 24. Its principal shortcoming is that before the enrichment process can be started, some data must be collected. 25. It is quite true that a great deal is done with very little information, but this should not blind one to the fact that the method still embodies the raw-data-flaw. 26. The ultimate objective, complete freedom from the inconvenience and embarrassment of experimental results, still lies un-attained before us.

Steps in the conduct of an investigation

1. A clear and precise statement of problem
2. Formulation of hypotheses
3. Evaluation of problem and hypotheses
4. Design of the investigation
5. Calibration and standardization of measuring instruments
6. Conduct of the investigation
7. Summarization and interpretation of data
8. Final report

From the above complete questions 57 to 62.

57. Sentences 1, 2, 4, and 5 refer to step number
   a) 1   b) 2   c) 3   d) 4   e) 5

58. Sentences 3 and 26 refer to step number
   a) 1   b) 2   c) 3   d) 4   e) 5

59. Sentence 6 refers to step number
   a) 3   b) 4   c) 5   d) 6   e) 7

60. Sentences 11, 12, 13, and 14-21 refer to step number
   a) 3   b) 4   c) 5   d) 6   e) 7

61. Sentence 9 refers to step number
   a) 3   b) 4   c) 5   d) 6   e) 7

62. Sentence 10 refers to step number
   a) 3   b) 4   c) 5   d) 6   e) 7

Questions 73 to 85 pertain to the survey described below:
    The following is a version of a Gannett News Service article about a survey, which appeared in the Ithaca Journal on February 14, 1978. It has been re-written somewhat in order to fit the purposes of an examination question. Sentences 7 and 9-20 were copied verbatim from the article.

Sentences are numbered and you are to associate a sentence(s) associated with each of the steps in conducting a survey.

N.Y. Poll: NYC Backed; Upstate Cool to Idea

1. The New York poll covered the entire state electorate. 2. Lists of registered voters with a telephone listing were used for a telephone survey. 3. 1,000 interviews of people who would answer the questions were to be obtained. 4. Any registered voter who answered the telephone was to be interviewed. 5. If the telephone was not answered or if the person answering was not a registered voter, the next number on a supplementary list was selected. 6. Registered voters, representing a cross-section of the statewide electorate, were telephoned. 7. The main question asked of those polled was: "If New York City is faced with bankruptcy and financial help isn't available from any other source, would you favor or oppose the state government giving whatever financial aid is necessary for the city to avoid bankruptcy?" 8. Other questions asked related to political party affiliation, race, what would their answer be if aid to NYC meant higher state income taxes, etc.. 9. State aid for the city, if all else fails, was favored by a margin of 64-27 of the registered voters contacted for the New York Poll. 10. This margin dropped to 51-39, however, when those who favored it were asked if they would be willing to pay higher taxes to finance the aid. 11. While upstaters were deadlocked, 43-43, on the question of aid to the nation's largest city, people who live in the boroughs and the New York City suburbs favored last-ditch state aid by a wide margin. 12. Not surprisingly, New York City dwellers endorsed it overwhelmingly, 83-12. 13. In the northern suburbs, including Westchester, Rockland and Dutchess, it was backed, 66-30. 14. This was somewhat lower than Long Island's endorsement of the aid, 75-14. 15. Close to 40 percent of New Yorkers wouldn't favor a state bailout of New York City if it would cost them more in state taxes. 16. Those who labelled themselves "liberal" gave the aid question a 70-20 vote of support, a margin which dropped to 60-32 among those who called themselves "conservative". 17. Blacks, who supported it 85-11, dropped off sharply, however, when asked if they'd go along with higher taxes to pay for the extra state aid. 18. On the second question, blacks backed it, 60-30. 19. Support for the aid proposal went across political party lines, but Democrats, 73-19, gave it more support than Republicans, 54-37. 20. It also was endorsed by all age groups, but those in the middle-age, 35 to 49, group gave it its lowest marks, 59-31. 21. Note that when the percentage of "unsure" is included, percentages would add to 100 percent.

Sentence Numbers

63. Definition of population or   a) none   b) 1   c) 2   d) 3   e) 4
    universe to be surveyed

64. Definition of sampling frame
    list or description of indivi-   a) none   b) 1   c) 2   d) 3   e) 4
    duals in a population

65 Definition of information   a) none   b) 1,4   c) 2-3   d) 5   e) 6
   sought in precise terms

66. Determination of whether
    or not information sought   a) none   b) 1   c) 2   d) 3   e) 6-10
    is already available

67. Procurement of pertinent   a) none   b) 1-2   c) 3   d) 4-6   e) 7-8
    information only

68. Determination of sample size,
    observational unit, sampling   a) none   b) 1-2   c) 3   d) 4-6   e) 7
    unit, group of sampling units
    - stratum

69. Sample survey design   a) 1-2   b) 3   c) 4-5   d) 6   e) 7-8

70 Construction of question-   a) none   b) 4   c) 5   d) 6   e) 7-8
   naire or reporting form

71. Training of interviewers,   a) none   b) 7   c) 8   d) 9   e) 10
    recording of answers

72. Determination of what to do
    with not-at-homes ,   a) 3   b) 4,5   c) 6   d) 7   e) 8
    refusal-to-answer , "unable-to-answer", etc.

73. Conduct the survey   a) none   b) 10   c) 11   d) 12   e) 13

74. Summarize and   a) none   b) 9-10   c) 11-13   d) 14-17   e) 9-21
    interpret results
75. Reporting the results   a) none   b) 9-10   c) 11-12   d) 13-14   e) 15-16

76. For the survey for questions 35-39, what are the assignable sources of variation in the population?

77. For the survey related to questions 40-43, what are the assignable sources of variation in the population?

78. For the survey related to questions 44-51, what are the assignable sources of variation in the population?

79. For the survey related to questions 73-85, what are the assignable causes of sources of variation in the results?

80. From the following list of random numbers obtain (describe your procedure)
    (i) a random ordering of the ten numbers, 0,1,2,3,4,5,6,7,8,9.
    (ii) a random ordering of the 16 numbers 0,1,2, ... ,15.
           Set of random numbers

28391	05940	55583	81256	44164
05490	65974	11186	15357	64805
92457	89200	94696	11370	13517
69758	79701	29511	55968	19195
17264	82840	59556	37119	77862

SECOND PRELIM

Questions 1 to 40, inclusive, are multiple choice questions. Indicate the correct answer on the enclosed sheet for mark sense grading. Be certain to print your name and put your I.D. number on the sheet. Also, put your name on Prelim II in the space provided above. Questions 4 to 44 are to be answered in the space provided on Prelim II. Questions 1 to 40 each count 3/2 points, and questions 41 to 44 have the number of points indicated beside the problems. Question 45 is 10-point bonus problem.

A completely randomized design with a control treatment, number 4, and four spray treatments, numbers 1, 2, 3, and 5, with four experimental units per treatment, was conducted on cabbage plants. The response measured, Y_{ij}, was number of loopers on 50 cabbage plants. The data are:

Treatment	Response	Treatment totals $Y_{i.}$	Treatment means
1	11, 4, 4 5	24	24 / 4 = 6
2	6, 4 3 6	19	19 / 4
3	8, 6 4 11	29	29 / 4
4	14, 27, 8, 18, 0, 0	67	67 / 6
5	7, 4, 9, 14	34	34 / 4
Total		173	173 / 22 = mean $\bar{y}_{..}$

The estimated treatment effect for treatment 4, the control treatment, is:
  (A) 67 - 173   (B) 173 / 20 - 67 / 6   (C) 67 / 6 - 173 / 22
  (D) 14 - 27 - 8 + 18   (E) 67 / 4 - 173 / 22

The estimated random error effect associated with the last response for treatment 4, $Y_{46} = 0$, is computed as:
 (A) 173 / 22 - 0   (B) 0 - 173 / 22   (C) 0 - 4 - 9 + 14
         (D) 0 - 67 / 6   (E) 34 / 4 - 14

The number of replications for treatment four is:
  (A) 1   (B) unequal   (C) 6   (D) 22   (E) 4

The sources of variation in this experiment in terms of sample values are:

        (A) true population mean, bias, true treatment effect
        (B) estimated overall mean $\bar{y}_{..}$ and treatment effects
        (C) number of loopers on 50 cabbage plants
        (D) control treatment plus four sprays
        (E) estimated overall mean, treatment effects, and random error

Consider the following experiment on comparing two brands of wine and a standard wine. All samples were identical in color and cooled to the same temperature. Six judges were used and samples were presented to each judge in the order indicated. The judge's mouth was rinsed between each sample. Scores of one to five were given with 5 being the best.

Row = order of presentation	Judge (column)						Order	
	1	2	3	4	5	6	Total	Mean
1	A-3	B-5	A-4	C-5	C-5	B-5	27	4.5
2	C-4	A-2	B-4	A-4	B-5	C-5	24	4.0
3	B-5	C-5	C-4	B-5	A-3	A-5	27	4.5
Judge total	12	12	12	14	13	15	78	$\bar{y}_{..} = 13/3$
Judge mean	4.0	4.0	4.0	14/3	13/3	5.0		

$Y_{.A}$ = total for treatment A = 21 and $\bar{y}_{..A}$ = 21 / 6 = 7 / 2 = 3.5

$Y_{.B}$ = total for treatment B = 29 and $\bar{y}_{..B}$ = 29 / 6

$Y_{.C}$ = total for treatment C = 28 and $\bar{y}_{..C}$ = 28 / 6 = 14 / 3

5. The estimated random error effect for treatment C for Judge 2 in order 3 is e_{32C} equals:
(A) 5 - 4.5 - 4.0 + 13/3   (B) 5 - 4.5 + 13 / 3
(C) 5 - 4.5 - 4.0 - 14 / 3 + 2(13 / 3)
(D) 4 - 4.0 - 4.0 - 14 / 3 + 2(13 / 3)   (E) 2 - 4 - 5 + 4

6. Treatment B effect is: (A) 13 / 3 - 29 / 6   (B) 29 / 6 - 13 / 3
(C) 29 / 6 - 21 / 6   (D) 21 / 6 + 14 / 3 - 2(29 / 6)   (E) 29 - 78

7. A researcher has 16 relatively homogeneous plots of land (experimental units); he wishes to compare a new fertilizer with the one currently recommended. He randomly allots the new one to 8 of the 16 plots and puts the current fertilizer on the remaining. Select the correct answer for design, v = number of treatments and r = number of replicates:
(A) randomized complete block, v = 2, r = 8
(B) completely randomized, v = 8, r = 2
(C) completely randomized, v = 2, r = 8
(D) stratified sample, v = 8, r = 16 (E) stratified sample, v = 2, r = 8

8. A farmer wants his lambs to gain weight as fast as possible in order to have them ready for market sooner. He has two different feeding schemes he wishes to compare. He decides to run an experiment. Among his lot of

100 lambs, he has 10 sets of identical twin lambs. What experiment design
would be appropriate to recommend given that he wishes to use the 10 sets
of twins?
(A) a balanced incomplete block (B) a simple change-over
(C) a completely randomized   (D) a latin 2 × 10 rectangle
(E) a randomized complete block

9. Given that a randomized complete blocks design is to be used, and given
the following set of random numbers where the numbers in the first row
are used for block I, the numbers in the second row are used for block II,
the numbers in the third row are used for block III, and the numbers 1, 2,
3, and 4 are used for treatments A, B, C, and A, respectively. (The
random numbers 1, 2 3, 4 only were used. The numbers 0, 5, 6, 7, 8, 9
and any of the numbers 1, 2, 3, and 4 already used, were discarded.)

4855	27029	01542	72443	72302
5434	21124	91087	87800	34870
6800	16781	65937	65946	65728

I II III	I II III	I II III	I II III	I II III
A A A	A A A	A A A	A A A	A A B
B A A	A A A	B C A	B C C	A A C
A B C	B B B	A A B	A B A	B C A
C C B	C C C	C B C	C A B	C B A
(a)	(b)	(c)	(d)	(e)

The correct randomized plan using the set of random numbers described
above is (a)   (b)   (c)   (d)   (e)

Suppose a researcher was interested in five brands of shoes (m_1 m_2, m_3,
m_4, and m_5 = control). Twenty individuals, constituting a simple-random
sample from a specified population, were used in the experiment. Two
pairs of shoes were to be worn by an individual in one of two
arrangements. For example, for the pair m_1 and m_3, the two arrangements
possible on an individual were:

	1st period	2nd period		1st period	2nd period
left foot	m_1	m_3	left foot	m_3	m_1
right foot	m_3	m_1	right foot	m_1	m_3

One of these two arrangements was randomly selected. The brand of shoes worn for two months on each foot and amount of wear was recorded. The amount of wear on both feet was the response recorded for each treatment (brand) on an individual. The symbols m_1, m_2, m_3, m_4, and m_5 were randomly allocated to the five shoe brands, and every pair of brands of shoes appeared together on two of the 20 individuals. Any given pair of treatments was randomly allocated to an individual.

10. The experiment design is   (A) completely randomized
(B) randomized complete block   (C) 2 × 2 factorial
(D) balanced incomplete block   (E) simple change over.

11. The number of treatments v is (A) 2 (B) 5 (C) 20 (D) 4 (E) 10

12. The number of blocks b is (A) 20 (B) 10 (C) 5 (D) 2 (E) 4

13. The number of replicates r is (A) 5 (B) 10 (C) 8 (D) 4 (E) 20

14. The block size k is (A) 4 (B) 1 (C) 10 (D) 5 (E) 2

15. The number of times that every treatment appears with every other treatment in a block is (A) 0 (B) 2 (C) variable (D) 1 (E) 4

Complete the following plan in order to obtain a plan for a 7 row by 4 column Youden design (questions 16-19)

Stream number	year 1	2	3	4
1	2	1	3	5
2	5	4	6	1
3	1	7	2	4
4	7	6	1	3
5	3	2	6	
6	6	5		
7				

16. The correct number in year 4 and stream 7 is
     (A) 1,   (B) 2,   (C) 3,   (D) 5,   (E) 7

Treatments 1, 2, 3, 4, 5, 6, and 7 represent methods of sampling to determine number and kind of fish present in a stream. The experiment could not be conducted for additional years because of construction on two of the streams.

17. The number of times any given treatment occurs with each of the other four treatments in rows of the above design is
   (A) 9   (B) 1   (C) 2   (D) 3   (E) 4

18. Rows and columns are
(A) orthogonal only   (B) orthogonal and balanced   (C) balanced only,
(D) neither orthogonal nor balanced   (E) confounded

19. Rows and treatments are (select an answer from 18):
   (A)   (B)   (C)   (D)   (E)

20. $Y_{..}$ is equal to (for i = 1,2, ... ,r rows, j = 1,2, ... ,c columns)
(a) $\sum_{i=1 \text{ to } r} Y_{ij}$ (b) $\sum_{j=1 \text{ to } c} Y_{ij}$ (c) $\sum_{i=1 \text{ to } r}\sum_{j=1 \text{ to } c} Y_{ij}$
(d) $\prod_{i=1 \text{ to } r}\prod_{j=1 \text{ to } c} Y_{ij}$ (e) None of the preceding.

21. In order to obtain a reduction in experimental variation (i.e., to have more precision) for estimated treatment effects, it is necessary to have
(A) replication and treatment design   (B) randomization and confounding
(C) fairness and blocking   (D) replication and randomization
(E) replication and blocking.

22. Two main purposes of randomization in experimentation are to obtain
   (A) unbiased estimates of treatment effects and experimental variation
   (B) fairness and efficiency   (C) replication and blocking
   (D) simple random sample and completely randomized design
   (E) orthogonality and balancedness.

23. If more observations are necessary to obtain the same precision for one design as compared to a second, we say design one relative to design two is less:
(A) orthogonal   (B) confounded   (C) efficient   (D) balanced   (E) fair

24. There is no confounding of effects in a design when the following is present:
   (A) balance   (B) replication   (C) orthogonality
   (D) partial confounding   (E) randomization.

25. One or more controls or standards may be included in an experiment in order to have: (A) more treatments (B) a point of reference (C) enough treatments to run an experiment (D) a placebo (E) control treatment.

26. When the percent of fish eggs hatched is a function of temperature and species only we say that the independent factors present are: (A) fish eggs (B) fish eggs and temperature (C) fish eggs and species (D) fish eggs, temperature, and species (E) temperature and species.

An experimenter wishes to compare four treatments, A = no insecticide, B = insecticide B, C = insecticide C, D = mixture of B and C in equal amounts. A simple random sample of 24 trees from an apple orchard is obtained. Each of the four treatments is applied to a simple random sample of six from the 24 trees. Beginning six weeks before harvest and every two weeks thereafter, a simple random sample of 10 apples is collected from each tree and each apple is examined for number of worms (questions 27-30).

27. The experimental unit is (A) 24 trees (B) the orchard (C) a tree (D) sample of 10 apples (E) an apple.

28. The sampling unit is (select answer from 27).

29. The treatment design is called a (A) completely randomized (B) simple random sample (C) single factor (D) 2 × 2 factorial (E) insecticide.

30. The experiment design is called a (A) simple random sample (B) factorial (C) randomized complete block (D) completely randomized (E) stratified.

31. Which of the following statements DOES NOT pertain to a factorial treatment design: (A) involves two or more factors (B) involves two or more levels of each factor (C) involves only a subset of all possible combinations (D) levels of factors may be equally or unequally spaced (E) all possible combinations of levels of the factors involved.

32. In selecting a treatment design involving one or more factors which of
the following conditions ARE NOT pertinent:
(A) specification of factors
(B) specification of range over which factors are to be varied
(C) the number and spacings of levels of each factor
(D) the experiment design to be used
(E) the form of the material to be used in varying the levels of factors.

The following is a field plan for an experiment involving nitrogen (n),
phosphorus (p), and potash (k) fertilizers as they affect the yield (gms) of
wheat where letter means mineral applied and absence of letter means
mineral (n, p, or k) was not applied. (1) means no fertilizer was applied.
The size of the experimental unit was 3 feet × 10 feet.

								Sum	Mean
p	n	np	k	nk	(1)	npk	pk		
18.8	12.2	18.3	15.8	11.4	11.5	19.4	18.9	126.3	15.8
n	nk	pk	npk	p	k	np	(1)		
12.9	7.3	17.4	17.2	19.7	12.0	19.0	15.6	121.1	15.1
nk	np	n	p	(1)	npk	pk	k		
10.7	17.5	10.4	18.0	9.8	16.6	1.5	14.3	114.8	14.4
pk	k	npk	(1)	n	np	p	nk		
18.3	12.6	14.2	12.2	11.4	14.5	16.9	16.1	116.2	14.5
np	(1)	nk	n	pk	p	k	npk		
17.9	12.8	13.3	11.3	16.5	15.6	10.9	16.7	115.0	14.4
k	pk	(1)	np	npk	n	nk	p		
14.9	18.2	12.8	17.1	15.8	9.5	8.9	20.6	117.8	14.7
npk	p	k	pk	np	nk	(1)	n		
19.0	18.9	11.2	17.1	17.9	8.6	10.2	14.5	117.4	14.7
(1)	npk	p	nk	k	pk	n	np		
17.5	20.4	20.8	16.4	16.8	18.5	13.6	23.0	147.0	18.4

| Sum | 130.0 | 119.9 | 118.4 | 125.1 | 119.3 | 106.8 | 116.4 | 139.7 | 975.6 | |
| Mean | 16.2 | 15.0 | 14.8 | 15.6 | 14.9 | 13.4 | 14.6 | 17.5 | | 15.2 |

Treatment Sums (Means)

(1) 102.4 (12.8) p 149.3 (18.7) np 145.2 (18.2) pk 142.4 (17.83)
 n   95.8 (12.0) k 108.5 (13.6) nk 92.7 (11.6) npk 139.3 (17.4)

33. The treatment design for the above is a (A) randomized complete block   (B) 2 × 2 × 2 factorial   (C) fractional replicate (D) 2 × 3 factorial   (E) latin square.

34. The experiment design for the above is a
(A) factorial   (B) incomplete block   (C) diallel crossing,
(D) randomized complete block   (E) latin square.

35. The number of replicates for each treatment is
  (A) 6   (B) 2   (C) 3   (D) 8   (E) 5.

The following plan represents the actual randomized layout for an experiment in a greenhouse on celery plants. The letters f, n, p, and k stand for iron, nitrogen, phosphorous, and potash, respectively. If a letter is present the corresponding chemical was present in the nutrient solution, otherwise it was absent. The different solutions were randomly allotted to the experimental units within each block.

Block 1		Block 2		Block 3		Block 4	
fn	n	pk	f	np	n	nk	f
pk	fk	n	fk	nk	fk	fp	fk
np	fp	np	fp	pk	fp	fn	np
nk	f	nk	fn	fn	f	n	pk

36. The experiment design for the above is a
(A) balanced incomplete block   (B) factorial
(C) randomized complete block   (D) Youden   (E) latin rectangle.

37. The treatment design for the above is a (A) balanced incomplete block   (B) 4 × 8 factorial   (C) diallel crossing   (D) randomized complete block   (E) 1 / 2 replicate of a 2^4 factorial.

A committee studying the honor code on campus is interested in the proportion p of undergraduates who engage in cheating on examinations. To gather information, the committee selects a simple random sample of 300 undergraduates to interview. Each of the 300 students is given a fair die and two cards, which read:
Card I. Did you cheat on any examination you took at Cornell last semester?
Card II. Is the number showing on the die you threw odd?
[Note: even numbers are 2, 4, 6, etc.; odd numbers are 1, 3, 5, etc.]

The student is then told: test the die a few times so you are sure that any face can turn up; then put up the cardboard barrier so the interviewer cannot see the outcome, and throw the die; if the die shows 1, 2, or 3, take card I and respond truthfully "Yes" or "No"; if the die shows 4, 5, or 6, take card II and answer truthfully. Do not show or reveal which card you took (questions 38-40).

38. The above procedure is called
(A) two-questions   (B) indirect question   (C) randomized response
(D) locked-box   (E) block total response.

39. Out of 300 students, how many are expected to answer the question on card II? (A) 50 (B) 100 (C) 150 (D) 250p (E) 250(1 - p).

40. Sixty yeses were obtained, p is estimated to be
(A) 60 / 300 (B) 1 / 15 (C) 1 / 60 (D) 0.3333 (E) 0.1.

Points
(17) 41. Consider the following schematic plans which, when properly randomized, may be used as experiment designs:

Block

I II III	I II III	I II III	I II III
A A A	A A A	A B C	A A B
A A A	B B B	A B C	A A B
A A A	C C C	A B C	A C B
B B B	A B C	A B C	A C B
B B B	B C A	A B C	B C C
C C C	A B C	A B C	B C C
C C C	B C A	A B C	B C C
C C C	C A B	A B C	B C C

Given that
    v = number of treatments which are designated with capital letters,
    b = number of blocks which are designated by I, II, III,
    k = the block size or number of experimental units per block,
    r = number of times a treatment is replicated, and
    λ = number of times a given treatment pair occurs in the design,
answer the following questions (write your answers in the table or circle the correct answer in the last part of the table):

	(i)	(ii)	(iii)	(iv)
$v = ?$ $r_A, r_B, r_C = ?$ $b = ?$ $k = ?$ $\lambda_{AB}, \lambda_{AC}, \lambda_{BC} = ?$				
orthogonal ?	yes, no	yes, no	yes, no	yes, no
balanced ?	yes, no	yes, no	yes, no	yes, no
completely ?	yes, no	yes, no	yes, no	yes, no
confounded ?	yes, no	yes, no	yes, no	yes, no
complete ?	yes, no	yes, no	yes, no	yes, no

Circle the correct answer
The method of blocking for the four designs is   the same   is different.
The method of blocking controls   the same   different   amounts of heterogeneity in the experiment for the four designs.
The block size for the four designs is   the same   different .

(11) 42. The following table, rather than response equations has been prepared to indicate ASSIGNABLE sources of variation in a number of statistical designs. The effects are given in terms of estimated values rather than population parameters. Indicate with a check (x) the appropriate sources of ASSIGNABLE CAUSES for each of the designs listed:

Design	Sample mean	One-way grouping or blocking effects	Second category of blocking effects	Third category of blocking effects	Estimated random error effects	Estimated treatment effects Direct	Carry-over
1. Simple random sample							
2. Cluster or area sample							
3. Completely randomlzed							
4. Randomized							

complete blocks
5. Simple
changeover
6. Latin square
7. Youden
8. F-square
9. Balanced
incomplete blocks
10. Double reversal
or double-change-
over

----

(10) 43. Miss I. M. Fashion planned to conduct an investigation on eye
shadow cosmetics using four different colors (blue, orange, yellow, and
green) and two types of oil (lanolin and soybean oil) at two levels (zero and
1 gm. per 2 gms. of other materials). She planned to use 3 ages of
individuals (16 year old, 24 year old, and 40 year old, and to use all
possible combinations.
    What is the treatment design
    Write out the treatment design for Miss Fashion.

(2) 44. Write out a plan for an F-square design for a 6 × 6 square and for
treatments A, B, and C appearing three times, two times, and once,
respectively, in each row and in each column.

Bonus Problem [10 points-3 parts (a), (b), and (c)]
45. Given the following set of random numbers obtained from a Texas
Instrument SR-51 pocket calculator:

5 3 5 5 3 6 0 8 9 0 5 8 6 8 1 7 6 3 3 9 9 9 3 2 0 5 5 8 3 7 5 0 3 1 8 4 0 0 6
6 5 1 0 3 3 9 7 6 8 2 1 2 7 8 7 6 0 3 7 3 9 9 1 5 2 1 9 3 3 0 1 1 2 7 3 6 7 6
3 1 0 6 6 7 5 2 1 7 6 9 0 0 8 1 3 1 7 8 4 2 7 1 8 6 4 3 0 2 2 2 2 0 3 8 7 6 0
2 0 4 9 3 5 6 4 9 9 8 2 8 0 7 7 2 1 9 6 3 4 9 3 3 3 0 0 8 2 6 0 9 5 0 7 6 1 4
2 3 3 7 5 4 0 9 5 5 2 2

(a) Describe how to and obtain a randomized plan for a four row by five
column Youden design using four different diets for five different animals
in four different time periods. (It is known that these particular diet
effects do not extend beyond the period in which the diet is given.) Give

and describe a response equation in terms of estimated effects for this design of the form used in class.

(b) Describe how to and obtain a randomized plan, using the above set of random numbers, for the $v = 5$ diets on twin sheep, i.e., the block size is $k = 2$, in a balanced incomplete block design. Describe a response equation in terms of estimated effects.

(c) Consider the yields for the following 2×2 factorial treatment design:

Factor a	Factor b 0	1	Mean
level 0	$2 = Y_{00.}$	$4 = Y_{01.}$	3
level 1	$6 = Y_{10.}$	$? = Y_{11.}$	?
mean	4	?	

In order to have zero interaction, the yield of treatment combination $a_1 b_1$ must be

   (A) 0   (B) 2   (C) 4   (D) 6   (E) 8

To have an interaction effect ab_{11} equal to -2, the treatment combination $a_1 b_1$ should be equal to

   (A) 2   (B) 0   (C) 10   (D) 4   (E) 6

THIRD PRELIM

Questions 1 through 48 are multiple choice questions and are to be answered on the attached standard answer sheet. Questions 49, 50 and 51 are to be answered on the examination sheet. Questions 1 to 48 are each worth 1.5 points. Note that the last three questions (49, 50, 51) are worth 10, 10, 8 points, respectively. There are bonus problems at the end of the examination.

The following is a randomized plan involving 6 treatments:

```
1        1        5        2
4        6        4        5
3        2        1        4
5        5        3        3
6        3        2        1
2        4        6        6
```

Treatments 1 and 2 represent one level of factor a; treatments 3 and 6 represent a second level of factor a; treatments 4 and 5 represent a third level of factor a; treatments 1, 3, and 5 represent one level of factor d; treatments 2, 6, and 4 represent a second level of factor d.

1. The treatment design for the above is a
   (A) randomized complete block   (C) fractional replicate
   (B) 2 × 2 × 2 factorial   (D) 2 × 3 factorial
                                    (E) latin square

2. The experiment design for the above is a
   (A) factorial   (C) diallel crossing
   (B) incomplete block   (D) randomized complete block
                                    (E) latin square

3. The number of replicates for each treatment is
   (A) 6   (B) 2   (C) 3   (D) 8   (E) 4

Given the following data (means) for a 4 variety by 3 fertilizer level factorial

| Variety | Fertilizer level(bu / A) | | | Total | Mean |
	low	medium	high		
a	50	59	50	159	53.0
b	45	60	75	180	60.0
c	30	45	60	135	45.0
d	35	60	85	180	60.0
Total	160	224	270	654	
Mean	40.0	56.0	67.5		54.5

4. The main effect for variety a is computed as
   (A) 159 - 654   (B) 159 - 160   (C) 159 / 3 - 654 / 12

(D) 50 - 53.0 - 40.0 + 54.5   (E) 53.0 - 40.0
5. The interaction effect for variety b at the medium fertilizer level is
computed as   (A) 50 - 53 - 40 + 54.5
(B) 60 + 60 - 56 - 54. 5   (C) 60 - 60 - 56 + 54. 5
(D) 60 - 180 - 224 + 654   (E) 60.0 - 54.5

6. The estimated interaction effects for varieties b and d at the medium
fertilizer level are
(A) the same   (B) for d is larger   (C) not able to determine
(D) for b is larger

For the following data, the graph of the
7. A × C interaction is given by (A), (B), (C), (D), (E)
8. B × C interaction is given by (A), (B), (C), (D), (E)
9. A × B interaction is given by (A), (B), (C), (D), (E)
10. A × B × C interaction is given by (A), (B), (C), (D), (E)
11. Zero interaction is given by (A), (B), (C), (D), (E)

The following data were obtained from a 3 × 3 × 2 factorial experiment:

	c_0					c_1		
	b_0	b_1	b_2			b_0	b_1	b_2
a_0	5	15	20		a_0	5	5	5
a_1	25	10	20		a_1	5	30	25
a_2	25	10	20		a_2	5	30	25

From these data the following graphs were drawn:

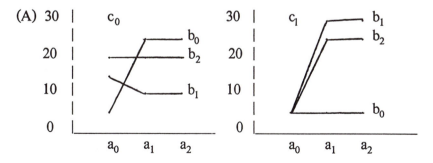

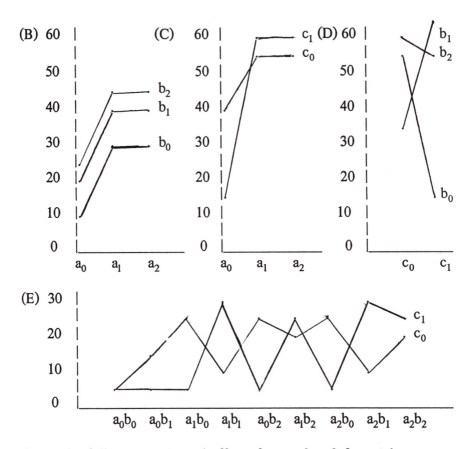

Given the following estimated effects from a 3 × 3 factorial treatment design, one can reconstruct the responses Y_{ij}, $i = 0, 1, 2$, and $j = 0, 1, 2$, from which these effects were estimated:

		level of factor b_j			
level of factor a_i	b_0	b_1	b_2		$b_0 = -6$
		(ab_{ij} effects)			$b_1 = -1$
					$b_2 = 7$
a_0	-3	1	2		$a_0 = -9$
a_1	3	1	-4		$a_1 = 9$
a_2	0	-2	2		$a_2 = 0$
					$\bar{y}_{..} = 20$

12. The one correct equation in the following five is:
   (A) $Y_{00} = 20 - (-6) - (-9) - (-3)$
   (B) $Y_{10} = 20 + 9 - 6 - 3$
   (C) $Y_{11} = 20 - 9 - 1 + 1$
   (D) $Y_{01} = 20 - 9 - 1 + 1$
   (E) $Y_{22} = 20 + 0 + 7 - 2$

13. A clinical trial is an experiment involving (A) humans (B) animals (C) clinics (D) medical treatment only (E) nutritional treatment only

14. The study of the distribution and determinant of health in a population of humans in order to elucidate and quantify the relationships between health and potential risk factor in order to prevent future health problems is known by the following name: (A) medical (B) environmental (C) longitudinal (D) clinical (E) epidemiology

Given the following data from a bioassay experiment:

Standard drug			New drug	
dose	response		dose	response
0	60		0	60
1 m / l	70		5 units / l	100
2 m / l	80		10 units / l	140

The following graphs were made from the data

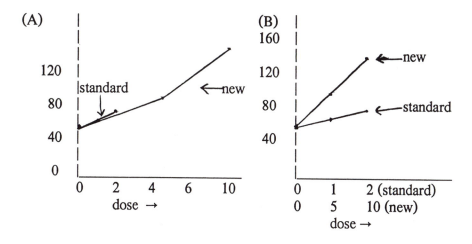

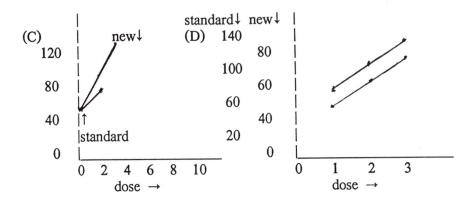

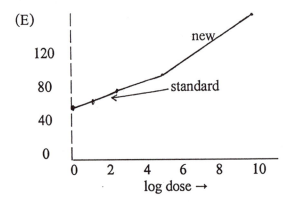

For the above bioassay data

15. The correct graphical representation is
        (A)   (B)   (C)   (D)   (E)

16. For the above, the type of bioassay treatment design is
    (A) parallel line (B) slope-ratio   (C) direct assay
    (D) analytic assay   (E) ratio of means

17. For the above, the relative potency R is computed as
    (A) {(140 - 60) / (10 - 0)} / {(80 - 60) / (2 - 0)}
    (B) (140 - 60) /(80 - 60)   (C) (2 - 0) / (10 - 0)
    (D) 140 / 80   (E) log(140 - 60)(2 - 0) - log(10 - 0)(80 - 60)

18. For a parallel line bioassay treatment design the response equations are

given by (S = standard, N = new)

(A) $Y_S = a + bX$ and $Y_N = a + RbX$

(B) $Y_S = \log(a + bX)$ and $Y_N = \log(a + RbX)$

(C) $Y_S = a + b \log X$ and $Y_N = a + b \log R + b \log X$

(D) $Y_S = abX$ and $Y_N = abRX$

(E) $\log Y_S = a + b \log 1 / X$ and $\log Y_N = a + b \log R / X$

19. The following is a treatment design for v entities, where c denotes that combination is present and blank denotes that it is absent:

		1	2	3	4	...	v
entity				entity				
1			c	c	c	...	c
2				c	c	...	c
3					c	...	c
4						...	c
...							
v - 1							c
v							

The above is NOT one of the following five. Indicate the incorrect one.

(A) matched pairs

(B) balanced incomplete block for v, $b = v(v - 1) / 2$, $k = 2$, and $r = v - 1$

(C) round-robin tournament (D) factorial (E) diallel-crossing plan

20 If a fair coin is flipped 4 times, 3 of the possible 16 sequences of outcomes are (i) T H H T, (ii) T T T T, (iii) H H H H. Given the condition that the coin must land in the order given above, which of the following statements is correct?

(A) Sequence (i) is more likely   (B) Sequence (ii) is more likely.

(C) Sequence (iii) is more likely

(D Sequences (i) and (ii) are not equally likely

(E) All three are equally likely

21. A nursery school class of 10 children is playing the game musical chairs. It is started with 8 chairs. In how many ways can 10 children fill these eight chairs?

(A) 90   (B) 45   (C) 8   (D) 10   (E) 10! / 8!

22. Given that a student knows the answers to question 1 to 21 on this examination but knows nothing about questions 22 to 24. He randomly selects one of five letters A, B, C, D, E, as the correct answer for these questions. The probability that he obtains a perfect score on these 24 questions is:

(A) $(1/5)^3 \, 21/24$ (B) $1 - (1/5)^3$

(C) $(1/5 + 1/5 + 1/5)(21/24)$ (D) $3/24$ (E) $1^{21}(1/5)^3$

23. Suppose that questions 1 to 4 on a test had 4 choices and questions 5 to 10 had three choices. Given that a student randomly selects an answer, the probability that he obtains all correct answers on these 10 questions is:

(A) $(1/4)^4(1/3)^6$ (B) $2(1/4)^4 3(1/3)^6$ (C) $(1/4)^4 + (1/3)^3$

(D) $1 - (1/4)^4 - (1/3)^6$ (E) $4(1/4) + 6(1/6)$

24. Suppose that student Zee is one who never really wakes up until two hours after getting out of bed. The color of socks is unknown to this student until after the first class attended. Suppose that the sock drawer is in disarray and contains only four red socks, three green socks, and two yellow socks arranged in random order. The student randomly draws two socks from the drawer. The student's probability of wearing two socks of the same color on a specified day is:

(A) $(2/72)(12/72)(6/72)$ (B) $4/9 + 3/9 + 2/9$

(C) $(4/9)(3/9)(2/9) = 8/243$

(D) $(4/9)(3/8) + (3/9)(2/8) + (2/9)(1/8) = 20/72 = 5/18$

(E) $1 - (4/9)(3/9)(2/9) = 235/243$

25. In problem 24, suppose that student Zee has selected a green sock on the first draw. The probability of wearing green socks that day is:

(A) $2/8$ (B) $(4/9)(3/9)$ (C) $3/8$

(D) $1/9$ (E) $(1/4)(4/9)(2/9)$

26. In problem 24, the total number of orderings of colors of two successively drawn socks is computed as:

(A) $9!/8!$ (B) $9!/7!(4)(3)(2)$ (C) $4! \, 3! \, 2!$

(D) 9   (E) 6

27. The word permutation has eleven letters, ten of which are different. How many orderings (permutations) of these eleven letters are there

(A) 11! / 6! 5!   (B) 11! / (11 - 11)!   (C) 11!
(D) 11! / 2!   (E) 11(10)

28. A balanced incomplete block design for v = 5 treatments in blocks of
k = 3 and with each pair of varieties occurring together λ = 3 times
requires the following number of blocks:
     (A) 5! / 3!   (B) 5(3)   (C) 5(3)(2)
     (D) 5! / 3!(5 - 3)!   (E) 5(4)

29. Which of the following is NOT a measure of central tendency
(location):
    (A) arithmetic mean   (B) average absolute deviation
    (C) median   (D) trimmed mean   (E) mode

30. In a uni-modal distribution with a long tail to the left, the location of
the arithmetic mean, the median, and the mode is:
    (A) at the same point on the abscissa
    (B) mode, median, and mean in increasing values on the abscissa
    (C) mode, median, mean in decreasing values on the abscissa
    (D) mode, mean, median in increasing values on the abscissa
    (E) mode, mean, median in decreasing values on the abscissa

Given the following form of a distribution of 180 observations

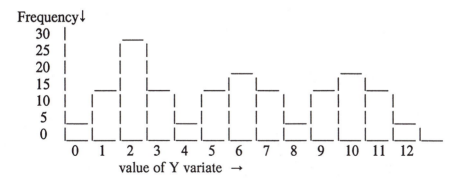

Cumulative 5   20   50   65   70   85   105 120 125 140 160 175 180
frequency

31. The median is located at
   (A) 2   (B) 4   (C) 7   (D) 10   (E) 6

32. The frequency of the modal value is
   (A) 15   (B) 30   (C) 2   (D) 3   (E) not given

33. The mode(s) is(are) located at
   (A) 2   (B) 30   (C) 6   (D) 10   (E) 6 and 10

34. The arithmetic mean lies in the interval
   (A) 6 to 7   (B) 5 to 6   (C) 0 to 2   (D) 7 to 9   (E) 9 to 12

35. The minimal and maximal values in the above distribution are
(A) 5 and 30   (B) 0 and 12   (C) 0, 4, 8, 12, and 2   (D) 2 and 10
(E) not given

36. The minimal and maximal frequencies in the above distribution are
(A) 5 and 15   (B) 0 and 12   (C) 15 and 30   (D) 2 and 5   (E) 5 and 30

Suppose that the variable Y can take on the following six values with
relative and cumulative frequencies as indicated:

	Frequency	
Y	Relative	Cumulative
0	5 / 100	5/ 100
1	10 / 100	15 / 100
2	15 / 100	30 / 100
3	20 / 100	50 / 100
4	25 / 100	75 / 100
5	25 / 100	100 / 100

A graphical representation of the above data (2nd column) is:
↓Relative frequency

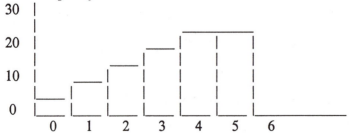

37. The mode(s) for these data is(are)
   (A) 25   (B) no mode   (C) 4 and 5   (D) 15   (E) 4

38. The median for these data is
   (A) 4   (B) 3   (C) 3.5   (D) 20   (E) 25, 25

39. If relative frequency is given as the interpretation of probability, the
probability that Y is at most 2 is computed as Pr(Y ≤ 2) =
   (A) 70 / 100
   (B) 1 - 15 / 100
   (C) 1 - 30 / 100
   (D) 5 / 100 + 10 / 100 = 15 / 100
   (E) 5 / 100 + 10 / 100 + 15 / 100 = 3 / 10

40. Relative to parameters for various distributions one of the following
statements is correct. Which one?
(A) The normal distribution is determined by knowing the variance σ^2 and
the sample size n.
(B) The Poisson distribution is determined by knowing the mean λ and the
sample size.
(C) The binomial distribution is determined by knowing the true
proportion π.
(D) The negative binomial distribution is determined by knowing the
sample size n when the κth success occurs and the proportion of successes
π, that is, the parameters are κ and π.
(E) None of the above.

41. Which of the following statements is correct?

(A) The binomial and negative binomial distributions are continuous-
valued.
(B) The binomial, Poisson, and negative binomial are discrete valued.
(C) The Poisson and normal are discrete valued.
(D) The Poisson and normal are continuous valued.
(E) None of the above.

The following graphs were constructed from the 1982 class data for eye
color (own data).

Eye color 1982 class

(A)	Black	Brown	Grey	Green	Blue	Total
No. of students	3	17	1	11	17	49
% of students	6	35	2	22	35	100

(B)

			(C)
blue	brown	brown	
35%	35%	blue	
	grey	green	
black	2%	black	
6%	green 22%	grey	

```
(C)
brown  |_____(17)
blue   |_____(17)
green  |_____(11)
black  |_____(3)
grey   |__(1)
       |_____
       0              10              20
```

42. Figure (A) is called a
    (A) three-way array   (B) graph   (C) one-way tabular array
    (D) per cent table   (E) pictograph

43. Figure (B) is called a
    (A) circular array   (B) pie chart or graph   (C) per cent circle
    (D) eye-color circle   (E) ideaograph

44. Figure (C) is called a
    (A) vertical line graph   (B) horizontal line graph
    (C) frequency polygon   (D) histogram   (E) color graph

Suppose the following data were available to construct a graph:

Annual Inflation Rates for Last 9 Years*

year	Y	ln Y	year	Y	ln Y
1973	3.0	1.10	1977	7.0	1.94
1974	3.5	1.25	1978	11.0	2.40
1975	4.0	1.37	1979	15.0	2.71
1976	5.5	1.70	1980	19.0	2.94
			1981	24.0	3.18

*ln equals natural logarithm, e.g., $e^{1.10} = 3.0$, $e^{1.25} = 3.5$, etc.

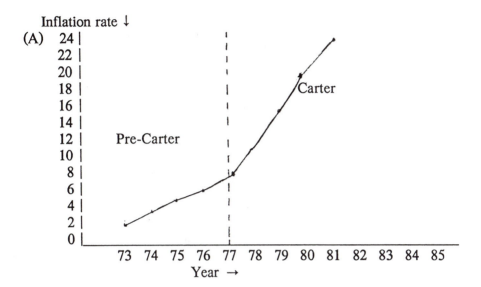

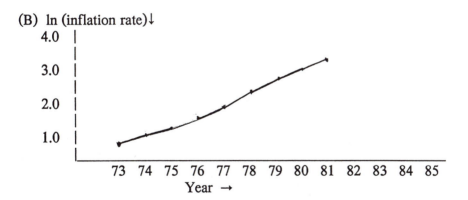

The following graph is constructed for Carter by a "supporter"

(C)

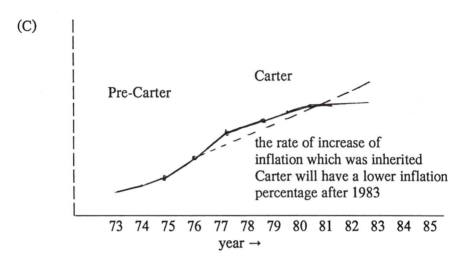

Pre-Carter

Carter

the rate of increase of
inflation which was inherited
Carter will have a lower inflation
percentage after 1983

73 74 75 76 77 78 79 80 81 82 83 84 85
year →

(D) Inflation rate (%)   (E) Inflation rate (%)

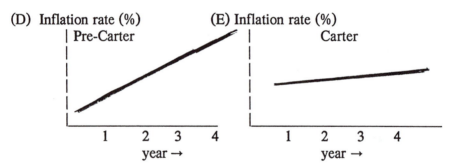

Pre-Carter   Carter

1   2   3   4   1   2   3   4
year →   year →

45. Figures (C), (D), and (E) above have
    (A) axes labelled correctly
    (B) clearly presented ideas
    (C) axes not clearly labelled or not labelled
    (D) useful ideas for science
    (E) ideas for truthful dissemination of information

46. To depict the actual percentages of pre-Carter and Carter inflation
rates use figure

    (A)   (B)   (C)   (D)   (E)

47. A change in slope of inflation rates is correctly depicted in figure
(A) (B) (C) (D) (E)

48. A bar chart and a histogram are identical when
(A) the abscissa is in scaled units of measurement
(B) the bars are horizontal   (C) the bars are vertical
(D) they are always the same
(E) the abscissa is not in scaled units of measurement

49. Completely randomized design (or stratified sample random sample survey design); data obtained for the response equation
Y_{ij} = mean + ith treatment effect + a random deviation of jth experimental unit from the treatment mean
= bias + μ + τ_i + ε_{ij}, or in terms of sample values
$Y_{ij} = \bar{y}_{..} + t_i + e_{ij}$, i = A, B, C, D, j = 1, 2, ..., r_i, r_A = 2, r_B = 4, r_C = 5, r_D = 5, were as follows:

Treatment

	A		B		C		D	
	Y_{Aj}	e_{Aj}	Y_{Bj}	e_{Bj}	Y_{Cj}	e_{Cj}	Y_{Dj}	e_{Dj}
	Y_{A1}=11	2	Y_{B1}=1	-3	Y_{C1}=10	0	Y_{D1}= 6	?
	Y_{A2}=7	-2	Y_{B2}=3	-1	Y_{C2}=10	0	Y_{D2}= 4	?
			Y_{B3}=7	3	Y_{C3}=10	0	Y_{D3}= 6	?
			Y_{B4}=5	1	Y_{C4}=12	2	Y_{D4}= 7	?
					Y_{C5}=8	-2	Y_{D5}=17	?
Total	$Y_{A.}$=18	0	$Y_{B.}$=16	0	$Y_{C.}$=50	0	$Y_{D.}$=40	?
Mean	$\bar{y}_{A.}$= 9		$\bar{y}_{B.}$= 4		$\bar{y}_{C.}$=10		?	

Mean for treatment D is computed as: $\bar{y}_D$ =_____

The overall mean $\bar{y}_{..}$ is computed as: $\bar{y}_{..}$ =_____
Median for treatment D is equal to_____
The mode(s) for all the data is(are)_____
Show how to compute range for treatment D above_____
Show how to compute the residuals for treatment D above where question-marks appear: _____
The degrees of freedom for a pooled or average variance =_____

Show how to compute the average (or pooled) sample error variance s_e^2 for the above data: _____

Given that the value of s_e^2 = 13.4, show how to compute the following:

Standard deviation = s_e = _____

Standard error of a mean of 5 observations for treatment D =

$s_{\bar{y}_{D.}}$ _____

Standard error of a difference for means $\bar{y}_{A.}$ = 9 and $\bar{y}_{C.}$ = 10 =

$s_{\bar{y}_{A.}-\bar{y}_{C.}}$ = _____

Average coefficient of variation is computed as:_____

50. Randomized Complete Block Design: the data obtained were as follows for a response equation in terms of sample values equal to Y_{ij} = $\bar{y}_{..}$ + $(\bar{y}_{i.} - \bar{y}_{..} = t_i) + (\bar{y}_{.j} - \bar{y}_{..} = b_j) + e_{ij}$ where i = A, B, or C and j = 1,2,3,4 (effect of washing and removing excess moisture from turnip greens on ascorbic acid content):

			Block				Treatment	
Treatment	1	2	3	4	5	Totals	Means=$\bar{y}_{i.}$	Effects=t_i
	mgm/100 gms ascorbic acid

			Block				Treatment	
Treatment	1	2	3	4	5	Totals	Means=$\bar{y}_{i.}$	Effects=t_i
	mgm/100 gms ascorbic acid							
Control C	150	87	97	50	175	559	112	?
Blotted B	57	389	118	168	109	841	168	?
Air dried A	117	272	175	130	154	848	170	?
Block total $Y_{.j}$	324	748	390	348	438	2248		0
Block mean $\bar{y}_{.j}$	108	249	130	116	146	-	150 = $\bar{y}_{..}$	
Block effect b_j		?	?				0	

Show how to compute the b_j values (block effects) and the t_i (treatment effects) for the above data (where question marks appear):_____

The range of yields for treatment A is _____

The median for treatment A is_____

The table of estimated error deviations or residuals has been partially completed. Show how to compute the deviations e_{B2} and e_{C4}:

_____

_____

Treatments	Blocks (e_{ij} values)						Sum within rounding errors
	1	2	3	4	5		
C	80		5		67		0
B	-69		-30		-55		0
A	-11	3	25	-6	-12		0
Sum	0	0	0	0	0		

The number of degrees of freedom for the sample variance for residuals =

_____

The pooled or average sample variance is computed as s_e^2 = _____

Given that the numerical value of s_e^2 is equal to 6591 the sample standard deviation is computed as _____

The sample standard error of a difference between two treatment means is computed as $s_{\bar{y}_{i.}-\bar{y}_{i'.}}$ _____

A summary table of differences between all possible pairs of treatment means is:

	$\bar{y}_{A.} = 170$	$\bar{y}_{B.} = 168$	$\bar{y}_{C.} = 112$
$\bar{y}_{C.} = 112$	58	?	-
$\bar{y}_{B.} = 168$	2	-	-

How does one compute the difference for the remaining pair:_____
Show how to compute the average coefficient of variation._____

51. Field lay-out showing yields in kilograms per experimental unit of 4 wheat varieties (A,B,C,D) arranged in a 4 × 4 latin square design:

Row	Column				Total	Mean
	1	2	3	4		
1	C = 10.5	D = 7.7	B = 12.0	A = 13.2	43.4	10.85
2	B = 11.1	A = 12.0	C = 10.3	D = 7.5	40. 9	10. 225
3	D = 5.8	C = 12.2	A = 11.2	B = 13.7	42.9	10.725
4	A = 11.6	B = 12.3	D = 5.9	C = 10.2	40.0	10
Total	39.0	44.2	39.4	44.6	167.2	
Col mean	9.75	11.05	9.85	11.15	-	10.45

		Variety totals and means			
		A	B	C	D
Total		48.0	49.1	43.2	26.9
Mean		12.0	12.275	10.8	6.725

The estimated residuals are:

			Column				
Row		1	2	3	4		Sum
1		0	-0.025	-0.075	0.10		0
2		-0.25	-0.375	0.325	0.30		0
3		-0.50	0.525	-0.475	0.45		0
4		0.75	-0.125	0.225	-0.85		0
Sum		0	0	0	0		0

Sum of squared residuals $\{0^2 + (-0.25)^2 + (-0.5)^2 + 0.75^2 + ... + (-0.85)^2\} = 2.72$

Show how to compute e_{44C} =_____

Show how to compute the mean for treatment A, using the original yields

---

Show how to compute:

Degrees of freedom for error variance:_____

Sample error variance s_e^2 =_____

Standard error of a difference between any pair of means, say for

treatments D and A means, say. $\bar{y}_{A.}$ = 12.0 and $\bar{y}_{D.}$ = 6.725, that is

$s_{\bar{y}_{..A} - \bar{y}_{..D}}$ = _____

Standard error of a treatment mean $s_{\bar{y}_{.j}}$ =_____

Coefficient of variation =_____

Bonus Problem (3 points)

52. A, B, C, D, and E are playing poker with a 32 card deck composed of four suits (hearts, diamonds, spades, and clubs) with cards 7, 8, 9, 10 Jack, Queen, King and Ace in each suit. Show how to compute

(A) Probability that A gets one pair

(B) Probability that A gets 5 cards of the same suit (a flush)

(C) Probability that A gets 4 Aces and B gets 4 Kings

Bonus Problem (2 points)

53. The following data are for the 1982 class and are weights in pounds. Construct a stem and leaf diagram for the data.

141	143	122	170	145	180	130	120	155
135	198	127	110	159	165	132	148	185
187	110	123	155	130	112	160	165	155
140	150	161	160	115	140	165	148	157
153	125	165	155	162	127	133	150	178
119	171	181	142					

FINAL EXAMINATION

Points as indicated. There are 200 points.
(Points)
(29) 1. Indicate whether or not the following statements are true (T) or false (F).
_____An absolute experiment and a comparative experiment differ in that the former is for one phenomenon or treatment while the latter is for two or more treatments.
_____A treatment design and an experiment design are the same in the respect that both involve treatments.
_____A theory and a hypothesis differ in that the former is a statement of fact whereas the latter relates to a fairly well-documented postulate.
_____Empirical and analytic investigation differ in that the latter has to do with data whereas the former does not.
_____Deductive and inductive inference are two terms that may be used interchangeably.
_____The use of "phony" statistics and the phony use of statistics always deal with using soundly based statistics in a false manner.

_____An assignable cause is of interest to an experimenter whereas an un-assignable cause of variation is not.

_____Precision and accuracy have nothing to do with bias.

_____A census involves a complete enumeration of a population with respect to some characteristic.

_____Populations are always infinite.

_____An important reason for sampling arises from the fact that frequently it is too expensive to carry out a census.

_____Quota sampling is a probability sampling method.

_____One important aspect of probability sampling is that it allows the calculation of reliability of the estimate from the sample itself.

_____In constructing strata for stratified samples one should make the items within a stratum as unlike as possible.

_____If a universe is grouped into sub-groups and a sample of items from some of the sub-groups is collected, the sample design is known as a stratified sample.

_____If all items in a universe have an equal chance of being selected, the resulting sample is known as a simple random sample.

_____Simple random sampling is a method of sampling based on expert judgment.

_____An important objective of statistics is to draw conclusions about the population from information obtained from a sample.

_____A competent statistician is able to draw meaningful conclusions from any kind of data that is presented to him.

_____In a table of random numbers each of the integers 0, 1, 2, 3, 4, 5, 6, 7, 8, 9 occurs about 1 / 10 of the time.

_____Drawing numbers out of a hat and selecting numbers from a table of random digits are roughly equivalent operations.

_____It is impossible to combine the technique of stratified sampling with that of cluster sampling.

_____In area sampling the clusters are always geographical units.

_____A census is a partial enumeration of the population while a sample is a complete enumeration.

_____A statistic is an estimate of a population parameter.

_____A parameter is calculated from all sampling units in a population.

_____In a cluster sampling procedure, we have to select samples from each of the selected clusters.

_____The smallest unit used in the selection process of the sample is called a sampling unit.

_____Judgment sampling is an unbiased method of sampling.

(9) 2. Complete the following statements:

Selection of the most readily accessible portion of the population, or the portion most easily enumerated, is known as_____

Based on experience and judgment, a sample of items is selected to represent the population. This is known as_____

A procedure which gives each of the possible samples an equal chance of being selected is known as a_____

A sampling procedure for which the probability of selection of each possible sample is known is termed a_____

If the population is divided into sub-populations and if a simple random sample is drawn from each of the sub-populations or groups, the sample design is a_____

If a simple random sample is made of the sub-populations or groups in a population and then a simple random sample, or a complete enumeration, is made in each of the selected sub-populations, the sample design is a_____

If the elements of a population have a natural ordering and if every kth ordered element is selected following a random start within the first k items, the sample design is an_____

If the sub-populations or sub-groups of a population are areas and if a simple random sample of these areas is obtained, the sample design is known as_____

If the fraction of items sampled in a sub-group relative to the sub-group size remains constant for all sub-groups sampled, this is known as _____
_____ allocation.

(15) 3. A large bakery delivers loaves of baked bread to 210 supermarkets on a daily basis. The owner wants to know
    (i) what type of display of the bread is used in stores (response Y_1),
    (ii) the condition of loaves of bread on the shelf at the end of a
    shopping day (Y_2), and
    (iii) the time that deliveries are actually made at stores (Y_3).

The owner of the bakery decides to take a simple random sample of 10 stores to determine Y_1 and Y_3. The owner takes every 20th loaf of bread after placing the loaves in order and after taking a random starting point between 1 and 20 to measure response Y_2 in the selected 10 stores.

From the above description fill in the appropriate design number and type
of allocation (equal = E, proportional = P, or neither = N) used for the
three responses Y_1, Y_2 and Y_3.

Response	No. of design	Type of allocation	Design
Y_1			1. Simple random sample
			2. Randomized complete block
			3. Cluster-simple random sample
Y_2			4. Stratified-simple random sample
			5. Cluster-every 20th item after a random start
Y_3			6. Stratified-every 20th item after a random start
			7. Convenience
			8. Judgement
			9. Quota

Suppose that in each store the owner obtained response Y_2 from the first
ten loaves of bread most readily accessible. The survey design is a _____
_____design.

Suppose instead the owner decided to sample 5 loaves of rye, 5 of wheat,
and 5 of white bread in each of the seven randomly selected stores. The
sample survey design is a _____ design.

What are the assignable and non-assignable causes of variation in response
Y_2 in terms of sample statistics?

(6) 4. For the situation described below, indicate whether or not the
statement is true by circling either yes or no for statements (a) to (f):

A student in a statistics class drew 40 numbers from a random numbers
table. He then paired the first and second numbers, the third and fourth
numbers, etc. to obtain 20 pairs. He called the first number of a pair X
and the second one Y. The Y values were plotted against the X values to
obtain the following graph:

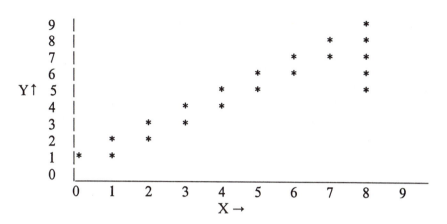

After connecting the points with line segments, the graph appears to indicate a relationship between Y and X. Should the student conclude that:

(a) The apparent relationship is a real phenomenon   yes   no

(b) He should never use a random numbers table because it is misleading?   yes   no

(c) Occasionally extreme samples are obtained which indicate a relationship when none is present?   yes   no

(d) Relationship studies and sampling should never be combined?   yes   no

(e) It is a good idea to understand the nature of the experiment when interpreting experimental data?   yes   no

(f) One should investigate the student's method of pairing and plotting observations before drawing conclusions?   yes   no

(27) 5. Indicate whether the following statements are true (T) or false (F):

____If the relative proportions of the number of times the v treatments occur together in a block remains constant for all blocks, the treatment effects are said to be orthogonal to the block effects.

____If the number of times each and every pair of treatments occurs together in r of the b blocks is constant, the treatments are said to form a balanced arrangement in the b blocks.

____We can liken the completely randomized design to simple random sampling and the randomized complete block design to stratified simple random sampling before the treatments are applied.

____Blocking is an effective device to reduce experimental error, according to one of the Fisherian principles of experimentation.

_____A treatment effect is the increase or decrease in the size of the response over what would have been observed had the treatment not been applied.

_____An experimental unit is always the smallest unit of observation.

_____In a randomized complete block design all treatments must be included an equal number of times.

_____The randomized complete block and the balanced incomplete block designs are experiment designs with two-way control of heterogeneity in the experimental material.

_____The simple change-over design and the Youden design are Latin rectangle designs.

_____The Latin square design is an experiment design with three-way elimination of assignable causes of variation due to heterogeneity of material.

_____Randomization is one method of being fair in making comparisons among a set of treatments.

_____The RCB design is a special case of a BIB design.

_____The paired design is a special case of a RCB.

_____Latin rectangle designs have row, column, and treatment effects as assignable causes of variation.

_____The completely randomized design requires the same number of observations for each treatment.

_____The simple change-over design is an orthogonal design.

_____The principle of random allocation of treatments to experimental units within the framework of the design is essential in obtaining an unbiased estimate of experimental variation.

_____Replication is essential for finding an unbiased estimate of a treatment mean.

_____Experimental variation can often be reduced by blocking or grouping of the experimental units.

_____Insofar as possible, statistical objectives as related to treatment design should be planned before actually collecting the data.

_____Bioassay refers to the assessment of potency of a drug by means of responses produced when the treatments are applied to a living organism.

_____In a regression line model we assume that the levels of X may be chosen by the experimenter and are not subject to measurement error.

_____Multiple regression cannot be considered as a fractional replicate of a complete factorial.

_____Any n-way classification with missing combinations may be treated as a fractional replicate of a complete factorial.

_____By using the Kiefer-Wolfowitz procedure, we can estimate the LD_{50} point.

_____By using the Robbins-Munro procedure, we can estimate the LD_{50} point.

_____Bioassay applies to all biological experiments.

(11) 6. Given the following data for 3 levels of factor a by 2 levels of factor b factorial treatment design,

level of factor a	level of factor b b_0	b_1		mean
a_0	$18 = Y_{00}$	$14 = Y_{01}$		$16 = \bar{y}_{0.}$
a_1	$10 = Y_{10}$	$10 = Y_{11}$		$10 = \bar{y}_{1.}$
a_2	$14 = Y_{20}$	$12 = Y_{21}$		$13 = \bar{y}_{2.}$
mean	$14 = \bar{y}_{.0}$	$12 = \bar{y}_{.1}$		$13 = \bar{y}_{..}$

Show how to compute the main effects and the interaction effects.

(9) 7. Suppose that the following main effects and interaction effects were available from a 3×3 factorial:

main effects for factor a: $a_0 = -5$, $a_1 = 3$, $a_2 = 2$
main effects for factor b: $b_1 = -4$, $b_2 = 0$, $b_3 = 4$
interaction effects: $ab_{01} = -4$, $ab_{02} = 2$, $ab_{03} = 2$
                            $ab_{11} = 1$, $ab_{12} = -1$, $ab_{13} = 0$
                            $ab_{21} = 3$, $ab_{22} = -1$, $ab_{23} = -2$.

The overall arithmetic mean $\bar{y}_{..} = 10$.

Show how to reconstruct the yields for the 3^2 combinations Y_{ij} .

(4) 8. Indicate whether or not the following statements are true (T) or false (F):

_____A curve-line graph can always be used without any danger whatsoever for predictions beyond the range of observations.

_____A histogram is a vertical bar graph which measures the relative area or proportion of occurrence in a given class interval measured on the abscissa in scaled units of measurement.

____A vertical bar graph is always a histogram.

____Two-way tables and graphs may be used to summarize data.

(14) 9. Using the 90% confidence interval for the binomial distribution as given on the appended chart, construct the 90% confidence intervals for the following pairs of values:

Estimated proportion equals x / n	Sample size	Lower limit	Upper limit
p = 0.4	n = 30		
p = 0.6	n = 30		
p = 0.5	n = 10		
p = 0.5	n = 100		
p = 0.5	n = 1000		

Suppose that the data from which p = 0.4 was derived were from 100 pea plants and 40 were white-flowered and 60 were red-flowered. Suppose that the investigator believed that the theoretical genetic ratio of white to red should be 1:3. Does the estimated interval contain the theoretical ratio 1 / 4? _____. Are the confidence interval limits for p = 0.6 equal to one minus those for p = 0.4?_____ Will the 95% confidence interval length be larger than given above for n = 10 and p = 0.5? _____ (Insert table for 90% confidence interval as given in W. J. Dixon and F. J. Massey, Jr., (1957),.*Introduction to Statistical Analysis, 2nd edition.* McGraw-Hill Book Co., Inc., New York, page 414, e. g.)

(20) 10. Following are 5 experiment designs listed as 1, 2, 3, 4, 5. Then, a set of 5 response equations is given. Match the response equation for each design. Then, a set of 5 residual equations is given. Again match the number of the equations with the design. Fill in the blanks for "total experimental or sampling units" and for "error degrees of freedom" in the pooled error variance s_e^2. Note: The blanks have been filled in for design number 5 to explain what is wanted.

Design	Response equation	Estimated residual	Total experimental or sampling units	Error degrees of freedom
1. Simple random sample				
2. Completely randomized or cluster-simple random sample (v treatments or clusters and r e.u.'s or s.u.s on each)				
3. Randomized complete block (v treatments, once in each of r blocks)				
4. Latin square or simple change-over (k rows by b columns and k = v treatments)				
5. Balanced incomplete block (b blocks of size k each with v > k treatments)	3	4	$bk = vr$	$bk-v-b+1$

Response equations:

1. $Y_i = \text{mean} + \text{random error} = \bar{y}_. + e_i$, $i = 1, 2, \ldots, n$.

2. $Y_{hij} = \text{mean} + \text{row h effect} + \text{column i effect} + \text{treatment j effect} + \text{random error} = \bar{y}_{..} + r_h + c_i + t_j + e_{hij}$

3. $Y_{hi} = \text{mean} + \text{hth block effect} + \text{ith treatment effect} + \text{random error} = \bar{y}_{..} + b_h + t_i + e_{hi}$

4. $Y_{hi} = \bar{y}_{..} + (\bar{y}_{h.} - \bar{y}_{..}) + (\bar{y}_{i.} - \bar{y}_{..}) + e_{hi} = \text{mean} + \text{estimated hth block effect} + \text{estimated ith treatment effect} + \text{estimated random error.}$

5. $Y_{ij} = \text{mean} + \text{ith group effect} + \text{random error} = \bar{y}_{..} + t_i + e_{ij} = \mu + \tau_i + e_{ij}$.

Estimated residuals:

1. $Y_{hi} - \bar{y}_{h.} - \bar{y}_{.i} + \bar{y}_{..} = e_{hi}$

2. $Y_{hij} - \bar{y}_{h..} - \bar{y}_{.i.} - \bar{y}_{..j} + 2\bar{y}_{..} = e_{hij}$

3. $Y_i - \bar{y}_. = e_i$

4. $Y_{hi} - \bar{y}_{..} - b_h - t_i = e_{hi}$

5. $Y_{ij} - \bar{y}_{i.} = e_{ij}$

The pooled error variance s_e^2 for all of the above designs is computed as (use words instead of symbols):_____

The variance of a treatment mean, using the pooled error variance s_e^2 is:

_____

The variance a difference between any pair of treatment means in all of the above designs except the simple random sample (1) and the balanced incomplete block design (5) and using the pooled error variance s_e^2, is obtained as: _____

Using tabulated values of the t-statistic and for normally distributed random errors, the $(1 - \alpha)$ % confidence interval for the difference between any two treatment means, say $\bar{y}_{1.} - \bar{y}_{2.}$, from designs (2), (3), and (4) is computed as: _____

11. The following is a field plan for an experiment involving the following treatments which are six corn double-crosses (the cross of two single-crosses and a single-cross is the cross of two lines (inbreds) of corn). There were three single-crosses (1, 2, 3) and two forms of one other cross involving male sterility, I, and male fertility, II. The six double-crosses were designated A $= 1 \times$ I, B $= 2 \times$ I, C $= 3 \times$ I, D $= 1 \times$ II, E $= 2 \times$ II, and F $= 3 \times$ II. The data are number of plants growing in a 6 foot $\times$ 10 foot area.

							$Y_{h..}$ Totals	$\bar{y}_{h..}$ Means	Row effects r_h
B	C	D	E	F	A				
16	15	16	19	21	14		101	16.83	-0.25
A	B	C	D	E	F				
18	16	18	14	15	17		98	16.33	-0.75
D	F	B	A	C	E				
18	18	20	18	22	17		113	18.83	1.75
C	E	A	F	B	D				
15	16	16	19	15	17		98	16.33	-0.75
F	A	E	C	D	B				
18	20	19	17	17	15		106	17.67	?
E	D	F	B	A	C				
15	18	19	16	16	15		99	16.50	?

$Y_{.i.}$ 100   103   108   103   106   95 | 615   -   0

$\bar{y}_{.i.}$ 16.67 17.17 18.00 17.17 17.67 15.83 | -   17.08

c_i =   -0.41 0.09 0.92 0.09 0.59 -1.25 | 0   -   -

Treatment Totals:	$Y_{..A}$ = 102	$Y_{..B}$ = 98	$Y_{..C}$ = 102
	$Y_{..D}$ = 100	$Y_{..E}$ = 101	$Y_{..F}$ = 112
Treatment Means:	$\bar{y}_{..A}$ = 17.00	$\bar{y}_{..B}$ = 16.33	$\bar{y}_{..C}$ = 17.00
	$\bar{y}_{..D}$ = 16.67	$\bar{y}_{..E}$ = 16.83	$\bar{y}_{..F}$ = 18.67;
Treatment effects:	t_A = -0.08	t_B = -0.75	t_C = -0.08
	t_D = -0.41	t_E = -0.25	t_F = 1.59

Sum of squares of estimated random errors is equal to 67.667
Sum of squares of estimated row effects is equal to 29.583
Sum of squares of estimated column effects is equal to 17.583
Sum of squares of estimated treatment effects is equal to 19.917

The number of replications for any of the treatments A, B, C, D, E, or F is
_____

The experiment design is a _____
The treatment design is a _____
The number of degrees of freedom for estimated random error sum of squares = 67.667 is _____
Show how to compute the estimated error variance for this design: s_e^2 =
_____

The estimated variance of a difference between 2 means, $\bar{y}_{..j}$ - $\bar{y}_{..j'}$, is computed as: $s_{\bar{y}_{..j} - \bar{y}_{..j'}}$ = _____

Given the following table of tabulated t-values

Degrees of freedom	α = proportion of time a larger value of $|t|$ is observed				
	0.3	0.2	0.1	0.05	0.01
2	1.39	1.89	2.92	4.30	9.92
8	1.11	1.40	1.86	2.31	3.36
9	1.10	1.38	1.83	2.26	3.25
10	1.09	1.37	1.81	2.23	3.17
18	1.07	1.33	1.73	2.10	2.88
20	1.06	1.32	1.72	2.09	2.84

Show how to compute the 80% confidence intervals for differences
between any pair of treatment means, e.g., $\bar{y}_{..A} - \bar{y}_{..C}$ _____
How would the computations change to compute the 95% confidence
interval for differences between pairs of treatment means? _____
The sources of variation for the above experiment are: _____
What assignable effects are orthogonal to each other? _____
The coefficient of variation is computed as: _____
Show how to compute row effects r_5 = _____
and r_6 = _____
Show how to compute the residual (estimated random error) effects for the
the 6th row, 6th column, and treatment C
    e_{66C} = _____
Show how to compute the sum of squares for treatments using the values
given above: _____
Complete the analysis of variance table for this experiment inserting the
sum of squares given with the data.

Source of variation	Degrees of freedom	Sum of squares	Variance
Total	36	---	---
Due to mean	1	10,506.25	

Set up a table showing how to obtain all possible differences between pairs
of treatment means:
Construct a graph to show the interaction effects for the above treatment
design: _____
Construct a stem and leaf diagram for the 36 observations. _____

(12) 12. A researcher has four different kinds of paint (rubber base = R,
lead base = L, aluminum base = A, and zinc base = Z); he wishes to
compare their relative "length of life" when the paint is exposed to
weathering conditions. The experimental materials available to him are
many pieces of white pine 4" × 4" × 24". The minimum acceptable size of

experimental unit is a surface 4" × 12", although larger experimental units are acceptable. Design a randomized complete block design with 3 replications (blocks) and 4 treatments (paints), and specify the following:
  (a) the size and nature of the experimental unit you used,
  (b) the randomization procedure using the following segment of a random numbers table:
  1428 3343 0132 5839 1954 5657 2358 2487 7726 2097
  3541 1789 8704 2832 1345 5903 9108 6924 8444 4283
  0789 3687 9873 7764 7519 0561 1164 3175 4938 9660
  2759 1558 1968 9547 2569 1190 2619 0740 8359 9095
  9598 4552 2735 8681 1629 3760 3935 0524 4900 2907

  (c) the experimental plan for all three blocks,
  (d) the sources of variation in terms of the sample statistics.

(6) 13. From the English alphabet of 26 letters, what is the probability that
  (i) a randomly selected letter is a vowel (a, e, i, o, u)?
  (ii) a randomly selected letter occurs in one of the last 5 positions?
  (iii) a randomly selected letter appears between "a" and "i" and is a consonant?
  (iv) two randomly selected letters are "a" and "b"?
(4) 14. Given the proper noun Statistics with 10 letters, how many permutations of letters are there?
  How many combinations of the 10 letters are there?
(2) 15. In playing a game of 5-card stud poker with a 52-card deck, how many different hands could you be dealt?
(2) 16. What is the probability that you would have a 5-card hand composed of the ace, king, queen, jack, and 10 of diamonds from a 52-card deck?

(6) 17. Match the four terms on the left with the statements on the right:
  i) regression   _____ is used when responses are numbers in a category
  ii) correlation   _____ to compare two treatment means when residuals are normally distributed
  iii) chi-square test _____ to measure association of two variables such as height and weight
  iv) t-test   _____ to explain the variation in one variable accounted for by a second variable

INDEX

acceptance sampling, 268
accuracy, 38
analysis of variance, 440 *et seq.*
anonymous response, 111-9
 -BIBD block total, 114-6
 -"black box", 112
 -partially balanced block total, 116
 -randomized block total, 118-9
 -randomized response, 112-4
 -supplemented block total, 116-8
area sample survey design, 107-8
Arthur, 378
assay, 257-9
 -analytical, 257
 -biological, 257
 -parallel line, 257
 -relative potency, 257
 -slope ratio, 257
 -standard preparation for an, 257
 -test preparation for an, 257
average
 -arithmetic average or mean, 6, 309, 328 *et seq.*
 -geometric mean, 309, 332, 454
 -harmonic mean, 332
 -median, 6, 327 *et seq.*
 -mode, 327 *et seq.*
 -orthogonal designs, 334-41
 -percentage, 308, 333
 -proportion, 333, 372
 -trimmed mean, 348

Bacon, F., 368
bias

-definition, 36
-estimate of, 104
-example of, 37, 94 *et seq.*, 167
blocking, 157, 170 *et seq.*
Bortkewitch, 401
box plots, 346-9
 -fences, 347-8
 -interquartile range, 347
 -major outlier, 348
 -minor outlier, 348
 -scale factor, 347
Bronfenbrenner, U., 80, 316
Brothers, Joyce, 53
Brunk, M. E., 196, 198, 229

Carroll, Lewis, 68, 83
catastrophic events, 150-1
categorical scale, 20
census, 5, 86
central limit theorem, 404
chance allotment, 99-103
class
 -center, 284
 -definition, 284
 -interval, 284, 308
 -modal, 285
class survey, 31-4, 44, 86, 285 *et seq.*, 333, 349, 458 *et seq.*
clinical trial, 260
Clopper and Pearson charts, 415-6
cluster sample survey, 107-8, 334
combinations, 381-6
computing
 computers, 12-3, 357-9
 -packages, 358-9
 -programs, 397-400
confidence coefficient, 415-6